Ham Radio

for
dummies®
A Wiley Brand

Ham Radio

4th Edition

by H. Ward Silver

Ham Radio For Dummies®, 4th Edition

Published by: **John Wiley & Sons, Inc.**, 111 River Street, Hoboken, NJ 07030-5774, www.wiley.com

Copyright © 2021 by John Wiley & Sons, Inc., Hoboken, New Jersey

Published simultaneously in Canada

For general information on our other products and services, please contact our Customer Care Department within the U.S. at 877-762-2974, outside the U.S. at 317-572-3993, or fax 317-572-4002. For technical support, please visit https://hub.wiley.com/community/support/dummies.

Wiley publishes in a variety of print and electronic formats and by print-on-demand. Some material included with standard print versions of this book may not be included in e-books or in print-on-demand. If this book refers to media such as a CD or DVD that is not included in the version you purchased, you may download this material at http://booksupport.wiley.com. For more information about Wiley products, visit www.wiley.com.

Library of Congress Control Number: 2021934046

ISBN 978-1-119-69560-8 (pbk); ISBN 978-1-119-69576-9 (ebk); ISBN 978-1-119-69561-5 (ebk)

Manufactured in the United States of America

SKY10068669_030224

Contents at a Glance

Table of Contents

Introduction

You may have come across ham radio in any number of ways. Did you browse a ham radio website, see a social media post about ham radio, or watch a YouTube video? Maybe you have a teacher, friend, or relative who enjoys ham radio. You could have seen hams on your newsfeed providing communication after natural disasters like hurricanes or during wildfires. Maybe you saw them helping out with a parade or race or you encountered a Field Day setup, ham radio's nation-wide "open house." Maybe you saw someone operating in a park or on a mountain-top trail. Wherever you find it, ham radio has room for an amazing number of activities and lots of hams just like you!

The traditional image of ham radio is of a room full of vacuum tube radios, flicking needles, Morse code keys, and enormous microphones, but today's hams have many more options to try. Ham radio has been changing rapidly! Although the traditional shortwave bands are certainly crowded with ham signals hopping around the planet, hams use the Internet, lasers, and microwave transmitters and traveling to unusual places high and low to make contact, even to and from the International Space Station and bouncing signals off the moon!

Simply stated, ham radio provides the broadest and most powerful wireless communications capability available to any private citizen anywhere in the world. Because the world's citizens are craving ever-closer contact and hands-on experiences with technology of all sorts, ham radio is attracting attention from people like you. The hobby has never had more to offer and shows no sign of slowing its expansion into new wireless technologies. (Did I say wireless? Think *extreme* wireless!)

About This Book

Ham Radio For Dummies, 4th Edition, is meant to get you started in ham radio and answer some of your many questions. If you've just become interested in ham radio, you'll find plenty of information here on what the hobby is all about. I will explain how to go about joining the fun by discovering the basics and getting a license. Many resources on ham radio's technical and operating specialties are available, but this book introduces them briefly so you can get up to speed as quickly as possible. It is true that a ham radio license is really a license to learn!

Some readers have asked for more license exam study information. That would make this a very thick book! There are plenty of great study guides out there, both online and in print, for all three license classes so I don't overlap with them. Material is available with lots of background information or in question-and-answer formats. There are flash cards, too! Just do an online search for "ham radio exam study guide" and you'll find lots of choices.

If you've already received your license, congratulations! This book helps you change from a listener to a doer. Any new hobby, particularly a technical one, can be intimidating to newcomers. By keeping *Ham Radio For Dummies* handy in your station, you'll be able to quickly understand what you receive on the airwaves. I cover the basics of getting equipment connected properly and the fundamentals of on-the-air behavior. Use this book as your bookshelf ham radio mentor, and soon, you'll be making contacts with confidence.

You can read this book in any order. Feel free to browse and flip through the pages to any section that catches your interest. The sidebars and icons are there to support the main story of each chapter, but you can skip them and come back to them later.

The book has five parts. Parts 1 and 2 are for readers who are getting interested in ham radio and preparing to get a license. Parts 3 and 4 explain how to set up a station, get on the air, and make contact with other hams. Part 5 is the Part of Tens (familiar to all *For Dummies* readers), which presents some tips and suggestions for you to get the most out of ham radio. In the online website for this book there is an extensive glossary and a handy supplement to help you with some of the basic math ham radio uses.

Within this book, you may note that some web addresses break across two lines of text. If you're reading this book in print and want to visit one of these web pages, simply enter the address exactly as it's noted in the text, pretending that the line break doesn't exist. If you're reading this book as an e-book, you've got it easy; just click the web address to be taken directly to the web page.

My Assumptions about You

In writing this book, I made some assumptions about you. You don't have to know a single thing about ham radio or its technology to enjoy *Ham Radio For Dummies*, 4th Edition, and you definitely don't need to be an electronics expert to enjoy this book.

But I ask two things of you:

>> You have an interest in ham radio.

>> You can get online.

Due to the broad nature of ham radio, I couldn't include everything in this book. (Also, if I'd done that, you wouldn't be able to lift it.) But I steer you in the direction of additional resources, that will help you get more out of this book with current information and more explanations.

Icons Used in This Book

While you're reading, you'll notice icons that point out special information. Here are the icons I use and what they mean.

TIP

This icon points out easier, shorter, or more direct ways of doing something. Tips also let you know about topics that are covered on the license exam.

REMEMBER

This icon goes with information that helps you operate effectively and avoid technical bumps in the road.

TECHNICAL STUFF

This icon signals when I show my techie side. If you don't want to know the technical details, skip paragraphs marked with this icon.

WARNING

This icon lets you know that some regulatory, safety, or performance issues are associated with the topic of discussion. Watch for this icon to avoid common gotchas.

Beyond the Book

In addition to what you're reading right now, this book also offers free access-anywhere information at www.dummies.com. This includes two appendixes: a comprehensive glossary and some tutorials on "radio math" that are part of ham radio. There is a long list of short entries and tips on topics like tuning, troubleshooting, ways to operate, suggestions for building gear, and many more. You can access these at www.dummies.com/go/hamradiofd4e.

The website also includes a handy Cheat Sheet that includes a summary of your Technician (and soon-to-be General) class license privileges, common Q-signals and repeater channel info, a list of Go Kit gear, and some handy online resources for you. Just search for *Ham Radio For Dummies Cheat Sheet* in the Search box.

Where to Go from Here

If you're not yet a ham, I highly recommend you find your most comfortable chair and read Parts 1 and 2 to discover the basics about ham radio and solidify your interest. If you're a licensed ham, browse through Parts 3 and 4 to find the topics that interest you most. For a bit of fun and interest, turn the pages of Part 5 for tips and bits of know-how that will help you along your way.

For all my readers, welcome to *Ham Radio For Dummies*, 4th Edition. I hope to meet you on the air someday!

1

Getting Started with Ham Radio

Chapter **1**

Getting Acquainted with Ham Radio

H am radio invokes a wide range of visions. Ham radio operators (or *hams*) are a varied lot — from go-getter emergency communicators and radiosport competitors to casual chatters to workshop tinkerers. Everyone has a place, and you do too.

Hams employ all sorts of radios and antennas using a wide variety of signals to communicate with other hams across town and around the world. They use ham radio for personal enjoyment, for keeping in touch with friends and family, for public service, and for experimenting with radios and radio equipment. They communicate by using microphones, computers, cameras, lasers, Morse keys, and even their own satellites.

Hams meet on the air, online, and in person, in groups devoted to every conceivable purpose. Hams run special flea markets and host conventions large and small. Some hams are as young as 6 years old; others are centenarians. Some have a technical background, but most do not; it's not necessary to enjoy ham radio. One thing that all these diverse people share, however, is an interest in radio that can express itself in many ways.

This chapter gives you an overview of the world of ham radio and shows you how to become part of it.

HAM: NOT JUST FOR SANDWICHES ANYMORE

Everyone wants to know the meaning of the word *ham*, but as with many slang words, the origin is murky. Theories abound, ranging from the initials of an early radio club's operators to the use of a meat tin as a natural sound amplifier. Of the many possibilities, the following theory seems to be the most believable.

"*Ham:* a poor operator" was used in telegraphy even before radio. The first wireless operators were landline telegraphers who brought with them their language and the traditions of their much older profession. Government stations, ships, coastal stations, and the increasingly numerous amateur operators all competed for signal supremacy in one another's receivers. Many of the amateur stations were very powerful and could effectively jam all the other operators in the area. When this logjam happened, frustrated commercial operators would send the message "THOSE HAMS ARE JAMMING YOU." Amateurs, possibly unfamiliar with the real meaning of the term, picked it up and wore it with pride. As the years advanced, the original meaning completely disappeared.

Exploring Ham Radio around the World

Although the United States has a large population of hams, the amateur population in Europe is growing by leaps and bounds, and Japan has an even larger amateur population. With more than 3 million hams worldwide, very few countries are without an amateur (see the nearby sidebar "Where the hams are"). Ham radio is alive and well around the world. Listen to the ham radio frequency bands on a busy weekend and you'll see what I mean!

Hams are required to have licenses, no matter where they operate. (I cover all things licensing in Part 2 of this book.) The international agency that manages radio activity is the International Telecommunication Union (ITU; www.itu.int/en). Each member country is required to have its own government agency in charge of licensing inside its borders. In the United States, hams are part of the Amateur Radio Service (www.fcc.gov/wireless/bureau-divisions/mobility-division/amateur-radio-service), which is regulated and licensed by the Federal Communications Commission (FCC). Outside the United States, amateur radio is governed by similar rules and regulations.

REMEMBER

Amateur radio licenses in America are granted by the FCC, but the licensing exams are administered by other hams acting as volunteer examiners (VEs). (I discuss VEs in detail in Chapter 4.) Classes and testing programs are often available through local clubs (see the section "Clubs and online groups," later in this chapter).

Because radio signals know no boundaries, hams have always been in touch across political borders. Even during the Cold War, U.S. and Soviet hams made regular contact, fostering long personal friendships and international goodwill. Although the Internet makes global communications easy, chatting over the airwaves with someone in another country or participating in a planet-wide competition is exciting and creates a unique personal connection.

TECHNICAL STUFF

Since the adoption of international licensing regulations, hams have operated in many countries with minimal paperwork. For example, CEPT, the international treaty that enables countries to recognize one another's amateur licenses, allows hams licensed in their home countries to operate within any other CEPT country. The ARRL provides a lot of useful material about international operating at www.arrl.org/international-regulatory.

Tuning into Ham Radio

Your interest in ham radio may be technical, you may want to use ham radio for public service or personal communications, or you may just want to join the fun. These are all perfectly valid reasons for getting a ham radio license.

Using electronics and technology

Ham radio lets you work closely with electronics and technology (see Chapter 2). Transmitting and receiving radio signals can be as much of an electronics-intensive endeavor as you like. By digging into the technology of ham radio, you're gaining experience with everything from basic electronics to cutting-edge wireless techniques. Everything from analog electronics to the latest in digital signal processing and computing technology is applied in ham radio. Whatever part of electronic and computing technology you enjoy most, it's all used in ham radio somewhere . . . and sometimes, all at the same time!

In this section, I give you a quick look at what you can do with technology.

REMEMBER

You don't have to know everything that there is to know. I've been a ham for almost 50 years, and I've never met anyone who's an expert on everything. A ham radio license is a license to learn!

Design and build

Just as an audiophile might, you can design and build your own equipment or assemble a station from factory-built components. All the components you need are widely available. Some of the original do-it-yourself (DIY) makers, hams delight in *homebrewing*, helping one another build and maintain stations. In software-defined radio (SDR) equipment, computer code is the new component, and I encourage you to experiment as much as you wish.

Experiment with radio waves

Besides being students of equipment and computers, hams are students of *propagation*, which is the means by which radio signals bounce around from place to place. Hams take an interest in solar cycles and sunspots and in the ways they affect the Earth's *ionosphere*, that uppermost region of the atmosphere that reflects shortwave radio signals back to Earth. For hams, weather takes on new importance, too — microwave radio signals can travel long distances along storm fronts or reflect from rain or snow.

Create your own antennas

Antenna experimentation and computer modeling is a hotbed of activity in ham radio. New designs are created every day, and hams have contributed many advances and refinements to the antenna designer's art. Antenna systems range from small patches of printed circuit-board material to multiple towers festooned with large rotating arrays. All you need to start growing your own antenna farm are some wire or tubing, a feed line, and some basic tools. I give you the full picture in Chapter 12.

Invent networks and signals

You can write software to create brand-new types of signals. Hams also develop systems that are novel hybrids of radio and the Internet. Hams developed packet radio, for example, by adapting data communication protocols used in computer networks to operate over amateur radio links. Packet radio is now part of many commercial applications, including your mobile phone.

The combination of GPS technology with the web and amateur mobile radios resulted in the Automatic Packet Reporting System (APRS), which is now used around the world. For more information about these neat systems, see Parts 3 and 4 of this book.

Code yourself a radio

Modern radios are based on software-defined radio (SDR) technology, which allows the radio to adapt to new conditions or perform new functions, as I discuss in Part 4. Hams using design tools like *GNU Radio* (www.gnuradio.org) can experiment with all sorts of techniques to improve and customize their equipment. Digital signal processing (DSP) is a big part of ham radio and has created some very innovative designs, such as the FlexRadio Systems Maestro operating console shown in Figure 1-1.

FIGURE 1-1:
The FlexRadio Systems Maestro combines traditional operating controls with the latest in software defined radio (SDR) design and display techniques.

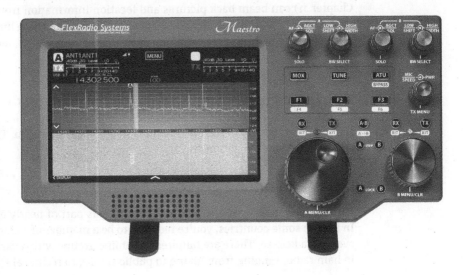

Digitize your radio

Voice communication is still the most popular way that hams use to talk to one another, but computer-based digital operation is gaining fast. New modes (methods of communication) enable world-wide contacts without requiring high power signals. The most common home station today is a combination of computer and radio. Hams also repurpose commercial network equipment to create their own microwave data networks.

Operate wherever you are

You can operate a remote-controlled station via a tablet or phone from anywhere in the world. All it takes is access to the Internet and some hosting software at the station. Most new radios are designed to support operation over an Internet connection. You can operate your home station while you're away or, if you can't have a station at home, use the Internet to access a station and keep on having fun!

Enhance other hobbies

Hams use radio technology in support of hobbies such as flying drones, model rocketry, and high-altitude ballooning. Hams have special frequencies for radio-controlled (R/C) model operation in their "6 meter" band, away from the crowded unlicensed R/C frequencies. Miniature ham radio video transmitters (described in Chapter 11) can beam back pictures and location information from robots, model craft, or portable stations carried by operators. Ham radio data links are also used in support of astronomy, aviation, auto racing and rallies, and many other pastimes.

Joining the ham radio community

Hams like to meet in person and online as well as on the radio. This section discusses a few ways to get involved.

Clubs and online groups

Participating in at least one radio club or group is part of nearly every ham's life. In fact, in some countries, you're required to be a member of a club before you can even get a license. There are hundreds of online groups with a variety of interests in ham radio, ranging from hiking to public service to technical specialties.

TIP

Chapter 3 shows you how to find and participate in ham radio groups, which are great sources of information, training, and assistance for new hams.

THE HAM RADIO BUSINESS CARD

A ham radio tradition is to exchange postcards called *QSLs* (ham shorthand for *received and understood*) with their call signs, information about their stations, and (often) colorful graphics or photos. If you are a stamp collector, you can exchange QSLs directly with the other station. There are online equivalents, too. Whether you prefer paper or electronic form, your QSL is your "ham radio business card" just like Ria N2RJ's in the figure below. You can find more sample QSL cards and information about the practice of QSLing in Chapter 14.

Hamfests and conventions

Two other popular types of gatherings are hamfests and conventions. The traditional *hamfest* is a ham radio flea market where hams bring their electronic treasures for sale or trade. Today's hamfests often include training or instruction opportunities along with commercial vendors for a complete "one-stop shop." Some hamfests are small get-togethers held in parking lots on Saturday mornings; others attract thousands of hams from all over the world and last for days, an in-person complement to eBay and Amazon.

Hams also hold conventions with a variety of themes, ranging from public service to DX (see "Radiosport — Competing with Ham Radio," later in this chapter) to technical interests. An increasingly popular feature is the all-day "university" format focused on a certain type of operating, a technical specialty, operating training — or all three combined!

Hams travel all over the world to attend conventions where they might meet friends formerly known only as voices and call signs over the radio. There is no better way to enjoy travel than being assisted by local hospitality!

Emergency teams

Hams don't need a lot of infrastructure to communicate. As a result, they bounce back quickly when a natural disaster or other emergency makes communications over normal channels impossible. Hams organize as local and regional teams that practice responding to a variety of emergency needs. They support relief organizations such as the American Red Cross and the Salvation Army, as well as police and fire departments.

TIP

Summer and fall are hurricane and wildfire seasons in North America so ham emergency teams gear up for these potentially devastating events. Hams staff an amateur station at the National Hurricane Center in Florida (w4ehw.fiu.edu) and keep The Hurricane Watch Net (an on-the-air response group) busy on 14.325 MHz (www.hwn.org) when storms are active. During wildfire season, hams deploy in support of fire crews across the West. They staff camp communications and set up radio gear in the field, often in remote and difficult locations. Many hams also act as NOAA SKYWARN (www.weather.gov/skywarn) severe weather spotters in their local communities, assisting the National Weather Service.

After disasters of all types, hams are some of the first volunteers to help out, on the job providing communications at emergency operations centers and in the field. Hams trained as emergency response teams help government agencies by handling health-and-welfare messages, performing damage assessments, and providing point-to-point communications until normal systems come back to life. Ham radio also provides the hams themselves with personal communications in and out of the affected area. To find out more about providing emergency communications and public service, see Chapter 10.

Community events

Hams provide assistance for more than just emergencies. Wherever you find a parade, festival, marathon, or other opportunity to provide communications services, you may find ham radio operators helping out. In fact, volunteering for community events is great training for emergencies!

FIELD DAY — HAM RADIO'S OPEN HOUSE

On the last full weekend of June, hams across the United States engage in an annual emergency-operations exercise called Field Day, which allows hams to practice operating away from their regular stations. You can find Field Day setups in parks, at schools, near athletics venues, even in fields! An amateur emergency team or station probably operates in your town or county; go visit them! The American Radio Relay League (ARRL), the national association for amateur radio, provides a Field Day Station Locator web page (www.arrl.org/field-day-locator) that shows you how to find the team or station nearest you.

Radiosport — Competing with Ham Radio

Just like playing a sport or exercising, hams like engaging in challenging activities to build their skills and station capabilities. Called *radiosport*, these encourage continuous improvement of both the operator and the station. Competition provides training and that pays off for public service or in emergencies! Here are a few of the most popular radiosport activities:

>> **DX:** In the world of ham radio, *DX* stands for *distance,* and the allure of making contacts ever more distant from one's home station has always been part of ham radio. Hams compete on the shortwave bands to contact faraway stations and to log contacts with every country. They especially enjoy the thrill of contacting exotic locations, such as "DXpeditions" to uninhabited islands and remote territories. On higher frequencies, even the microwave bands, hams guide their signals along weather systems and ionospheric features, even bouncing signals off the moon, to make contacts far beyond the "radio horizon." When conditions are right and the band is full of faraway stations, succumbing to the lure of DX is easy.

>> **Contests:** Contests are ham radio's version of a contact sport. The point is to make as many contacts as possible during the contest period— sometimes thousands — by exchanging short messages. These exchanges are related to the purpose of the contest: to contact a specific area, use a certain band, find a special station, or just contact the most stations.

>> **Awards:** Thousands of awards are available for various operating accomplishments, such as contacting different countries or states. There are award programs for contacting islands, summits, parks and trails, counties — almost any type of station or location. Awards are great incentives for improving your station and your operating skills.

>> **Special-event stations:** These temporary stations are on the air for a short time to commemorate or celebrate an event or location, often with a special or collectible call sign. Each December, for example, the Marconi Cape Cod Radio Club sets up a special temporary station at the location of Marconi's Wellfleet transatlantic operations. Find out more on the club's Facebook page, *KM1CC - Marconi Cape Cod Radio Club* (www.facebook.com/KM1CC).

If you enjoy the thrill of the chase and the feel of a good workout, go to Chapter 11 to find out more about all these activities.

Communicating through Ham Radio Contacts

If you were to tune a radio across the ham bands, what would you hear hams doing? They're talking to other hams, of course. These chats, called *contacts*, consist of everything from simple conversations to on-the-air meetings to contesting (discussed later in this chapter). I discuss contacts in depth in Chapter 8.

Though you make contacts for different purposes, most contacts follow the same structure:

1. You make a call to someone or respond to someone else's call by transmitting your call sign over the air.

2. You and the other operator exchange names, information about where you're located, and the quality of your signals for an understanding of conditions between your stations.

3. If the purpose of the contact is to chat, proceed to chat.

   You might talk about how you constructed your station, what you do for a living, your family, and your job, for example.

REMEMBER A *call sign*, often shortened to just *call*, is a ham's "radio name." (The term *call letters* is only used by broadcast radio and TV stations.) Call signs have two parts; a *prefix* of letters and a number, such as KE7 or W5, and a *suffix*, which is all letters. The prefix tells you what country licensed the ham and the suffix tells you which ham it is. My call, NØAX, says "N" (so I'm American), "Ø" (so I was licensed in the tenth call district), and "AX" (that's me!). Chapter 7 covers call signs in detail.

Except for the fact that you and the other ham take turns transmitting, and except that this information is converted to radio waves that bounce off the ionosphere or are retransmitted by powerful relay stations, making a contact is just like talking to someone you meet at a party or convention. You can hold the conversation by voice, by keyboard (using a computer connected to the radios), or by Morse code. The average contact satisfies a desire to meet another ham and see where your radio signal can be heard.

REMEMBER

A question that I'm frequently asked about ham radio is "How do you know where to tune for a certain station?" Usually, my answer is "You don't!" Ham radio operators don't have specific frequency assignments or channel numbers. This situation is a good news/bad news situation. The good news is that ham radio gives you unparalleled flexibility to make and maintain communications under continually changing circumstances. The bad news is that making contact with one specific station requires you to know when and on what frequency to call. As you see in Chapter 11, hams have found many ways to solve this problem, however; the result is an extraordinarily powerful and adaptive communications service.

Ragchews

By far the most common type of activity for hams is casual conversation, called *chewing the rag*. Such contacts are *ragchews*. Ragchews take place via voice or keyboards or Morse code across continents or across town. You don't have to know another ham to have a great ragchew; ham radio is a friendly hobby with little class snobbery or distinctions. Just make contact, and start talking! Find out more about ragchews in Chapters 8 and 9.

TIP

The origins of the word *ragchew* are fairly clear. The phrase *chewing the rag* was well known even in the late Middle Ages. *Chew* was slang for *talk*, and *rag*, derived from *fat*, was a reference to the tongue. Thus, people began to use *chewing the rag* to describe conversations, frequently those that took place during meals. Later, telegraph operators picked up that use, and hams picked it up from telegraphers. Because most of ham radio is in fact conversation, ragchewing has been part of radio since its earliest days.

Nets

Nets (an abbreviation of *networks*) are organized on-the-air meetings scheduled for hams who have a shared interest or purpose. Your club or public service team probably has a regular net on a weekly basis. These are great practice for new hams! Here are some of the types of nets you can find:

>> **Public service:** Under normal circumstances, these nets meet for training and practice. When disasters or other emergencies strike, hams organize using these nets to provide crucial communications into and around the stricken

areas until normal services are restored. The nets are also used to provide non-emergency assistance to public events, like parades or foot races.

>> **Technical specialties:** These nets are like radio call-in programs; stations call in with specific questions or problems. The net control station may help, but more frequently, one of the listening stations contributes the answer. Many technical-assistance nets are designed specifically to assist new hams.

>> **Mobile and boating:** Hams operate while on the road or on the water, fresh or salt. They like to stay in touch during their travels and other hams like to contact them as they visit unusual locations. If there are mechanical problems, the station has a ready group of helpers. When there's no phone service or Internet, such as at sea or in remote locations, the net can relay messages and status reports.

>> **Digital networks:**

- *Messaging:* Ham radio was the original "text messaging" system and we're still pretty good at it! Not only do hams exchange messages directly between stations, they have built relay networks, such as APRS (discussed earlier in this chapter), D-RATS, WSPR, and many more.

- *Email:* If you could listen to Internet systems make contact and exchange data, a "mailbox" station might be what they'd sound like. Mailbox stations monitor a single frequency all the time so that others can connect to it and send or retrieve messages via the ham radio Winlink system (www.winlink.org).

- *High-speed data:* Hams share access to frequencies used by WiFi and similar services. By reprogramming common routers and other network equipment, hams have created their own high-speed networks, such as *HSMM-MESH* and the Amateur Radio Emergency Data Network (AREDN). The repurposed routers listen for other routers nearby and connect to them, forming an "ad hoc" network. These flexible network can also connect to the Internet and are a valuable public service tool, especially in remote areas without reliable mobile phone service.

Citizen Science and HamSCI

Hams have supported "real science" since the earliest days of wireless when *everyone* was an experimenter. One of the best examples is the series of "listening tests" conducted during the early 1920s, in which hams supplied many observations that helped establish the existence of the ionosphere. Amateur radio and science have gone hand-in-hand ever since. The ARRL publication *A History of QST*,

Volume 1: Amateur Radio Technology describes the 100-year story of collaboration between hams and scientists, discovering and inventing technologies at the foundation of our present-day wireless world. Ham radio is also a great activity for STEM (science, technology, engineering, and math) students in secondary or university-level education. The hands-on nature of ham radio makes the equations and graphs "real." Ham radio provides practical experience with all sorts of know-how that is offered by few other activities.

Today, there are opportunities for hams to participate in scientific research. These are just a few of the opportunities hams have to make real contributions:

>> **High-altitude ballooning:** Student teams and individuals launch weather balloons with APRS equipment (described earlier in this chapter) to track the balloon position and altitude. Data and images are either transmitted back to the ground or stored on a memory card and recovered along with the balloon. Find out more about amateur high-altitude ballooning at www.arhab.org.

>> **CubeSats:** Working with universities and government space programs, teams of students and researchers build micro-scale satellites (www.cubesat.org) that beam telemetry data from on-board experiments back to Earth. Some satellites also have simple repeater or translator stations on-board that hams can use for point-to-point communication.

>> **Society of Amateur Radio Astronomers (SARA):** If keeping an eye on the sky sounds interesting, check out the SARA website (radio-astronomy.org). It can help you build your own equipment, find kits, or purchase preassembled gear.

>> **WSPRnet:** A special digital protocol, WSPR was designed to make measurements of propagation using extremely low power. To collect the observations, WSPRnet was created (wsprnet.org) with stations reporting in world-wide, 24 hours a day. The data is used for modeling the ionosphere, examining the effect of solar activity, and making propagation predictions.

In August of 2017, the United States was treated to a coast-to-coast total solar eclipse. Hams realized this would have a big effect on the ionosphere as the solar shadow traveled west to east. Because the ionosphere reflects shortwave signals, those signals could be used to measure the effect of the eclipse. HamSCI (www.hamsci.org) was created to conduct the experiment, inviting hams to participate in a large-scale experiment to characterize the ionosphere's response to the eclipse and other open scientific questions. Hundreds of hams helped out by getting on the air during the Solar Eclipse QSO Party (SEQP — see Figure 1-2), a contest-like operating event designed to generate data for studying the eclipse.

FIGURE 1-2:
Students from the New Jersey Institute of Technology club (K2MFF) operating during the eclipse.

The success of the SEQP both in the number of observations and their high quality, led the group to create today's forum for academic and other professional researchers to engage the ham radio community. Today's HamSCI features a wide range of researchers and interested amateurs. Check out the group's projects and events if you're interested in using ham radio to advance science — any interested person is welcome. And there's another total solar eclipse traversing the United States from south to north in 2024!

Chapter **2**

Getting a Handle on Ham Radio Technology

Ham radio covers a lot of technological territory — one of its most attractive features. To get the most out of ham radio, you need to have a general understanding of the technology that makes ham radio work.

In this chapter, I cover the most common terms and ideas that form the foundation of ham radio. If you want, skip ahead to read about what hams do and how we operate our radios; then come back to this chapter when you need to explore a technical idea.

Getting to Know Basic Ham Radio Gear

For some hams, their entire station consists of a small handheld radio or two. Other hams operate on the go in a vehicle. Most hams also have a spot somewhere at home that they claim for a ham radio and associated gadgets. I discuss building and operating your own station in Part 4 of this book. You can see examples of several stations, including mine, in Chapter 13. For now, though, here's a list of the usual gear that makes up a ham's station:

» **The radio:** The modern radio *transceiver,* also referred to as a "rig," combines a receiver and transmitter in a single compact package about the size of an average home entertainment receiver. Transceivers usually have a large tuning knob that controls the frequency, but computer-style displays and screens have replaced the dials and meters of older gear. Today's software-based radios use a PC or tablet for controls and displays.

» **Computer:** Most hams today have at least one computer in the station. Computers can control many of a radio's functions. Using most digital signals simply wouldn't be possible without them. Hams often use more than one computer at a time to perform different functions. Accessories and gadgets using Arduino and Raspberry Pi computers are quite common. Software allows you to control the operating frequency and many other radio functions from a keyboard. Computers can also keep your *log,* a record of your contacts. Computers can send and receive Morse code, too.

» **Handheld radio:** Popular with new hams, a handheld transceiver is a convenient way to get started making contacts through local *repeaters.* Battery-powered, these radios can be used in a vehicle or any time you are away from your home station.

» **Mobile/base radio:** These radios are used for regular mobile operation and in home stations. They produce much more output power than the small handheld models and are about the size of a mid-sized hardcover book.

» **Microphones, keys, and headphones:** Depending on the station owner's preferences, you'll see a couple (or more) of these important gadgets, the radio's original user interface. Microphones and keys range from imposing and chrome-plated to miniaturized and hidden. The old Bakelite headphones, or *cans,* are also a distant memory (which is good; they hurt my ears!), replaced by lightweight, comfortable, high-fidelity designs.

» **Antennas:** In the station, you'll find switches and controllers for antennas that live outside. A ham station tends to sprout antennas ranging from thin whips the size of pencils to wire antennas stretched through the trees and super-sized directional *beam* antennas held high in the air on steel towers. See Chapter 12 for more info on antennas.

» **Cables and feed lines:** Look behind, around, or under the equipment and you find wires. Lots of them. The radio signals are piped through thick, round *coaxial cables,* or *coax.* Power is supplied by wires not terribly different in size from those that power a car stereo. I cover cables and feed lines in detail in Chapter 12.

Building a Basic Ham Radio Station

Although the occasional vintage vacuum-tube radio still glows in a ham's station, today's ham radios are sleek, microprocessor-controlled communications centers, as you see in this section.

REMEMBER

Ham stations are often referred to as a *radio shack*. That's an old term from the very early days of radio when noisy and somewhat dangerous equipment was used. As a result, the station was often in a small shed of its own — literally a shack! Today, your "shack" is simply the place you keep your radio and other ham equipment. The days of bulbous vacuum tubes, jumping meters, and two-handed control knobs are largely in the past.

Basic stations

The stations shown here are very versatile — they can be used as a home or base station, a mobile station in a vehicle, or as a portable station. Equipment is available for very lightweight, low-power operation or heavy-duty full-power stations. Once you get a look at *QST* or *CQ* magazine or the many ham radio websites and social media, you'll start to get a feel for the wide range of opportunities out there!

When you get your entry-class Technician license, you'll probably set up a station like those in Figure 2-1. Many hams install a *mobile* rig in their vehicle, powering it from the battery. They often use a *mag-mount* antenna on the roof or trunk held on with a large magnet. You can also use these radios and antennas at home with an AC power supply and the antenna on a metal surface.

Many hams also have a small handheld radio. Figure 2-1 shows you some of the common accessories that are available. There are all sorts of batteries and battery chargers. For better range, you can use an adapter to connect a mag-mount antenna, replacing the flexible "rubber duck" antenna supplied with the radio.

Figure 2-2 shows the equipment in a station intended to operate on the "HF" or traditional "shortwave" bands. Radios are sold with a handheld microphone and an internal speaker, but you'll have to provide your own headphones or external speaker. If you prefer *Morse code* (also referred to as *CW* for *continuous wave*), you can use the traditional *straight key* (an old-fashioned Morse code sending device), but more commonly, you use a *paddle* and *keyer*, which are much faster to use than straight keys and require less effort. (Morse code operating is discussed in Chapter 8.)

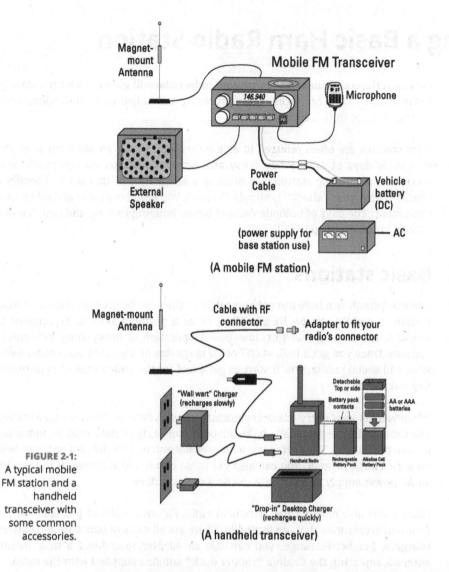

Mobile FM Transceiver

Magnet-mount Antenna

146.940

Microphone

External Speaker

Power Cable

Vehicle battery (DC)

(power supply for base station use)

AC

(A mobile FM station)

Magnet-mount Antenna

Cable with RF connector

Adapter to fit your radio's connector

"Wall wart" Charger (recharges slowly)

Detachable Top or side

Battery pack contacts

AA or AAA batteries

Handheld Radio

Rechargeable Battery Pack

Alkaline Cell Battery Pack

"Drop-in" Desktop Charger (recharges quickly)

(A handheld transceiver)

FIGURE 2-1:
A typical mobile FM station and a handheld transceiver with some common accessories.

The radio is connected with a *feed line* to one or more antennas. Three popular antenna types — dipole, beam, and vertical — are shown. A *dipole* is an antenna made from wire and typically connected to its feed line in the middle. Dipoles can be held up by poles or trees. A *beam* antenna sends and receives radio waves in a preferred direction; it's often mounted on a mast or tower with a *rotator* that can point it in different directions. *Antenna switches* allow the operator to select one of several antennas. An *antenna tuner* sits between the antenna/feed line combination and the transmitter, like a vehicle's transmission, to make the transmitter operate at peak efficiency. If the radio doesn't use AC line power directly, a *power supply* provides the DC voltage and current.

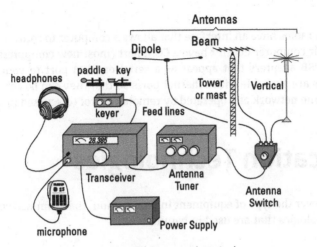

(A typical HF (shortwave) station)

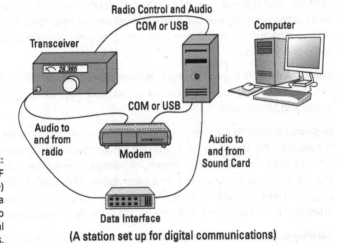

FIGURE 2-2:
A typical HF (shortwave) station and a station set up to use digital communications.

(A station set up for digital communications)

A computer can be used to exchange data over the air using a *digital mode*. Figure 2-2 shows typical digital mode setups. Some radios have USB ports that include support for digital audio along with control of the radio. A *data interface* passes audio signals between the radio's speech circuits and computer. For some types of data, a computer can't do the necessary processing, so a *modem* is used. The computer talks to the modem through a COM or USB port.

Many radios are designed to be easy to use over an Internet connection as a *remote station*. Operating this way is called *remote control*. This is a perfect solution if you can't set up a station at home. Some hams share a remote station they can all operate. There are even remote station services that provide access to excellent stations! You can get a taste of remote operating by trying some of the online receivers at www.websdr.org.

TECHNICAL STUFF

Many radios have an interface that allows a computer to control the radio directly. If your computer doesn't have a COM port (most new computers do not), you can use USB adapters that appear as a serial or COM port to your computer. Some radios are available with Ethernet ports so that they can be connected to a router or home network and operated by remote control (discussed in Chapter 12).

Communication Technologies

TIP

Whatever the type of equipment in your station, the exam will cover the following technologies that are used in ham radio:

>> **Modulation/demodulation:** *Modulation* is the process of adding information to a radio signal so that the information can be transmitted over the air. *Demodulation* is the process of recovering information from a received signal. Ham radios primarily use two kinds of modulation: amplitude modulation (AM) and frequency modulation (FM), similar to what you receive on your car radio or home stereo.

>> **Modes:** A *mode* is a specific combination of modulation and information. You can choose among several modes when transmitting, including voice, data, video, and Morse code.

>> **Repeaters:** *Repeaters* are relay stations that listen on one frequency and retransmit what they hear on a different frequency. Because repeaters are often located on tall buildings, towers, or hilltops, they enable hams to use low-power radios to converse over a wide region. They can be linked by radio or the Internet to extend communication around the world. Repeaters can listen and transmit at the same time — a feature called *duplex* operation.

>> **Satellites:** Just like the military and commercial services, hams construct and use their own satellites. (We piggyback on commercial satellite launches; we don't build our own rockets!) Some amateur satellites act like repeaters in the sky; others make scientific measurements, and some relay digital messages and data.

>> **Software-Defined Radio (SDR):** Microprocessors perform many of the radio's functions, generating and receiving signals by processing them as digital data. Some radios consist almost completely of software running on a PC. SDR technology is extremely flexible and enables hams a wide variety of operating choices and features.

>> **Networks:** Hams have constructed radio-linked computer networks and a worldwide system of email servers accessed by radio. Some of the most popular voice modes consist of digitized speech routed between repeater stations over the Internet.

HAM RADIOS, CB RADIOS, AND MOBILE PHONES

Radios abound — enough to boggle your mind. Here are the differences between your ham radio and other radio services:

- **Citizens Band (CB):** CB radio uses 40 channels near the 28 MHz ham band. CB radios are low-power and useful for local communications only, although the radio waves sometimes travel long distances. You don't need a license to operate a CB radio. This lightly regulated service is plagued by illegal operation, which diminishes its usefulness.

- **Family Radio Service (FRS) and General Mobile Radio Service (GMRS):** These popular radios, such as the Motorola Talkabout models, are designed for short-range communications between family members. Both operate with low power on UHF frequencies. FRS operation is unlicensed, but using the GMRS channels (see the radio's operating manual or guide) does require a license.

- **Broadcasting:** Although hams are often said to be broadcasting, that is incorrect. Hams are barred from doing any one-way broadcasting of programs the way that AM, FM, and TV stations do. Broadcasting without the appropriate license attracts a lot of attention from a certain government agency whose initials are FCC.

- **Public-safety and commercial mobile radio:** The handheld and mobile radios used by police officers, firefighters, construction workers, and delivery-company couriers are similar in many ways to VHF and UHF ham radios. In fact, the bands are so close in frequency that hams often convert and use surplus equipment. Commercial and public-safety radios require a license to operate.

- **Mobile or cell phones:** Obviously, you don't need a license to use a mobile phone, but you can communicate only through a licensed service provider on one of the mobile phone allocations from 700 MHz through 2 GHz. (The new 5G services go much higher in frequency.) Although the phones are actually small UHF and microwave radios, they generally don't communicate with other phones directly and are completely dependent on the mobile phone network to operate.

- **WiFi:** Your wireless network router, gateway, or access point is really a radio transceiver operating on the 2.4- or 5.6-GHz bands. That's what those little moveable antennas are for! Your phone or tablet has small antennas and a WiFi transceiver inside, too.

TECHNICAL STUFF

Hams have always been interested in pushing the envelope when it comes to applying and developing radio technology — one of the fundamental reasons why ham radio exists as a licensed service. Today, ham inventions include such things as creating novel hybrids of radio and other technologies, such as the Internet or

GPS radio location. Ham Mesh networks, for example, consists of wireless local area network technology adapted to ham radio. Ham radio is also a hotbed of innovation in antenna design and construction — in short, techie heaven!

Understanding the Fundamentals of Radio Waves

Getting the most out of ham radio (or any type of radio) is greatly improved by having a general understanding of the purpose of radio: to send and receive information by using radio waves.

Radio waves are another form of light that travels at the same speed: 186,000 miles per second. Radio waves can get to the Moon and back in 2½ seconds or circle the Earth in ⅐ second.

TIP

For the exam, you'll need to know that the energy in a radio wave is *electromagnetic*. That is, the waves are made up of both electric and magnetic *fields*. (A *field* is just a way of storing energy in space, like a gravitational field that makes you experience weight.) The radio wave's field makes charged particles — such as the electrons in a wire — move in sync with the radio wave. These moving electrons are a *current*, just like in an AC power cord except that they form a radio frequency current that your receiver turns into, say, audible speech.

This process works in reverse to create radio waves. *Transmitters* cause electrons to move so that they, in turn, create the radio waves. *Antennas* are just structures in which the electrons move to create and launch radio waves into space. The electrons in an antenna also move in response to radio waves from other antennas. In this way, energy is transferred from moving electrons at one station to radio waves and back to moving electrons at the other station.

TIP

You will see radio waves referred to as *electromagnetic radiation*. Don't let the word *radiation* alarm you (or your neighbors). This is just a general term for any kind of electromagnetic energy flying around. There isn't nearly enough energy in a radio wave to cause the same kind of concern as nuclear or *ionizing* radiation. It's not even close! Radio waves are *non-ionizing* radiation and can't cause the genetic effects or other damage associated with radioactivity.

Frequency and wavelength

TIP

The fields of a radio wave aren't just one strength all the time; they *oscillate* (vary in direction back and forth) the way a vibrating guitar string moves above and below its stationary position. The exam asks about the time a field's strength takes to go through one complete set of values — it's called a *cycle*. The number of cycles in one second is the *frequency* of the wave, measured in *hertz* (abbreviated Hz).

The wave is also moving at the speed of light, which is constant. If you could watch the wave oscillate as it moved, you'd see that the wave always moves the same distance — one *wavelength* — during one cycle (see Figure 2-3). The higher the wave's frequency, the faster a cycle completes and the less time it has to move during one cycle. High-frequency waves have short wavelengths, and low-frequency waves have long wavelengths.

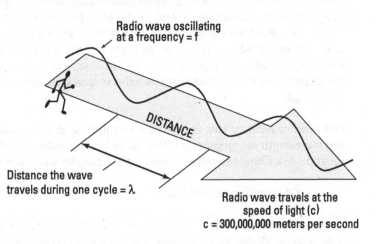

FIGURE 2-3:
As a radio wave travels, its fields oscillate at the frequency of the signal. The distance covered by the wave during one complete cycle is its wavelength.

Radio wave oscillating at a frequency = f

DISTANCE

Distance the wave travels during one cycle = λ

Radio wave travels at the speed of light (c)
c = 300,000,000 meters per second

$\lambda = c/f = 300 / f$ in MHz

Courtesy American Radio Relay League

If you know a radio wave's frequency, you can figure out the wavelength because the speed of light is always the same. Here's how:

> Wavelength = Speed of light / Frequency of the wave

> Wavelength in meters = 300,000,000 / Frequency in hertz

Similarly, if you know how far the wave moves in one cycle (the wavelength), you also know how fast it oscillates because the speed of light is fixed:

> Frequency in hertz = 300,000,000 / Wavelength in meters

Frequency is abbreviated as *f*, the speed of light as *c*, and wavelength as the Greek letter lambda (λ), leading to the following simple equations:

$$f = c / \lambda \text{ and } \lambda = c / f$$

The higher the frequency, the shorter the wavelength, and vice versa.

TIP

If you need some help with the math in this book (although I've used very little) there is a handy Radio Math supplement on this book's web page (see the Introduction). The supplement also lists a number of online references for even more help!

TIP

Radio waves oscillate at frequencies between the upper end of human hearing at about 20 kilohertz, or kHz (*kilo* is the metric abbreviation meaning 1,000), on up to 1,000 gigahertz, or GHz (*giga* is the metric abbreviation meaning 1 billion). They have corresponding wavelengths from hundreds of meters at the low frequencies to a fraction of a millimeter (mm) at the high frequencies. As an example, AM broadcast waves have frequencies of about 1 MHz and wavelengths of 300 meters or so. FM broadcast radio has a much higher frequency, around 100 MHz, so the wavelength is shorter, about 3 meters. WiFi waves (WiFi is a radio system, too!) are about 1/8 meter long. The exam includes several questions about frequency and wavelength.

The most convenient two units to use in thinking of radio wave frequency (RF) and wavelength are megahertz (MHz; *mega* means 1 million) and meters (m). The equation describing the relationship is much simpler when you use MHz and m:

$$f = 300 / \lambda \text{ in m and } \lambda = 300 / f \text{ in MHz}$$

For example, a wave with a frequency of 3.75 MHz has a wavelength of 300 / 3.75 = 80 meters. Similarly, a wavelength of 2 meters corresponds to a frequency of 300 / 2 = 150 MHz.

REMEMBER

If you aren't comfortable with memorizing equations, an easy way to convert frequency and wavelength is to memorize just one combination, such as 300 MHz and 1 meter or 10 meters and 30 MHz. Then use factors of ten to move in either direction, making frequency larger and wavelength smaller as you go.

The radio spectrum

The range, or *spectrum*, of radio waves is very broad (see Figure 2-4). Tuning a radio receiver to different frequencies, you hear radio waves carrying all kinds of different information. These radio waves are called *signals*. Signals are grouped by the type of information they carry in different ranges of frequencies, called *bands*.

FM broadcast-band stations, for example, transmit signals with frequencies between 88 and 108 MHz. That's what the numbers on a radio display mean — 88 for 88 MHz and 108 for 108 MHz, for example. Bands help you find the type of signals you want without having to hunt for them over a wide range.

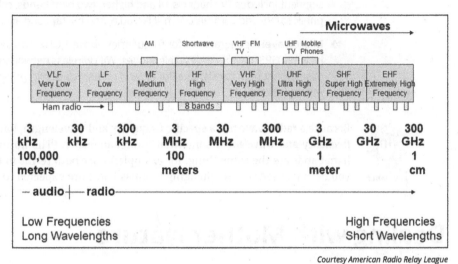

FIGURE 2-4: The radio spectrum extends over a wide range of frequencies and wavelengths.

Courtesy American Radio Relay League

The different users of the radio spectrum are called *services*, such as the Broadcasting Service or the Amateur Radio Service. Each service gets a certain amount of spectrum to use, called a *frequency allocation*. Amateur radio, or ham radio, has quite a number of allocations sprinkled throughout the radio spectrum. Hams have access to many small bands; I get into the exact frequencies of the ham radio bands in Chapter 8.

Radio waves at different frequencies act differently in the way they travel, and they require different techniques to transmit and receive. Because waves of similar frequencies tend to have similar properties, the radio spectrum hams use is divided into five segments:

>> **Low Frequency (LF) and Medium Frequency (MF):** Frequencies from 30 kHz to 300 kHz and from 300 kHz to 3 MHz. This segment includes AM broadcasting, radionavigation transmitters, and two ham bands.

>> **Shortwave or High Frequency (HF):** Frequencies from 3 to 30 MHz. This segment — the traditional shortwave band — includes shortwave broadcasting; eight ham radio bands; and ship-to-shore, ship-to-ship, military, and Citizens Band users.

>> **Very High Frequency (VHF):** Frequencies from 30 MHz to 300 MHz. This segment includes TV channels 2 through 13, FM broadcasting, three ham bands, public safety and commercial mobile radio, and military and aviation users.

>> **Ultra High Frequency (UHF):** Frequencies from 300 MHz to 3 GHz. This segment includes TV channels 14 and higher, two ham bands, cellular phones, public safety and commercial mobile radio, and military and aviation users.

>> **Microwave:** A general term for frequencies above 1 GHz. This segment includes GPS; digital wireless telephones; WiFi wireless networking; microwave ovens; eight ham bands; satellite TV; and numerous public, private, and military users.

REMEMBER

Because a radio wave has a specific frequency and wavelength, hams use the terms *frequency* and *wavelength* somewhat interchangeably. (The 40 meter and 7 MHz ham bands are the same thing, for example.) I use both terms in this book so that you become used to interchanging them as hams are expected to do.

Dealing with Mother Nature

Ham radio offers a whole new way of interacting with the natural world around us. The movement or *propagation* of radio waves is affected by the Sun, the characteristics of the atmosphere, and even the properties of ground and water. We may not be able to see these effects with our usual senses, but by using ham radio, we can detect, study, and use them.

TIP

Did you know you can hear the world turning through your radio? It's true! Because of daily, seasonal, and year-to-year changes in the Sun's activity and the way sunlight falls on the rotating Earth, radio wave propagation changes, too. As you listen, you can hear the bands "open" and "close" as signals bounce from place to place. It is one of the most fascinating things about ham radio!

Experiencing nature affecting radio waves

TIP

On their way from Point A to Point B, radio waves journey around the Earth and through its atmosphere, in a variety of ways. The exam covers several of these effects:

>> **Ground wave propagation:** For local contacts, the radio wave journey along the surface of the Earth is called *ground wave propagation*. Ground wave propagation can support communication up to 100 miles but varies greatly with the frequency being used.

>> **Sky wave propagation:** For longer-range contacts, the radio waves must travel through the atmosphere. At HF and sometimes at VHF (refer to "The radio spectrum," in the previous section), the very highest layers of the atmosphere, called the *ionosphere,* bend the waves back to Earth. This is called *sky wave propagation* or "skip." Depending on the angle at which the signal is reflected, a sky wave "hop" can be as long as 2,000 miles. HF signals often bounce between the Earth's surface and the ionosphere several times so that contacts are made worldwide.

>> **Tropospheric propagation:** Apart from the ionosphere, the atmosphere itself can direct radio waves. *Tropospheric propagation,* or *tropo,* occurs along weather fronts, temperature inversions, and other large-scale features in the atmosphere. Tropo is common at frequencies in the VHF and UHF range, often supporting contacts over 1,000 miles or more.

>> **Aurora:** When the aurora is strong, it absorbs HF signals but reflects VHF and UHF signals while adding a characteristic rasp or buzz. Hams who are active on those bands know to point their antennas north to see whether the aurora can support an unusual contact.

>> **Meteor trails:** Meteor trails are very hot from the friction of the meteoroid's passage through the atmosphere — so hot that the gases become electrically conductive and reflect signals until they cool. For a few seconds, a radio mirror floats high above the Earth's surface. Meteor showers are popular times to try meteor-scatter propagation (see Chapter 11).

Overcoming radio noise

One limiting factor for all wireless communication is noise. Certainly, trying to use a radio in a noisy environment such as a car presents some challenges, but I'm talking here about electrical noise, created by natural sources such as lightning, the aurora, and even the sun. Other types of noise are human-made, such as arcs and sparks from machinery and power lines. Even home appliances make noise — lots of it. When noise overpowers the signal, radio communication becomes very difficult.

Radio engineers have been fighting noise since the early days of AM radio. FM was invented and used for broadcasting because of its noise-rejection properties. Even so, there are practical limits to what transmitters and receivers can do, which is where digital technology comes in. By using sophisticated methods of turning speech and data into digital codes, digital technology strips away layers of noise, leaving only the desired signal.

Hams have been in the forefront of applying noise-fighting digital techniques to wireless. The noise-canceling technology in most mobile phones was pioneered in part by Phil Karn, an engineer and scientist for Qualcomm and amateur operator KA9Q. Recently, powerful noise-fighting coding and decoding techniques have been applied to amateur signals by Nobel Prize laureate Dr. Joe Taylor, also known by his ham radio call sign K1JT. Using Taylor's special software, known as *WSJT-X*, hams can communicate with signals hundreds of times weaker than the natural noise level, even bouncing their signals off the moon with simple equipment. You can download *WSJT-X* for free at physics.princeton.edu/pulsar/K1JT.

IN THIS CHAPTER

» **Finding mentors and clubs**

» **Checking out online communities**

» **Becoming a member of the ARRL**

» **Finding a specialty organization**

» **Going to hamfests and conventions**

Chapter **3**

Finding Other Hams: Your Support Group

O ne of the foundations of ham radio is helping newcomers. After all, a ham radio license is mostly a license to learn! Hams are great at providing a little guidance or assistance. You can make your start at ham radio operating much easier and more successful by taking advantage of those helping hands. This chapter shows you how to find them.

HAM RADIO — MEETING ONLINE

As this edition is being written, Covid-19 continues to require social distancing, so many face-to-face ham radio events have been canceled. For now, the hamfests, meetings, conventions, and conferences described in this chapter are still happening, but only online. Many hams are taking advantage of the situation to watch the many presentations and seminars now available to an online audience at little or no cost. The success of this format means you can expect to find many "virtual" events available in the future.

Finding and Being a Mentor

A mentor is very useful in helping you over the rough spots that every newcomer encounters. A good place to start your search for a mentor is to search for ham radio clubs in your area (refer to "Finding and choosing a club," later in this chapter). You might start on the clubs page of the QRZ.com website (go directly to www.qrz.com/clubs or use the Database menu), for example.

As your interests widen, you'll need additional help. Luckily, hundreds of potential mentors, known in ham radio as *Elmers*, are available around the world.

REMEMBER

Using the word *Elmer* to mean *mentor* is unique to ham radio. Rick Lindquist (WW1ME) traces the origin of the term *Elmer* to the March 1971 issue of *QST* magazine; the term appeared in a "How's DX" column by Rod Newkirk (W9BRD). Rod's mentor was a ham named Elmer and the message was that every new ham should have an Elmer to help them. The name stuck and since then, "Elmering" has meant "helping." Every ham has at least one Elmer at some point. You will, too, and if someone refers to you as "my Elmer," you can be proud.

Entering *ham radio elmer* or *ham radio mentor* into a search engine turns up lots of resources. Some specialize in helping you study for the exam. Some are organized in a frequently asked questions (FAQ) format. And a few are online forums where you can ask a specific question. You may want to join one of the groups set up specifically to answer questions and offer help to new hams and those studying for the license exam.

TIP

When looking for answers in an online forum or email list, check the website's archives first. It's likely that others will have had similar questions and you can find your answer right away. This is just good "netiquette." For example, in the eham.net "Elmer" forum (look in Community > Forums > Elmers) searching for *tuning dipole* turns up dozens of helpful responses. Along with answering your original question, you are likely to find a lot more information in the archives!

Can you be a mentor? Although you may not think you are ready to mentor someone, you might be the perfect person! Since you are learning about ham radio, you understand very well how other new hams might feel and what questions they might have. Don't hesitate to take someone else along for the ham radio ride. If you are both studying for the license exam at the same time, you can even mentor each other! Studying together is a great way to learn.

After you succeed in getting your license (and you will!) you are in a great position to help someone else learn and understand the material. The things that were confusing to you might also be confusing to him or her, and you can relate how you figured things out. Advice given by another newcomer can feel less intimidating than from a seasoned "old timer."

As you progress with ham radio, you'll acquire some equipment, learn about using it, and have more than a few *aha!* moments. You might not think you can act as a teacher, but why not give it a try by answering someone's question. Offer to help a new ham at an operating event or pair up when performing public service. As is often said (and demonstrated), the best way to learn something is to teach it! You were once brand new, too, so don't hesitate to reach out.

Interacting in Online Communities

Just like every other human activity, ham radio has online communities in which members discuss the various aspects of the hobby, provide resources, and offer support 24 hours a day. Will these communities replace ham radio? Not likely; the magic of radio is too strong. By their presence, though, they make ham radio stronger by distributing information, cementing relationships, and adding structure.

The number and type of online ham radio resources is increasing every day. There are too many to list here, so your best strategy for finding them is to use a search engine. Of course, you have to know what to look for! The online list of forums and articles under the Community tab at www.eham.net cover a lot of topics.

TIP

If you are searching for ham radio information online, be sure to use both *ham radio* and *amateur radio* in the search window. Both terms are used interchangeably. By using both, you'll see a full selection of links and pages.

Social media and blogs

Everything has a presence online, and ham radio is no different. Just search for *ham radio* on Facebook, for example, and you'll find dozens of possibilities, ranging from general-interest clubs to emergency communications to license-exam practice to contesting — and more.

TIP

As with all online communities, not everyone behaves perfectly, but I recommend that you give a few of these communities a try and see which of them suits your personal style.

Here are some popular streams of information about amateur radio:

>> **Twitter:** twitter.com/amateurradio (@Amateurradio) or search for ham radio topics at twitter.com/search?q=ham+radio

>> **Reddit:** www.reddit.com/r/amateurradio and www.reddit.com/r/hamradio

>> **Instagram:** #hamradio or #amateurradio

New platforms are being introduced every day — Slack and Discord both have a lot of ham radio content, for example. Use the platform's search tools to look for *ham radio* or *amateur radio*. Most posts and threads will treat those search terms the same so there will be a lot of overlap, but searching for both will find just about everything they have to offer.

WARNING

The Internet is full of misinformation and, ham radio being a technical hobby, it can be hard to tell the helpful from the inaccurate. If something seems a little too simple for a complicated question, or if you just don't understand the claims, get some second opinions. Asking a mentor often leads to learning something completely new. This is where an experienced mentor can be really helpful!

As a beginner, blogs and individual websites can be very valuable in finding answers to common questions. One blog that's very helpful for newcomers to amateur radio is KB6NU's Ham Radio Blog (www.kb6nu.com). Run by Dan Romanchik (KB6NU), it offers study guides and news. Dan writes books and guides for newcomers, too.

Videos, podcasts, and webinars

There's nothing quite like a demonstration to find out how to do something, such as put on a connector, make a contact, tune an antenna, or assemble a kit. Many video and photo websites are available to speed you on your way to ham radio success; YouTube (www.youtube.com) and Instructables (www.instructables.com/howto/ham+radio/) are just two of the options. Instagram (www.instagram.com) and Vimeo (www.vimeo.com) have quite a few ham radio videos, too. The YouTube channel of Randy Hall (K7AGE — www.youtube.com/user/K7AGE) has quite a few good instructional videos, for example.

Also available are several nicely produced talk show–style programs that have large followings. Here are a few of my favorites:

>> *Ham Nation* on TWiT.TV (twit.tv/shows) covers operating and technical topics in an informal and fast-paced format. A new show airs every week.

>> *Ham Radio Now TV* (www.hamradionow.tv), hosted by David Goldenberg (WØDHG), is a weekly podcast that tackles all sorts of interesting topics. Pearce's web page lists many other audio and video programs.

>> *Ham Radio Newsline* (www.arnewsline.org) is a podcast structured like a news program with the latest ham radio-related stories. It is produced by Neil Rapp (WB9VPG).

>> *QSO Today* (www.qsotoday.com), produced by Eric Guth (4Z1UG and WA6IGR) includes podcasts, videos (mostly interviews), and blog entries. This is a great way to meet some of ham radio's leaders and learn about specific topics, too.

TIP

It's hard to use a dictionary to look up a word you don't know how to spell! It can be the same with ham radio. If you don't know the right name for something, your online search can be pretty frustrating. You can use the Google Images service (images.google.com) to help, though. If you have a mental picture of what you're looking for, describe it in the images search window and click on some images that look right. Then follow the links to the web page the image came from.

Email reflectors

The first online communities for hams were email lists, known as *reflectors*. Reflectors are mailing lists that take email from one mailbox and rebroadcast it to all members. With some list memberships numbering in the thousands, reflectors get information spread around pretty rapidly. Every ham radio interest has a reflector.

Focused online communities like those on Groups.io (groups.io) offer much more than just email distribution. They also offer file storage, photo- and video-display, messaging, and excellent member management. To take advantage of these groups.io services, create a personal account; then search the service (use "Find a Group") for *amateur radio* or *ham radio*.

Table 3-1 lists several of the largest websites that serve as hosts for reflectors. You can browse the directories and decide which list suits your interests. (Be careful, though, that you don't wind up spending all your time on the reflectors and none on the air.)

TABLE 3-1   ## Hosts and Directories for Ham Radio Reflectors

Website	Topics
www.qth.com	Radios, bands, operating, and awards
www.contesting.com	TowerTalk, CQ-Contest, Amps, Top Band (160 meters), RTTY (digital modes) — look under "Contest Lists" and "Other Lists"
www.dxzone.com/catalog/Internet_and_Radio/Mailing_Lists and www.ac6v.com/mail.php	Directories of reflectors and forums hosted on other sites

Because my main interests are operating on the HF bands, contesting, and making DX (long-distance) contacts, for example, I subscribe to the TowerTalk reflector, the CQ-Contest reflector, a couple of the DX reflectors, and the Top Band reflector about 160 meter operating techniques and antennas. To make things a little easier on my email inbox, I subscribe in digest format so that I get one or two bundles of email every day instead of many individual messages. Most reflectors are lightly moderated and usually closed to any posts that aren't from subscribed members — in other words, spam.

TIP

As soon as you settle into an on-the-air routine, subscribe to one or two reflectors or groups. They are great ways to find out about new equipment and techniques before you take the plunge and try them yourself.

Online training and instruction

Webinars (online video seminars hosted by an instructor) are very common. Many of these events are archived, such as those hosted by the World Wide Radio Operator's Foundation (www.wwrof.org) webinar is the next-best thing to your mentor being there in the room with you. Many conferences are also recording their sessions and posting them online for you to learn from. Contest University (CTU — contestuniversity.com) is a good example of a day-long event with many presentations that are recorded and available online. Chances are, any conference or hamfest that presents speakers and training sessions will record and publish the videos for you to watch any time.

If you can take an in-person class to study for your FCC license exam, that is a good way to learn the material in-depth. You may prefer to study on your own, though. If that's the way you choose, there are study books and online resources for you — I discuss them in Chapter 5. Some books are combined with online practice exams, such as the ARRL's license manuals and the *ARRL Exam Review for Ham Radio* software.

Before you get started on getting a license, you should browse the available materials by doing an online search for *ham radio license study*. You'll find quite a number of choices ranging from simple flash-cards to interactive videos. Some are free, and none are terribly expensive. Take a look at what's available so you'll be ready to go when you decide to dive in and get started!

Many new hams get their licenses to support public service activities. Does that sound like you! If so, you might be interested in getting some training on how hams organize themselves to provide that service. Many ham radio public service groups use the Federal Emergency Management Agency (FEMA) model, called NIMS for the National Incident Management System. The FEMA training website (the Emergency Management Institute — training.fema.gov/is/crslist.aspx) lists numerous free courses, some of which may be required to become part of a public service team. Start with the current version of course ICS-100 to learn about this common method of managing public service activities.

Web portals

Portals provide a comprehensive set of services and function as ham radio home pages. They feature news, informative articles, radio buy-and-sell pages, links to databases, reflectors, and many other useful services to hams. Here are three popular portals:

>> **QRZ.com** (the ham radio abbreviation for "Who is calling me?") evolved from a call sign lookup service — what used to be a printed book known as a *callbook* — to the comprehensive site (www.qrz.com) that you see today. The call sign search features are incredibly useful, and the site offers a variety of call sign management functions.

>> **eHam.net** (www.eham.net) provides forums, articles, reviews, and classified ads for equipment sales. You will also find real-time links to a DX-station spotting system (frequencies of distant stations that are currently on the air) and the latest solar and ionospheric data that affects radio propagation.

>> **DXcoffee.com** (www.facebook.com/DxCoffee) is typical of a site with a theme. This site is all about the fun of DXing or trying to contact distant stations. There are hams traveling to exotic locations all the time. By watching a site like this, you'll know when they're going and their plans for operating. (DXing is discussed in more detail in Chapter 11.)

Joining Radio Clubs

To get in touch with other hams, find your local radio club! Although online help is convenient, there's no substitute for in-person contact and making friends. The following are true for most hams and clubs:

>> Most hams belong to a general-interest club as well as one or two specialty groups.

>> Most local or regional clubs have in-person meetings, because membership is drawn largely from a single area.

>> Almost all clubs have a website or social media presence, some kind of newsletter, and usually an email distribution list or Twitter feed.

>> Specialty clubs focus on activities. Activities such as contesting, low-power operating, and high-altitude ballooning may have a much wider (even international) membership. See "Taking Part in Specialty Groups," later in this chapter, for more information.

TIP

Clubs are great resources for assistance and mentorship. As you get started in ham radio, you'll find that you need answers to a lot of basic questions and maybe some in-person help. I recommend you start by joining a general-interest club (see the next section). If you can find one that emphasizes assistance to new hams, so much the better. You'll find the road to enjoying ham radio a lot smoother in the company of others, and you'll find other new hams to share the experience.

Before you start, remember that you don't have to immediately join a club "for life." You can attend a few meetings as a visitor to get a feel for the group. If you decide to join, most memberships are for a year and you can decide to renew or not. It's very common to be a member of a few clubs before you find one or two that are right for you. And if you change your ham radio style later, you can join a different club!

Finding and choosing a club

Here's one way to find ham radio clubs in your area:

1. Go to www.qrz.com/clubs then use the "List clubs in" menu to select your area, or enter a club name.

2. From the list of clubs in the database, click on the club name to go to their website.

   For an example club listing, see the nearby sidebar "Checking out a club."

TIP

The ARRL, covered later in this chapter, also has a directory of affiliated clubs at www.arrl.org/find-a-club. You can look up clubs by name or a keyword, by city/state/province/Zip Code, or by ARRL Section or Division.

If more than one club is available in your area, how do you make a choice? Consider these points when making a decision:

>> **Which club has meetings that are most convenient for you?** Check out the meeting times and places for each club.

>> **Which club includes programs that include your interests?** On the club's website or newsletter, review recent programs and activities to see if they sound interesting. Are there any activities coming up you could take part in?

>> **Which club has activities for new hams?** General-interest and service clubs often have activities designed specifically to educate, train, and welcome new hams. These are good starter clubs for you.

>> **Which club has informal "meet and greet" activities?** Many clubs have an informal breakfast or after-work every so often. It's a great way to introduce yourself to some of the members and ask questions without going to a full-blown meeting.

>> **Which club feels most comfortable to you?** Don't be afraid to attend a meeting or two to find out what different clubs are like.

You'll quickly find out that the problem isn't finding clubs, but choosing among them. Unless a club has a strong personal-participation aspect, such as a public-service club, you can join as many as you want just to find out about that part of ham radio.

CHECKING OUT A CLUB

I found this listing for one of the largest clubs in western Washington state through the ARRL website:

Mike & Key Amateur Radio Club

City: Renton, WA

Call Sign: K7LED

Specialties: Contest, Digital Modes, DX, General Interest, Public Service/Emergency, Repeaters, VHF/UHF

Services Offered: Club Newsletter, Entry-Level License Classes, General or Higher License Classes, Hamfest, License Test Sessions, Mentor, Repeater

Section: WWA

Links: www.mikeandkey.org

This club is well suited to a new ham. You'll find yourself in the company of other new license holders, so you won't feel self-conscious about asking questions. The club offers educational programs, activities, and opportunities for you to contribute.

Are you a college student or looking for a college radio club? Check out the ARRL Collegiate Amateur Radio Initiative on Facebook (www.facebook.com/groups/ ARRLCARI) to find college clubs and see what they are up to.

Participating in meetings

After you pick a general-interest club, show up for meetings, and make a few friends right away, your next step is to start participating. But how?

Obviously, you won't start your ham club career by running for president at your second meeting, but ham clubs are pretty much like all other hobby groups, so you can become an insider by following a few easy first steps. You're the new guy or gal, which means you have to show that you want to belong. Here are some ways to get acquainted and fit in:

>> Right at the start, introduce yourself to a club officer as a visitor or new member as soon as you get to the meeting. If a "stand up and identify yourself" routine occurs at the beginning of the meeting, be sure to identify yourself as a new member or visitor. If other people also identify themselves as new, introduce yourself to them later.

>> Show up early to help set up, make coffee, hang the club banner, help figure out the projector, and so on. Stay late and help clean up, too.

>> Be sure to sign in, sign on, or sign up if you have an opportunity to do so, especially at your first meeting.

>> Wear a name tag or other identification that announces your name and call sign in easy-to-read letters.

>> Introduce yourself to whomever you sit next to.

>> After you've been to two or three meetings, you'll probably know a little about some of the club's committees and activities. If one of them sounds interesting, introduce yourself to whoever spoke about it and offer to help.

>> Show up at as many club activities and work parties as possible.

>> Comb your hair. Brush your teeth. Sit up straight. (Yes, Mom!)

These magic tips aren't just for ham radio clubs; they're for just about any club. Like all clubs, ham clubs have their own personalities, varying from wildly welcoming to tightly knit, seemingly impenetrable groups. After you break the ice with them, though, hams seem to bond for life.

When you're a club elder yourself, be sure to extend a hand to new members. They'll appreciate it just as much as you did when you were in their shoes.

Getting more involved

Now that you're a regular, how can you get more involved? This section gives you some pointers.

Volunteering your services

In just about every ham club, someone always needs help with the following events and activities:

>> **Field Day:** Planners and organizers can always use a hand with getting ready for this June operating event (see Chapter 1). Offer to help with generators, tents, and food, and find out about everything else as you go.

TIP

Helping out with Field Day — the annual continent-wide combination of club picnic and operating exercise — is a great way to meet the most active members of the club. Field Day offers a little bit of everything ham radio has to offer.

>> **Conventions or hamfests:** If the club hosts a regular event, its organizers probably need almost every kind of help. If you have any organizational or management expertise, so much the better. (I discuss hamfests and conventions in detail later in this chapter.)

>> **Club "stuff":** Managing sales of club logo items is a great job for a new member. You can keep records, take orders, and make sales at club meetings.

TIP

If you have a flair for graphic art, don't be afraid to make suggestions about designing these items or new ones the club might enjoy.

>> **Shared equipment:** Many clubs maintain a supply loaner equipment. All you have to do is keep track of everything and make it available to other members.

>> **Club station:** If your club is fortunate enough to have its own radio station or repeater, somebody always needs to do maintenance work, such as working on antennas, changing batteries, tuning and testing radios, or just cleaning. Buddy up with the station manager, and you can become familiar with the equipment very quickly. You don't have to be technical — just willing.

>> **Website and newsletter:** If you can write, edit, or maintain a website, don't hesitate to volunteer your services to the club newsletter editor or webmaster. Chances are that this person has several projects backlogged and would be delighted to have your help. You'll also become informed about everything happening with the club.

Find out who's currently in charge of these areas and offer your help. You'll discover a new aspect of ham radio, gain a friend, and make a contribution.

Taking part in activities

Along with holding ongoing committee meetings and other business, most clubs sponsor several activities throughout the year. Some clubs are organized around one major activity; others seem to have one or two going on every month. Here are a few common club activities:

>> **Public service:** This activity usually entails providing communication services during a local sporting or civic event, such as a parade or festival. Events like these are great ways for you to hone your operating skills.

>> **Contests and challenges:** Operating events are great fun, and many clubs enter on-the-air contests as a team or club. Sometimes, clubs challenge each other to see which can generate the most points. You can either get on the air yourself or join a multiple-operator station. (For more on contests, see Chapter 11.)

>> **Work parties:** What's a club for if not to help its own members? Raising a tower or doing antenna work at the club's station or that of another member is a great way to meet active hams and gain experience with station building.

>> **Construction projects:** Building your own equipment and antennas is a lot of fun, so clubs occasionally hold "show and tell" or "homebrew" meetings. Some sponsor group construction projects in which everyone builds a particular item at the same time. Building your own equipment saves money and lets everyone work together to solve problems. If you like building things or have technical skills, taking part in construction projects is a great way to help out.

TIP Take part in activities for newcomers even if you think you already understand the topic or technique. You'll get some practice and may learn a thing or two that you had overlooked. If you do have it down pat, lend a hand to another new ham who needs a little help. That's mentoring!

REMEMBER Supporting your club by participating in activities and committees is important. For one thing, you can acknowledge the help you get from the other members. You also start to become a mentor to other new members. By being active within the club, you strengthen the organization, your friendships with others, and the hobby in general.

Exploring the ARRL

The *American Radio Relay League* (ARRL; www.arrl.org) is one of the oldest continuously functioning amateur radio organization in the world. It provides services to hams around the world and plays a key part in representing the ham radio cause to the public and governments. That ham radio could survive for more than 100 years without a strong leadership organization is hard to imagine, and ARRL has filled that role. I devote a whole section of this chapter to ARRL simply because it's such a large presence within the hobby for U.S. hams (and for those in Canada who belong to its sister organization, Radio Amateurs of Canada).

ARRL is a volunteer-based, membership-oriented organization. Rest assured that even as a new ham, you can make a meaningful contribution as a volunteer. To find out how to join, go to www.arrl.org/membership.

ARRL benefits to you

The most visible benefits of ARRL membership are the magazines that you receive month (see Figure 3-1). The largest, oldest, and most widely read ham radio magazine, QST includes feature articles on technical and operating topics, reports on regulatory information affecting the hobby, the results of ARRL-sponsored competitions, and columns on a wide variety of topics.

FIGURE 3-1:
QST, On The Air, QEX, and NCJ cover nearly every aspect of ham radio plus product reviews and ads from almost every ham radio vendor.

Courtesy American Radio Relay League

The magazine *On The Air* is intended for the new ham with articles and projects to help you get going. Two other publications, *QEX* for technical experimenters and *National Contest Journal* or *NCJ* for competitive radio, are also available with your membership. These four publications cover everything from current events to state-of-the-art technology.

Along with the magazines, ARRL maintains an active and substantial website, providing current news and general-interest stories; the Technical Information Service, which allows you to search technical documents and articles online; and several free email bulletins, online newsletters, and social media services.

TIP

Why does W1AW transmit bulletins over the air in this day and age of broadband connectivity? Bulletins may sound old-fashioned but they offer an opportunity to test equipment, assess radio propagation, and practice copying live Morse code on the air.

ARRL also manages the Amateur Radio Emergency Service (ARES), which helps hams organize at the local level to support local government and public-safety functions with emergency communication services. These teams also perform public service by providing support and communications services for parades, sporting events, and similar events. You can find out more about ARES and public service activities in Chapter 10.

In addition, ARRL is the largest single sponsor of operating activities for hams, offering numerous contests, award programs, and technical and emergency exercises.

ARRL benefits to the hobby

By far the most visible aspect of ARRL on the ham bands is its headquarters station, W1AW (see Figure 3-2). Carrying the call sign of ARRL founder Hiram Percy Maxim, the powerful station beams bulletins and Morse code practice sessions to hams around the planet every day. Visiting hams can even operate the W1AW station themselves (as long as they remember to bring a license). Most hams think that being at the controls of one of the most famous and storied ham stations in the world is the thrill of a lifetime.

FIGURE 3-2: The world-famous W1AW station in Newington, Connecticut.

The ARRL is a volunteer examiner coordinator (VEC) organization. You may take your licensing test at an ARRL-VEC exam session. (See Part 2 for information about the licensing process.) With the largest number of volunteer examiners (VEs), the ARRL-VEC helps thousands of new and active hams take their licensing exams, obtain vanity and special call signs, renew their licenses, and update their license information of record. When the Federal Communications Commission (FCC) could no longer maintain the staff to administer licensing programs, ARRL and other ham organizations stepped forward to create the largely self-regulated VEC programs that are instrumental to healthy ham radio.

One of the least visible of ARRL's functions, but arguably one of its most important, is its advocacy of amateur radio service to governments and regulatory bodies. In this telecommunications-driven age, the radio spectrum is valuable territory, and many commercial services would like to get access to amateur frequencies, regardless of the long-term effects. ARRL helps regulators and legislators understand the special nature and needs of amateur radio.

ARRL benefits to the public

Although it naturally focuses on its members, ARRL takes its mission to promote amateur radio seriously. To that end, its website is largely open to the public, as are all bulletins broadcast by W1AW (see the preceding section). The organization also provides these services:

>> **Facilitates emergency communications:** In conjunction with the field organization, ARES teams around the country provide thousands of hours of public service every year. While individual amateurs render valuable aid in times of emergency, the organization of these efforts multiplies the usefulness of that aid. ARRL staff members also help coordinate disaster response across the country.

>> **Publishes** *The ARRL Handbook for Radio Communications:* First published in 1926, "The Handbook" is used by telecommunications professionals and amateurs alike. Information about the current edition is available at www. arrl.org/arrl-handbook-reference, including a link where you can get your own copy.

>> **Provides technical references:** The league publishes numerous technical references and guides, including conference proceedings and standards.

>> **Promotes technical awareness and education:** ARRL is involved with the Radio merit badge for scouting and with Jamboree-on-the-Air programs. It also sponsors the Teachers Institute on Wireless Technology to train and license primary and secondary educators.

Taking Part in Specialty Groups

Ham radio is big, wide, and deep. The hobby has many communities that fill the airwaves with diverse activities. A *specialty club or organization* focuses on one aspect of ham radio that emphasizes certain technologies or types of operation. Many specialty organizations have worldwide membership.

Some clubs focus on particular operating interests, such as qualifying for awards or operating on a single band. An example of the latter is the 10-10 International Club (www.ten-ten.org), which is for operators who prefer the 10 meter band — a favorite of low-power and mobile stations and one of the HF bands open to entry-level Technician class licensees. The 10-10 club sponsors several contests every year and offers a set of awards for contacting its members. A similar group, Parks On the Air (POTA: parksontheair.com) sponsors awards for operating from and for contacting stations operating in local and national parks, monuments, recreation areas, trails, and all sorts of fun spots. Some operate casually and some are quite active in their quest to activate or contact as many "counters" as possible.

To find specialty clubs, search your favorite search engine for your area of interest and the phrase *radio club*. Using the search term *10 meter amateur radio club*, for example, turns up a bunch of ham clubs and forums about operating on the 10 meter band.

This section lists only a few examples of the specialized groups you'll find in ham radio; there are many, many more.

On the Air — IOTA, SOTA, and POTA

With the recent introduction of superb quality low-power radios and portable antennas, operation from the field has really become popular. The Islands On the Air program (IOTA; www.iota-world.org) encourages hams to operate from salt-water islands and the less-common rocks and reefs are highly sought after. The enthusiasm extends to dry land as hams clamber to the tops of mountains to put small stations on the air for Summits On the Air (SOTA; www.sota.org.uk). For the 2016 centennial of the U.S. National Parks, the ARRL sponsored the National Parks On the Air award. The program proved so popular that activity evolved into the Parks On the Air (POTA) program I discussed earlier. What do all these programs have in common? That "OTA" or "On the Air," which is what hams like doing best. Why not combine OTA with your favorite outdoor activity? (See Figure 3-3.)

Young Hams — YOTA

Another "on the air" group, YOTA stands for Youngsters On the Air. ("Youngsters" means younger adults and youth in Europe, home of the organization.) The members of YOTA (www.ham-yota.com) are mostly between the ages of 15 and 25, including everyone from high school students through young adults already starting their careers. Founded in Europe, YOTA hosts numerous on-the-air events, such as YOTA Month with special -YOTA call signs, and sponsors the Youth Contesting Program (YCP) where young operators get to operate at world-class contest stations. YOTA also helps radio clubs host gatherings like in Figure 3-4 that are open to young hams from around Europe, Africa, and world-wide. YOTA-like groups are growing here in the Americas and in Asia with a camp planned for the summer of 2021 by Youth On the Air (www.youthontheair.org).

REMEMBER

If you have an interest in scouting, the annual national Jamboree also hosts a ham radio station and radio merit badge programs. Watch for Jamboree On the Air (JOTA: www.scouting.org/international/jota-joti/jota) to learn how to contact the scouts during the event.

Competitive clubs

One type of specialty club is the contest club. Members enjoy participating in competitive on-the-air events known as contests or *radiosport* (see Chapter 11). These clubs challenge one another, sponsor awards and plaques, and generally encourage their members to build up their stations and techniques to become top contest operators.

FIGURE 3-4:
These young hams from Europe, Africa, and Asia gathered in Bulgaria for YOTA Camp 2019.

Contest clubs compete locally, regionally, or nationally depending on the rules of the competition. You can view an extensive list of clubs that compete in the ARRL club competition at www.arrl.org/contest-club-list.

No less competitive than contest operators are the long-distance communications specialists, or *DXers,* who specialize in contacts with places well off the beaten track. The quest to *work 'em all* (contact every country on every ham band) lasts a lifetime, so DXers form clubs to share operating experiences and host traveling hams, fostering international communications and goodwill along the way.

TIP

Many contesters are also DXers, and vice versa. Because of the international nature of DXing and contesting, clubs that specialize in these activities tend to have members sprinkled around the globe. You can find lists of these organizations at www.dailydx.com/clubs.htm.

Handiham

Ham radio provides excellent communication opportunities to people who otherwise find themselves constrained by physical limitations. Handiham (www.handiham.org), founded in 1967, is dedicated to providing tools that make ham radio accessible to people with disabilities of all sorts, helping them turn their disabilities into assets. The website provides links to an extensive set of resources.

Handiham not only helps hams with disabilities reach out to the rest of the world, but also helps its members link up with other members and helpful services.

Even if you're not disabled, Handiham may be a welcome referral for someone you know, or you may want to volunteer your services.

The CQ Communications family of print and digital magazines (see Figure 3-5) provides a lot of good information on ham specialties. CQ focuses on general-interest stories and news, product reviews, and columns on technical and operating interests. CQ Communications also publishes books on a wide range of topics and offers a good-looking yearly calendar each fall.

FIGURE 3-5: CQ Communications publications cover just about every style and interest in ham radio.

Magazine covers courtesy CQ Communications, Inc.

AMSAT

AMSAT (short for Radio AMateur SATellite Corporation, www.amsat.org) is an international organization that helps coordinate satellite launches and oversees the construction of its own satellites. Yes, Virginia, there really are amateur radio satellites whizzing through the heavens! The first one, launched in 1962, sent a Morse code beacon consisting of the letters *HI* (in Morse code speak, "di-di-di-dit, di-dit"), known as "the telegrapher's laugh." The first, OSCAR-1 (Orbiting Satellite Carrying Amateur Radio), was about the size of a briefcase.

The big news in amateur satellites these days are the nano-sized satellites known as *CubeSats* (www.cubesat.org). Pioneered by amateurs, hundreds of these satellites are launched every year by NASA, the ESA, and JAXA. These are often constructed by university student teams (see Figure 3-6) and use amateur radio as their means of communication to send data. Some CubeSats include a repeater or translator so that hams can communicate through them.

FIGURE 3-6: University of Washington students designed and built HuskySat-1, launched in 2019 with a ham radio transponder and science experiments.

Radio operation via satellite is a lot easier than you may think, however, as you can find out in Part 4 of this book. All you need to make contacts through — or with — satellites is some simple equipment. Figure 3-7 shows Sean Kutzko (KX9X) using a handheld radio and a hand-aimed antenna to make contacts through satellite AO-27.

TAPR

Tucson Amateur Packet Radio (TAPR; www.tapr.org) has been instrumental in bringing modern digital communications technology to ham radio. In return, TAPR members created several innovative communication technologies that are now commonplace beyond ham radio, such as the communications system known as *packet radio*, which is widely used in industry and public safety. Recently, TAPR members have been involved in developing a state-of-the-art package of scientific instrumentation called the Personal Space Weather Station. It will provide research quality data and be affordable so that hams can install it at their stations around the world.

FIGURE 3-7:
Sean Kutzko
(KX9X) shows that
satellite
operation can be
easy!

TAPR also sponsors conferences and publications, working with other organizations such as AMSAT and HamSCI (www.hamsci.org) to help develop state-of-the-art digital communications technology for amateur radio. If you have a strong computer or digital technology background, TAPR is likely to have activities that pique your interest.

YLRL

The Young Ladies' Radio League (YLRL; www.ylrl.org) is dedicated to promoting ham radio to women, encouraging them to be active on the air, promoting women's interests within the hobby, and providing a membership organization for female hams.

The organization has chapters in many countries, some of which host conventions, thereby creating opportunities for members to travel.

The YLRL's website provides a list of activities and member services. The organization also has a vigorous awards program; it sponsors on-the-air nets and on-the-air competitions for members throughout the year.

QRP clubs

QRP is ham radio shorthand for *low-power operating*, in which hams use just a few watts of power to span the oceans. Like bicyclists among motorists, QRP enthusiasts emphasize skill and technique, preferring to communicate by using minimal power. They're among the most active designers and builders of any group in ham radio. If you like building your own gear and operating with a minimum of power, check out these clubs and other groups of QRPers.

TIP

One way to find QRP clubs is to visit www.arrl.org/find-a-club and search for clubs with *QRP* in their name.

The largest U.S. QRP club is QRP Amateur Radio Club International, known as QRP ARCI (www.qrparci.org). Its magazine, *QRP Quarterly* (see Figure 3-8), is full of construction projects and operating tips. The club sponsors numerous low-power activities and achievement programs such as the 1000-Miles-Per-Watt award.

TECHNICAL STUFF

Many QRP clubs have worldwide membership. One of my favorites is the British club GQRP. (*G* is a call sign prefix used by stations in England.) You can find the GQRP Club website at www.gqrp.com.

FIGURE 3-8:
The QRP ARCI publishes this excellent quarterly magazine featuring articles on technical topics and operating.

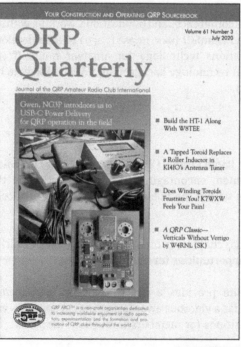

Courtesy QRP ARCI

Attending Hamfests and Conventions

Depending on how much you like collecting and bargaining, I may have saved the best for (almost) last. Despite online retail being everywhere, *hamfests* — ham radio flea markets — continue to be some of the most interesting events in ham radio. Imagine a bazaar crammed with technological artifacts spanning nearly a century, old and new, small and massive, tubes, transistors, computers, antennas, batteries . . . I'm worn out just thinking about it. (I love a good hamfest; can you tell?) Commercial vendors are often present, as well, for you to stock up on needed items without a special trip — or maybe a new radio!

Ham radio conventions have a much broader slate of activities than hamfests do; they may include seminars, speakers, licensing test sessions, and demonstrations of new gear. Some conventions host competitive activities such as foxhunts or direction finding, or they may include a hamfest flea market along with the rest of the functions. Conventions usually have a theme, such as emergency operations, QRP, or digital radio transmissions. Similarly, hamfests may also include training opportunities, license exams, and presentations along with the flea markets.

Finding and preparing for hamfests

In the United States, the best place to find hamfests is ARRL's Hamfests and Conventions Calendar (www.arrl.org/hamfests-and-conventions-calendar). Search for events by location or ARRL Section or Division. The calendar usually lists about 100 hamfests. Most metropolitan areas have several good-size hamfests every year, even in the dead of winter.

TIP

Although as this edition was written, many 2020 gatherings had been cancelled due to the pandemic, they will resume as soon as possible. In the meantime, lots of events are being conducted entirely online through Zoom or other multi-user platforms. This makes it hard to shop for used gear but it allows many more people to attend and take part in the programs. Some vendors are also taking advantage of the online format and creating a virtual "booth" for online sales and demos. Expect this format to continue even after the pandemic subsides.

After you have a hamfest in your sights, set your alarm for early morning, and get ready to be there at the opening bell. Although most are Saturday-only events, more and more are opening on Friday afternoon.

Be sure to bring the following things:

>> **An admission ticket:** You need a ticket, sold at the gate or by advance order through a website or email.

>> **Money:** Take cash. Commercial vendors can take credit cards but most individual sellers don't.

>> **Something to carry your purchases in:** Take along a sturdy cloth sack, backpack, or another type of bag that can tolerate a little grime or dust.

>> **A handheld or mobile radio:** Most hamfests have a talk-in frequency, which is almost always a VHF or UHF repeater. The talk-in operator can give directions and updates on weather or parking.

If you attend with a friend, and both of you take handheld radios, you can share tips about the stuff you find while walking the aisles.

TIP

>> **Water and food:** Don't count on food being available, but the largest hamfests almost always have a food concession. Gourmet food is rarely available; expect the same level of quality that you'd find at a ballpark. Taking along a full water bottle is a good idea.

Buying equipment at hamfests

After parking, waiting, and shuffling along in line, you finally make it inside the gates and you're ready to bargain. No two hamfests are alike, of course, but here are some general guidelines to live by, particularly for hamfest newcomers:

>> If you're new to ham radio, buddy up with a more experienced ham who can steer you around hamfest pitfalls.

>> Most prices are negotiable, especially as the sellers begin to think about packing up, but good deals go quickly. Don't be afraid to make a reasonable offer and bargain!

>> Most vendors aren't interested in trades unless they post a sign that they are willing to consider them.

>> Hamfests are good places to buy accessories for your radio, often for a fraction of the manufacturer's price if they're sold separately from the radio. Commercial vendors of new batteries often have good deals on spare battery packs.

>> Many hamfests have electricity available so that vendors can demonstrate equipment and maybe even a radio test bench. If a seller refuses to demonstrate a supposedly functional piece of gear or won't open a piece of equipment for inspection, you may want to move along.

Unless you really know what you're doing, avoid antique radios. They often have quirks that can make using them a pain or that require impossible-to-get repair parts.

WARNING

>> Don't be afraid to ask what something is. Most of the time, the ham behind the table enjoys telling you about his or her wares, and even if you don't buy it, the discussion may attract another buyer.

Be familiar with the smell of burned or overheated electronics, especially transformers and sealed components. Direct replacements may be difficult to obtain.

>> If you know exactly what you're looking for, check auction and radio swap sites such as ebay.com, eham.net, and qrz.com before and while attending the hamfest with your smartphone. You can get an idea of the going price and average condition, so you're less likely to pay too much.

>> The commercial vendors will sell you accessories, tools, and parts on the spot, which saves you shipping charges.

Don't forget to look under the tables, where you can occasionally find some real treasures!

Finding conventions and conferences

Conventions tend to be more extravagant affairs, held in hotels or conventions centers, that are advertised in ham radio magazines as well as online. The main purposes are programs, speakers, and socializing.

The two largest ham radio conventions are the Hamvention (www.hamvention. com), held in Dayton and Xenia, Ohio in mid-May, and the Internationale Exhibition for Radio Amateurs (www.hamradio-friedrichshafen.de/ham-en), held in Friedrichshafen, Germany, in late June or early July. Dayton regularly draws more than 20,000 hams; Friedrichshafen, nearly that many. Both events have mammoth flea markets, an astounding array of programs, internationally known speakers, and more displays than you can possibly see.

ARRL National and Division conventions (listed on the ARRL website at www.arrl.org/hamfests-and-conventions-calendar) are held all over the United States. Most states and ARRL Sections also host a convention or hamfest. Radio Amateurs of Canada (www.rac.ca) also hosts a national convention every year. These conventions typically attract a few hundred to a few thousand people and are designed to be family friendly. They also provide a venue for specialty groups to host conferences within the overall event. These smaller conferences offer extensive programs on regional disaster and emergency communications, direction finding, QRP, county hunting, wireless networking on ham bands, and so on. Table 3-2 lists a number of the large regional conventions around the U.S.

TABLE 3-2   **Regional General-Interest Events**

Region or State	Name and Website
Florida	Hamcation — Orlando (www.hamcation.com)
California	Pacificon — San Francisco Bay Area (www.pacificon.org)
Oregon/Washington	SeaPac — Seaside, OR (www.seapac.org)
Colorado	HamCon — Denver (www.hamconcolorado.com)
New England	Northeast HamXposition — Boston, MA (www.hamxposition.org)
Missouri	Winterfest — Collinsville, IL (www.winterfest.slsrc.org)
Alabama	Huntsville Hamfest (www.hamfest.org)

Some conventions and conferences emphasize one of ham radio's many facets, such as DXing, VHF and UHF operating, or digital technology. If you're a fan of a certain mode or activity, treating yourself to a weekend convention is a great way to meet hams who share your tastes and to discover more about your interests. Table 3-3 lists a few of the themed conventions held each year. There are many more; just search for your theme and *ham radio conference* or *convention*.

TIP

As mentioned earlier, lots of events are being conducted entirely online through Zoom or other multi-user platforms. This has become so popular that some events are being redesigned to be completely online!

REMEMBER

Even if you can't attend, the "proceedings" (a compilation of material presented at the event) are often available at little or no cost for downloading as PDF or for watching as recorded video.

TIP

Even if the conference isn't a ham radio conference, it might have a group of hams hosting ham radio-themed events or presentations. For example, look for the Ham Radio Social at the IEEE International Microwave Symposium or HamCon at the Hackercon convention. Maker Faires (www.makerfaire.com) frequently feature one or more ham radio groups, too.

TABLE 3-3   Themed Conventions

Theme	Name and Website
DXing and Contesting	International DX Convention — Visalia, CA (www.dxconvention.com)
	IOTA Convention (www.facebook.com/islandsontheair/)
	Contest University (www.contestuniversity.com)
	DX University (www.dxuniversity.com)
	W9DXCC — Chicago, IL (www.w9dxcc.com)
	W4DXCC — Pigeon Forge, TN (www.w4dxcc.com)
VHF, UHF, and Microwave	Southeastern VHF Conference (SVHFS: www.svhfs.org)
	Eastern VHF Conference (NEWSVHF: www.newsvhf.com)
	Central States VHF Conference (CSVHF: www.csvhfs.org)
	Pacific Northwest VHF Conference (PNWVHF: www.pnwvhfs.org)
	Microwave Update (www.facebook.com/MicrowaveUpdate/)
	International EME Conference (search for *International EME Conference* online)
Technical	Digital Communications Conference — hosted by ARRL and TAPR (www.tapr.rg/conferences.html)
	Microhams — Bellvue, WA (www.microhams.com/mhdc)
	SDR Academy — Friedrichshafen, Germany (www.2020.sdra.io)
	GNU Radio (www.gnuradio.org)
QRP (Low Power)	QRP Four Days in May — held at the Hamvention (www.qrparci.org/ www.qrparci.org/fdim)
	OzarkCon (www.ozarkcon.com/index.php)

2

Wading through the Licensing Process

IN THIS PART . . .

Tour the Federal Communications Commission's amateur radio licensing system, see how call signs are structured, and find out how the call sign system works.

Find out about the exam itself, as well as how to study for it by yourself and with a mentor.

See how to locate and register for an exam session, and get ready to take the exam.

While you're waiting for your brand-new call sign, decide whether you want to customize your call.

Chapter **4**

Understanding the Licensing System

L
ike most people, you're probably familiar with the process of getting a license to drive your car, to fish, or to get married, but ham radio licensing is a little different. The process is easy to deal with when you know how it works, however. This chapter explains the FCC licensing system for amateur radio in the United States.

TIP

Hams call their license a "ticket" for good reason since it's a ticket to all of what ham radio has to offer.

Amateur radio is one of many types of *communication services* that use radio. Similar services include broadcast (AM and FM radio, television), public safety (police and fire departments), aviation, and even radar systems (radionavigation).

REMEMBER

When the name of a specific service is capitalized, such as Amateur Radio Service or Citizens Band, that's a formal reference to the set of Federal Communications Commission (FCC) rules for that service. Each service has a different set of rules for its type of operating and use.

To maintain order on the airwaves, the FCC requires that each signal must be transmitted by an authorized station. Stations in all the different services must abide by FCC regulations to obtain (and keep!) their licenses, which give them permission to transmit. That's what a ham license is: It authorizes you to transmit on the frequencies that licensees of the Amateur Radio Service are permitted to use.

Getting Acquainted with the Amateur Service

By international treaty, amateur radio in every country is a licensed service — that is, a government agency has to grant a license for a ham to transmit. Although regulation may seem to be a little quaint, given all the communications gadgets for sale these days, licensing is necessary for a couple of reasons:

>> It allows amateurs to communicate internationally and directly without using any kind of intermediate system that limits their activities.

>> Because of amateur radio's broad capabilities, hams need some technical and regulatory training. This allows hams to share the radio spectrum with other radio services, such as broadcasting.

Unlike some of the other types of radios available to the public, you can't transmit using a ham radio without a license — why? Hams are allowed more flexibility than other services which are restricted to use low power on only a few frequency channels. Other services are limited to specific types of equipment that can't be modified. Hams are encouraged to build our own equipment, operate on any of our many frequencies, and even invent new types of signals and ways of operating! This requires us to have a certain amount of know-how to operate legally and be good citizens on the air. Getting a license requires us to learn some of that know-how just as getting a driver's license requires you to show that you understand the basic rules of the road.

FCC rules

By requiring licensees demonstrate some basic knowledge of radio and the FCC rules, licensing helps ensure that the amateur service makes the best use of its unique citizen access to the airwaves. Licensing sets ham radio apart from the unlicensed services and is recognized in the FCC rules, Part 97. Summarized below, the very first rule states the basis and purpose for ham radio as Rule 97.1:

>> Recognition of ham radio's exceptional capability to provide emergency communications (Rule 97.1(a))

» Promote the amateur's proven ability to advance the state of the radio art (Rule 97.1(b))

» Encourage amateurs to improve their technical and communications skills (Rule 97.1(c))

» Expand the number of trained operators, technicians, and electronics experts (Rule 97.1(d))

» Promote the amateur's unique ability to enhance international goodwill (Rule 97.1(e))

Pretty heady stuff! Ham radio does all these good things in exchange for access to a lot of very useful radio spectrum. You can find all the pertinent rules at wireless at www.fcc.gov/wireless/bureau-divisions/technologies-systems-and-innovation-division/rules-regulations-title-47; click the Part 97 link for the amateur radio rules. (You can get to this page by searching for *fcc rules*, too.) Plain-English discussion of the rules is available in *FCC Rules and Regulations for the Amateur Radio Service*, published by the American Radio Relay League (ARRL; see Chapter 3). The ARRL website also includes an up-to-date copy of the Part 97 rules for amateur radio.

Ham radio frequency allocations

The International Telecommunication Union (ITU), part of the United Nations, provides a forum for countries to create and administer rules for the radio spectrum. This helps keep order between all the services around the world.

The ITU divides the spectrum into small ranges in which specific types of uses occur (see Figure 4-1). These ranges are *frequency allocations*, which hams call *bands*.

The world is divided into three regions, as follows:

» **Region 1:** Europe, Africa, and Russia and North Asia

» **Region 2:** North and South America

» **Region 3:** South Asia, Australia, and most of the Pacific

Within each region, each type of radio service — amateur, military, commercial, and government — is allocated a share of the available frequencies. Luckily for amateurs, most of their allocations are the same in all three regions, so they can talk to one another directly.

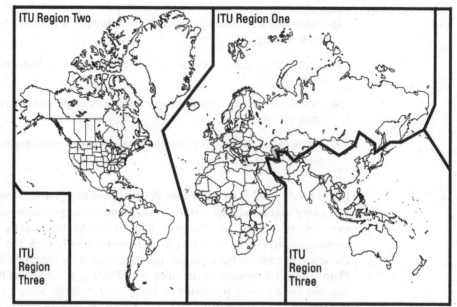

FIGURE 4-1: ITU region map showing the world's three administrative regions for telecommunication.

Figure 4-2 shows the high-frequency (HF) range frequencies (from 3 MHz to 30 MHz). This allocation is very important, particularly on the long-distance bands, where radio signals might propagate all the way around the Earth. Talking to someone in a foreign country is pretty difficult if you can't both use the same frequency.

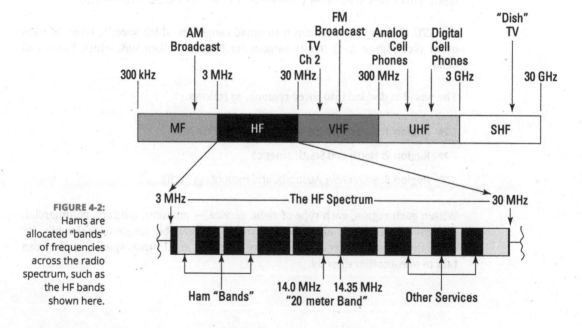

FIGURE 4-2: Hams are allocated "bands" of frequencies across the radio spectrum, such as the HF bands shown here.

To get an idea of the complexity of the allocations, browse to the Region 2 allocation chart from the National Telecommunications and Information Administration (NTIA) at www.ntia.doc.gov/files/ntia/publications/january_2016_spectrum_wall_chart.pdf. (If you have a PDF reader, you can download and display the chart in full color.) The individual colors represent different types of radio services. Each service has a small slice of the spectrum, including amateurs. (Can you find the amateur service on the chart? Hint: It's green.)

Amateurs have small allocations at numerous places in the radio spectrum, and access to those frequencies depends on the class of license you hold (see the next section). The higher your license class, the more frequencies you can use. The "ham bands" are shown in charts you can download at www.arrl.org/graphical-frequency-allocations. The US ham bands are also subdivided by types of signals; I discuss that later.

Learning about Types of Licenses

Three types of licenses are being granted today: Technician, General, and Amateur Extra.

By taking progressively more challenging exams, you gain access to more frequencies and operating privileges, as shown in Table 4-1. After you pass the test for one level of license, called an *element*, you have permanent credit for it as long as you keep your license renewed. This system allows you to progress at your own pace. Your license is good for ten years and you can renew it without taking an exam.

TABLE 4-1 **Privileges by License Class**

License Class	Privileges	Notes
Technician	All amateur privileges above 50 MHz; limited CW, Phone, and Data privileges below 30 MHz	See the nearby Tip about possible FCC rules changes.
General	Technician privileges plus most amateur HF privileges	
Amateur Extra	All amateur privileges	Small exclusive sub-bands are added on 80, 40, 20, and 15 meters.

TIP

At the time this edition was prepared, the FCC was considering several "Rulemaking Petitions" to change how amateur signals were regulated and the frequencies available to Technician licensees. The ARRL and other popular websites will announce any changes, if they occur.

Technician class

All hams start by obtaining a Technician class license, also known as a "Tech" license. A Technician licensee is allowed access to all ham bands with frequencies of 50 MHz or higher. These privileges include operation at the maximum legal power limit and using all types of communications. Tech licensees may also trans- mit using voice, code, and digital signals on part of the 10 meter band and Morse code on some of the HF bands below 30 MHz.

The Technician exam is 35 multiple-choice questions on regulations and technical radio topics. You have to get 26 or more correct to pass.

TIP

Morse code, or "CW" for "continuous wave," was once required for amateur operation below 30 MHz. At the time international treaties were adopted, code was used for a great deal of commercial and military radio traffic was — news, telegrams, ship-to-ship, and ship-to-shore messages. Emergency communications were often in Morse code, too. Back then, using Morse code was a standard radio skill. It's still a very effective part of ham radio. Its efficient use of transmitted power and spectrum space, as well as its musicality and rhythm, make it very popular with hams. It's easy and fun to use, too! Chapter 8 tells you all about Morse code.

General class

After earning the entry-level Technician license, many hams immediately start getting ready to upgrade to a General class license. When you obtain a General class license, you've reached a great milestone. General class licensees have full privileges on nearly all amateur frequencies, with only small portions of some HF bands remaining off limits.

The General class exam, which includes 35 questions (you have to get 26 right to pass), covers some of the same topics as the Technician exam, but in more detail. The exam also introduces new topics that an experienced ham is expected to understand to operate on the traditional shortwave or HF bands.

Amateur Extra class

General class licensees can't use every frequency; the lowest segments of several HF bands are for Amateur Extra class licensees only. These segments are considered to be prime operating territory. If you become interested in contesting, contacting rare foreign stations (*DXing*; see Chapter 11), or just having access to these choice frequencies, you want to get your Amateur Extra license — the top level.

The Amateur Extra exam consists of 50 multiple-choice questions, 37 of which you must answer correctly to pass. The exam covers additional rules and regulations associated with sophisticated operating and several advanced technical topics. Hams who pass the Amateur Extra exam consider their license to be a real achievement. Do you think you can climb to the top rung of the licensing ladder?

Grandfathered classes

The amateur service licensing rules have changed over the years, reducing the number of license classes. Hams who hold licenses in deleted classes may renew those licenses indefinitely, but no new licenses for those classes are being issued.

Two grandfathered license classes remain:

>> **Novice:** The Novice license was introduced in 1951 with a simple 20-question exam and 5-words-per-minute (WPM) code test. A ham with a General class (or higher) license administered the exam. Originally, the license was good for a single year, at which point the Novice upgraded or had to get off the air. These days, the Novice license, like other licenses, has a ten-year term and is renewable. Novices are restricted to segments of the 3.5, 7, 21, 28, 222, and 1296 MHz amateur bands.

>> **Advanced:** Advanced class licensees passed a written exam midway in difficulty between those for the General and Amateur Extra classes. They have access to frequencies beyond the General licensees.

Table 4-2 shows the relative populations of all types of U.S. license holders as of February 2021.

TABLE 4-2

Relative Populations of U.S. License Classes

License Class	Active Licenses	Share of Active Licensees
Tech	395,683	51.0%
General	183,283	23.7%
Amateur Extra	152,643	19.7%
Advanced	37,115	4.7%
Novice	7,254	0.9%
Total	**775,978**	**100%**

Source: www.arrl.org/fcc-license-counts

Getting Licensed

To pass the exam, you'll need to do a little studying and there are plenty of opportunities to practice. Then you'll take your exam, administered by volunteer hams who were also in your shoes once upon a time. After you pass, you'll receive a call sign that is yours and yours alone: your radio name. Ready? Let's go!

Studying the exam questions

ARRL (www.arrl.org) and other organizations publish study guides and manuals, some of which may be available through your local library. Also, online training and study guides are available. All of these use the actual questions that are on the test. Take advantage of these materials, and you'll be ready to pass the exam on test day.

WARNING

The exam questions, the *question pool*, changes every four years. Make sure that you have the current version of study materials, containing the correct questions and any recent changes in rules and regulations. Each license manual should clearly show its "expiration date" on the cover.

TIP

I go over some useful ways to study for the questions in Chapter 5.

Taking your license exam

In the Olden Days, hams took their licensing tests at the nearest FCC office, which could be hundreds of miles away. I vividly remember making long drives to a government office building to take my exams along with dozens of other hams.

Today, although the FCC still grants the licenses, it no longer administers amateur radio licensing examinations. In the United States, these exams are given by *volunteer examiners* (VEs); some VEs even file the results with the FCC. This process enables you to get your license and call sign much faster than in the days when the FCC handled everything on paper.

In-person exam sessions are usually available a short drive away at a club, a school, or even a private home. Beginning in 2020, "fully remote" exam sessions can be conducted online and monitored by teams of VEs. Digital signatures and other state-of-the-art tools are used to insure the quality of the exam process.

TIP

See Chapter 6 for full details on finding an exam session near you or online and taking your test.

Volunteer examiner coordinators

A *volunteer examiner coordinator* (VEC) organization takes responsibility for certifying and coordinating the volunteer examiners (VEs) who run the exam sessions. The VEC also processes FCC-required paperwork generated during the session. Each VEC maintains a list of VEs, upcoming exam sessions, and other resources for ham test-takers. It can also help you renew your license and change your address or name. There may be a small fee charged by the VEC to cover expenses of the volunteers and maintaining the exam program. As of mid-2020, the maximum fee was $15 to take an exam for any of the license tests.

The VEC with the most VEs is the group run by the American Radio Relay League (ARRL-VEC), but 13 other VECs are located around the United States. Some VECs, such as ARRL-VEC and W5YI-VEC, operate nationwide; others work in only a single region.

TIP

You can find VEC groups that conduct exams in your area at www.ncvec.org.

Volunteer examiners

VEs make the system run. Each exam requires three VEs to administer or *proctor* the session, certify that the test was conducted properly, and file the necessary information to process the exam results. VEs are responsible for all aspects of the testing process, including providing the meeting space and announcing the exam sessions. (For remote communities, exam sessions can be administered online by a VE team with local volunteers. This is discussed in Chapter 6.) If they incur any expenses, such as for supplies or facility rental, they're allowed to keep up to $7 per person of the test fee; any left-over fees go to the VEC to cover its expenses.

VEs are authorized to administer license exams for the same class of license they hold themselves or for lower classes. A General class VE, for example, can administer Technician and General exams but not Amateur Extra exams. Extra class licensees can give exams for Amateur Extra, as well.

TECHNICAL STUFF

General, Advanced, and Amateur Extra class licensees can become VEs by contacting one of the VEC organizations and completing whatever qualification process that VEC requires. The ARRL-VEC, for example, provides a booklet on the volunteer licensing system and requires applicants to pass a short exam. VE certification is permanent as long as it is renewed on time with the VEC.

REMEMBER

VEs are amateurs just like you; they do a real service to the amateur community by making the licensing system run smoothly and efficiently. Don't forget to say "Thanks!" at the conclusion of your test session, pass or fail. Better yet, become a VE yourself. It's fun and rewarding. As a VE, I've given dozens of exams to hams as young as 10 years old. You can be a VE for more than one VEC, too!

Receiving Your New Call Sign

Each license that the FCC grants comes with a very special thing: a unique call sign (*call* to hams). Your call sign certifies you have passed the licensing exam and gives you permission to construct and operate a station — a special privilege. If you're a new licensee, you'll get your call sign within a few days of taking your licensing exam.

Your call sign becomes your on-the-air identity, and if you're like most hams, you may change call signs once or twice before settling on the one you want to keep. Sometimes, your call sign starts taking over your off-the-air identity; you may become something like Ward NØAX, using your call sign in place of a last name. (I have to think really hard to remember the last names of some of my ham friends!)

TIP

Hams rarely use the term *handle* to refer to actual names; it's fallen out of favor in recent years. Similarly, they use the term *call letters* only to refer to broadcast-station licenses that have no numbers in them. Picky? Perhaps, but hams are proud of their hard-earned call signs.

Chapter 7 provides full coverage of call signs. In this section, I give you a brief overview.

Call-sign prefixes and suffixes

Each call sign is unique. Many call signs contain NØ or AX, for example, but only one call sign is NØAX. Each letter and number in a call sign is pronounced individually and not as a word — "N zero A X," for example, not "No-axe."

REMEMBER

Hams use the Ø (ALT-0216 on keyboards) symbol to represent the number 0, which is a tradition from the days of teleprinters and typewriters. It avoids confusion between capital-O and zero.

Ham radio call signs around the world are constructed of two parts:

>> **Prefix:** The prefix is composed of one or two letters and one numeral from Ø to 9. (The prefix in my call sign is NØ.) It identifies the country that issued your license and may also specify where you live within that country. For U.S. call signs, the numeral indicates the *call district* of where you lived when your license was issued. (Mine was issued when I lived in St. Louis, Missouri, which is part of the tenth, or Ø, district.)

>> **Suffix:** The *suffix* of a call sign, when added to the prefix, identifies you, the individual license holder. A suffix consists of one to three letters. No punctuation characters are allowed — just letters from A to Z. (The suffix in my call sign is AX.)

The ITU assigns each country a block of prefix character groups to create call signs for all its radio services. All U.S. licensees (not just hams) have call signs that begin with A, K, N, or W. Even broadcast stations have call signs such as KGO or WLS. Most Canadian call signs begin with VE. English call signs may begin with G, M, or 2. Germans use D (for *Deutschland*) followed by any letter; almost all call signs that begin with J are Japanese, and so on. You can find the complete list of ham radio prefix assignments at ac6v.com/prefixes.php.

Class and call sign

As a newly licensed ham, your license class will be reflected in your assigned call sign. When you get your first license, the FCC assigns you the next call sign in the heap for your license class, in much the same way that you're assigned a license plate at the department of motor vehicles. Just as you can request a specialty license plate, you can request a special vanity call sign — within the call sign rules, of course. The higher your license class, the shorter and more distinctive your chosen call sign can be.

Chapter 5

Preparing for Your License Exam

You've decided to take the plunge and get your ham radio license. Congratulations! Although you can't just run down to the store, buy your gear, and fire it up, becoming licensed isn't that hard. A lot of resources are available to prepare you for the ham radio exam. This chapter gives you some pointers on how best to prepare so that you will enjoy studying and do well at test time. (Exams and exam sessions are also referred to as *tests* or *test sessions*. I use mostly *exam* in this chapter.)

This chapter doesn't teach the answers to the exam's specific questions — that's the job of study guides and manuals. There are quite a few questions in the exam question pools! Use one of the several excellent print and online study resources that review each individual question. This book will help you through the process of studying and passing your exam.

Getting a Grip on the Technician Exam

To do the best job of studying, you need to know just what the exam consists of. The exams for all license classes are multiple-choice; you won't find any essay questions. Some questions refer to a simple diagram. No oral questions of any kind are used; no one asks you to recite the standard phonetic alphabet or sing a song about Ohm's Law.

The exam for each license class is called an *element*. The written exams for Technician, General, and Amateur Extra licenses (see Chapter 4) are Elements 2, 3, and 4, respectively. (Element 1 was the Morse code exam, which has been dropped.)

Your studies will focus on the *question pool*, the complete set of actual questions used on the exam. The questions are available to help you study. The exam that you'll take is made up of a selection of questions from that pool.

The exam covers four basic areas:

>> **Rules & Regulations:** Important rules of the road that you have to know to operate legally including some important definitions.

>> **Operating:** Basic procedures and conventions that hams follow on the air to be effective. In essence, you'll learn "radio manners."

>> **Basic Electronics:** Elementary concepts about radio waves and electronic components, with some basic math involved. There are some equipment questions, too.

>> **RF Safety:** Questions about how to transmit safely.

The exam must include a certain number of questions from each area; questions are selected randomly from those areas. The Technician and General exams have 35 questions; the Amateur Extra has 50. If you answer at least three-quarters of the questions correctly, you pass.

WARNING

Because the exam questions are public, you'll experience a strong temptation to memorize the questions and answers. Don't! Take the time to understand as much of the material as you can, memorizing only what you must. After you do get your license, you'll find that studying pays off when you start operating.

Finding Study Resources

If you're ready to start studying, what do you study? Fortunately for you, the aspiring ham, numerous study references are available to fit every taste and capability. Common study aids include classes, books, software, videos, and online help. Many aids are listed in the Getting Licensed area of the ARRL's website (www. arrl.org/getting-licensed) or you can try some of the resources listed below, including special resources for the disabled. If you need personal help, give the

ARRL's New Ham Help Desk a call at 1-800-32-NEWHAM or send email to newham@arrl.org.

REMEMBER

Exam questions and regulations change once every four years for each class of license. The latest changes in the Technician class questions, for example, took effect July 1, 2018, so the next set of questions will be apply starting on July 1, 2022. Be sure that any study materials you purchase support the latest updates. For the dates of the current question pools, see www.arrl.org/question-pools or www.ncvec.org.

Licensing classes

If you learn better with a group of other students, you'll find classes beneficial. You can find classes in several ways:

>> **Checking with your radio club:** You can take classes sponsored by the club. If you don't see the class you want, contact the club through its website or social media pages, and ask about classes. To find a club in your area, turn to Chapter 3.

>> **Looking for upcoming exams to be held in your area:** The American Radio Relay League (ARRL) has a search engine devoted to upcoming exam sessions at www.arrl.org/exam_sessions/search, as do the W5YI VEC (www.w5yi.org) and Laurel VEC (www.laurelvec.com). You can check other VEC sessions by going to the national VEC website (www.ncvec.org) and clicking the FCC Certified VECs link.

Get in touch with the exam's contact liaison and ask about licensing classes. Because exams are often given at the conclusion of class sessions, contact liaisons are frequently class instructors themselves.

>> **Asking at a ham radio or electronics store:** If a ham radio store is in your vicinity the staffers there usually know where classes are being held.

Businesses that sell electronics supplies or kits to individuals may also know about classes. In a pinch, you can do a web search for *ham radio class* or *radio licensing class* (or close variations) and your town or region.

TIP

Maker and robotics groups often include hams as members. Ask around and see if they can point you in the right direction.

>> **Community colleges:** It is increasingly common for local colleges to offer ham radio classes sponsored by a local club as part of an adult education or hobby instruction program. Even if the classes aren't being held right away, there will often be information to help you contact the previous instructor for more information about the schedule.

TIP

Because of the pandemic, in-person classes may have shifted to an online format. This means you can take classes from almost any organization since you don't have to travel to attend. Online classes are likely to continue since they have been very successful.

Other options for finding classes include local disaster-preparedness organizations such as CERT (Community Emergency Response Teams, sponsored by the Federal Emergency Management Agency, or FEMA); the National Weather Service's SKYWARN instructors; and public-safety agencies such as police and fire departments. By asking around, you can usually turn up a reference to someone who's involved with ham radio instruction.

Occasionally, classes are advertised that take you from interested party to successful exam-taker in a single weekend. The Technician exam is simple enough that a focused, concerted effort over a couple of days can cram enough material into your brain for you to pass. The good part about these sessions is that by committing a single weekend, you can pass your exam on Sunday and find your new call sign in the Federal Communications Commission (FCC) database right away. For busy folks or those who are in a hurry, this timesavings is a tremendous incentive.

WARNING

Remember when you crammed for a final exam overnight and the minute after you took the exam forgot everything that was on it? The same phenomenon applies to a weekend course. A lot of information that you memorize in a short period will fade quickly. In two days, you can't really absorb the material well enough to understand it. You'll use everything you learn in your studies later in real life. If you have time to take a weekly course, that's the better option.

TIP

After the license study course, the same sponsor may also offer "new ham training" courses or presentations that help you use your new license. These are well worth your time, will help you learn to operate effectively, and you'll have an opportunity to make new friends, as well!

Books, websites, and videos

You have a variety of options to help you study on your own. If you can, look at a few of the different resources to find one that seems right for your style. Here are a few of the more popular choices:

>> **Study guides:** The best-known guide for licensing studies is the ARRL's *Ham Radio License Manual*. Aimed at the person studying for a Technician exam, it goes well beyond presenting just the questions from the question pool;

it teaches the why and how of the material. A great companion to the manual is *ARRL's Tech Q&A* which gives the correct answer to each exam question. Both books are available in the ARRL Store (www.arrl.org) in the Licensing, Education & Training section and at numerous retail outlets.

- Gordon West (WB6NOA) has written a series of licensing guides and audio courses for all three license classes. These guides focus tightly on the question pool in question-and-answer format and are geared to students who want to pass the exam quickly without the more extensive background of the ARRL books. West's books are available at www.masterpublishing.com/radio.html, www.w5yi.org, and various retail outlets.

- Dan Romanchik (KB6NU) also writes a series of popular study guide books that present the material in a very focused format. Dan has also written a great guide to learning Morse code and blogs extensively on a variety of ham radio topics at www.kb6nu.com.

- A team led by Stu Turner (WØSTU) offers a two-license package of study material for the Technician and General licenses (www.hamradioschool.com). The book includes graphics and explanations for each question.

>> **Websites:** Here are a few good choices:

- HamTestOnline (www.hamradiolicenseexam.com/index.html) offers online tutoring and training material that you can access through your web browser.

- Hamstudy (hamstudy.org) is a practice guide in a flashcard format with apps for Android or iOS.

>> **Videos:** Most license study videos are on YouTube (www.youtube.com). Search for *ham radio technician class* or *ham radio technician study guide*. There are quite a few videos to choose from. The "crash courses" are designed to help you pass the exam quickly. Others are more in-depth. Almost all are free and many are given by clubs. After you've passed the exam, you may find the same instructors giving video lessons on other topics that you will find useful.

TIP

If you buddy up with a study partner, studying is much easier. Having a partner helps you both stick with it. Each of you will find different things are easy or difficult, so you can help each other get over the rough spots. Best of all, you can celebrate passing together.

TIP

When you find that a topic is difficult to understand, try different explanations from two or three different sources. Seeing the topic from a few different perspectives often leads to an *aha!* moment of understanding.

Online practice exams

Online practice exams can be particularly useful. When tutoring students, I urge them to practice the online exams repeatedly. Because the online exams use the actual test questions, they're almost like the real thing. Practicing with them reduces your nervousness and gets you used to the actual format.

The sites score your exams and let you know which of the study areas need more work. When you can pass the online exams by a comfortable margin every time, you'll do well in the actual session. You can find lots of online exams by doing an online search for *ham radio practice exams*. Practice with exams from two or three different sites to get a little practice with more than one style of exam. The ARRL provides a free online app, *Exam Review for Ham Radio* (arrlexamreview.appspot.com), which generates practice exams online.

TIP

How do you know when it's time to stop studying and take the actual exam? Take the practice exams until you consistently score 80 percent or higher. Also, make sure that you're practicing with a random selection of questions; you shouldn't see the same questions each time. Passing the practice exams with a little safety margin will give you the confidence to sign up for your exam session.

Locating Your Mentor

Studying for your license may take you on a journey into unfamiliar territory. You can easily get stuck at some point — maybe on a technical concept or on a rule that isn't easy to understand.

As in many similar situations, the best way to solve a problem is to call on a mentor — a more experienced person who can help you over the rough spots. They're called *Elmers*, as explained in Chapter 3, and having someone to fill that role is important at this stage. (I discuss my own mentoring experience in the sidebar at the end of this chapter.)

A lot of potential mentors are out there in Ham Radio Land. You can find them in the following places:

>> **Radio clubs:** Clubs welcome visitors and often have an introduction session during meetings. This session gives you an opportunity to say something like this: "Hi. My name is so-and-so. I don't have a license yet, but I'm studying and might need some help." Chances are that you'll get several offers of assistance and referrals to local experts or classes. (Find out more about clubs in Chapter 3.)

- >> **Online:** Although the best way to get assistance is in person, several popular ham radio websites have forums for asking questions. The eHam.net site, for example, has a licensing forum (from the home page, look in the Community tab and select Forums). So does QRZ.com in its Community Help Forum (go to the Forums index and look for Becoming an Amateur Radio Operator).

- >> **In your community:** Many of today's hams find mentors by looking around their own neighborhoods. A ham with a tower and antenna may live near you, or you might see a car with a ham radio call sign on the license plate. If you get the opportunity, introduce yourself, and explain that you're studying for a license. The person you're talking to probably also needed a mentor way back when and can give you a hand or help you find one.

REMEMBER

After you get your license, you're in an excellent position to help other newcomers, because you know exactly how *you* felt at the start of *your* journey. Even if you're just one step ahead of the person who's asking the questions, you can be a mentor. Some hams enjoy mentoring so much that they devote much of their ham radio time to the job. You won't find a higher compliment in ham radio than being called "my Elmer."

MY MENTOR EXPERIENCE

When I started in ham radio, my mentor was Bill (then WNØDYV, now KJ7PC), a fellow high-school student who had been licensed for a year or so. I wasn't having any trouble with the electronics but I sure needed a hand with the Morse code and some of the rules. I spent every Thursday over at Bill's house, practicing Morse code *(pounding brass)* and learning to recognize my personal-nemesis characters: D, U, G, and W. Without his help, my path to getting licensed would have been considerably longer. Thanks, Bill!

Since getting my license, I've required the assistance of several other mentors as I entered new aspects of ham radio. If you can count on the help of a mentor, your road to a license will be much smoother.

Chapter **6**

Taking the Exam

After your diligent studies, you find yourself easily passing the practice exams online by a comfortable margin. Now you're ready to — drumroll, please! — take the exam.

If you're part of a class or study group, the exam may already be part of the program. In this case, you're all set; just show up on time. Skip to the last section of this chapter.

If you're studying on your own, however, read on. This chapter tells you where and when you can take the exam.

REMEMBER

In December of 2020, the FCC released a notice that it would charge a $35 fee for new licenses, license renewals, and applications for a vanity call sign. There will be no fee for administrative updates, such as a change of mailing or email address. The starting date for collecting the fees has not been established as of February 2021 but it will be announced at least 30 days in advance. This fee is in addition to session fees charged by the sponsoring VEC.

Types of Exams

Until recently, nearly all exams were conducted in-person, using traditional paper-and-pencil forms, test booklets, and answer sheets. Then came the pandemic. Keeping lots of nervous test-takers and the test administrators in a small room just isn't safe! This was the incentive to create "remote sessions." You can now take the test from home via the Internet, supervised and monitored by VEs. More and more VECs are offering remote sessions, so be sure to check the VEC websites to see if a remote session would work for you.

Public in-person exams

Most exam sessions are open to the public and are held at schools, churches, and other public meeting places. Nearly all sessions are open to walk-ins — that is, you can just show up unannounced, pay your session fee, and take the exam — but some require an appointment or reservation.

After you find a session, contact the hosts or sponsors to let them know that you'll be attending the session and what elements you want to take. Many of them have online registration forms. If you require special assistance to take an exam, definitely contact the sponsor first.

TIP

Registering for an in-person session ahead of time isn't just good manners; it also can alert you to time or location changes. Contact the session's sponsor to confirm the date, get directions, and find out if there is a test fee.

Remote exams

A remote exam session is conducted using commercial online examination software and teleconferencing software like Zoom. Test administrators monitor you using your PC camera and microphone while you answer the test questions. The sponsor will provide all the necessary information for you to take the exam.

The Greater Los Angeles Amateur Radio Group (GLAARG: glaarg.org/ve_sessions) has a good description of the process on their website, including an extensive FAQ to answer common questions. If you can participate in regular teleconferencing, you can take the exam in a remote session!

Remote sessions are generally open to the public (some are prearranged for a specific group) but you have to register and provide some credentials to participate. Along with the camera and microphone, a web browser (usually Chrome) is needed for the exam software. Test fees are paid online, as well. The test administrators will need to view your photo ID to verify your identity. If you can't conduct the exam session from home, there are often in-person sessions that provide computers for you to take the exam online.

Exams at events

Exam sessions are common at public events such as hamfests and conventions. (See Chapter 3 for information on these events.) Whether conducted as in-person or remote sessions, they can attract dozens of examinees and often fill up quickly. Some exams are given more than once throughout the day, so you can take more than one exam or spend time enjoying the event.

Under FCC rules, you're not required to pay an attendance fee for the event if you're going just to take a license exam, but you may need to pay an entry fee or buy a ticket for the event. Don't be afraid to call ahead and ask, or check the event's website.

If you attend an event-sponsored exam that doesn't offer online registration, it's a good idea to get to the site early to register. Multiple sessions may be offered, and the exams for different elements may be given only at specific times.

Exam sessions in homes and online

Small sessions may be held in private residences, especially in rural areas and small towns. When I was a VE living in a small town, many new hams in my community passed their exams while sitting at my kitchen table.

You'll be a guest in someone's home, so act accordingly.

If you'll be taking your exam at a private residence, call ahead to ensure that there'll be room for you at the session and that the VE is prepared for the exam you want to take.

Finding an Exam Session

Fortunately for you, finding a schedule of exam sessions in your area is pretty easy. The Volunteer Examiner Coordinators (VECs) are listed on the National Council of VECs (NCVEC; www.ncvec.org) website. (For details on the connection between VECs and licensing exams, see Chapter 4.)

All of the VECs have websites showing how to contact them and find their exam sessions. Here are some of the larger VECs with the link to their exam session information:

>> **ARRL VEC exams:** The American Radio Relay League (ARRL) has a VEC that operates nationwide. Search for exams based on your zip code at www.arrl.org/find-an-amateur-radio-license-exam-session.

>> **W5YI VEC exams:** Like ARRL, W5YI (founded by Fred Maia (W5YI)) has a nationwide VEC. You can find a list of certified examiners to contact at www.w5yi-vec.org/exam_locations_ama.php.

>> **Greater Los Angeles Area Amateur Radio Group (GLAARG) exams** (glaarg.org/ve_sessions): This group is the first to conduct large-scale remote sessions using teleconferencing and online exam software (glaarg.org/remote-sessions). They also conduct in-person exams.

>> **Laurel VEC exams:** The Laurel Amateur Radio Club is based in Maryland but has teams of VEs all across the country. Their exam sessions are listed at www.laurelvec.com.

TIP

If you can't find an exam session that's convenient, contact the VECs and ask for help. The mission of VECs is to help prospective amateurs get licensed. No matter where you live, these organizations can put you in touch with volunteer examiners (VEs) so you can take your exam.

Registering with the Universal Licensing System (ULS)

The FCC's online licensee registration system is called the Universal Licensing System (ULS). The ULS allows you to process renewals, make address changes, and use other simple services to maintain and modify your license. Your license information is managed by the FCC's CORES database, which stands for COmmission REgistration System.

Before you take the exam, you must use the ULS to register with the FCC and create your own Federal Registration Number (FRN). Using your FRN on the license exam forms makes sure your successful test results are automatically credited to you along with your new call sign. You must have an FRN to access your stored license information.

You must have your FRN before you take the exam and submit your license application — it only takes a few minutes and there is no charge for registering. If you do not have an FRN, you may be able to get help with registration from the VEs administering your exam session.

REMEMBER

The FCC regularly updates its website, so the instructions in this section might not be up to date. (The instructions were current as of early 2021.) Check the VEC or ARRL websites if you need help using the ULS.

To register and get your FRN, follow these steps:

1. **Go to the ULS home page:** `wireless.fcc.gov/uls`.

    The ULS web page appears. Scroll down and click the New User Registration link.

2. **Click the Register button.**

    An options page appears.

3. **Select Register As an Individual and then click the Continue button.**

4. **Enter your name and address in the page that appears.**

    Any field marked with an asterisk is a required field. Your telephone number, fax number, and email address are optional.

    Everything you enter on this page *except* your Social Security number, phone number, and email address will be available for public inspection.

REMEMBER

5. **If you're registering for the first time, enter your Social Security number.**

    You're required to provide this information (or give a reason why you can't). Enter the number without any spaces, hyphens, or periods (*1234567890*).

TIP

Ignore any prompts or windows asking for a Sub-Group Identification Number (SGIN), which is used by managers of large communications services that have many call signs. You don't need an SGIN.

6. **At the bottom of the window, enter a password of 6 to 15 characters (or have the system pick one for you); then enter it again in the Re-Enter Password box.**

WARNING

Don't use your call sign (or any part of it) as your password, because an unauthorized person would try that immediately.

7. **In the Hint box, enter a password reminder.**

If you ever forget your password and want the FCC to tell you what it is, this hint verifies that you're you — not someone else. You can enter any word or words that fit in the box.

8. **Click the Submit button.**

The ULS system processes your entries and displays a page that lists any errors you made, such as omitting a required item or entering the wrong type of information in a particular field.

9. **If necessary, correct any errors and click Submit again.**

The system displays a confirmation form containing your licensee information and FRN.

REMEMBER

Print the confirmation page, add your password and hint, and keep a copy in a safe place in case you ever forget your password. You won't access the CORES database often, so it's a good idea to keep the confirmation page with a printed copy of your license.

You're now registered with the FCC! Have your FRN with you when you take the exam. By entering your FRN on the NCVEC Form 605, your FRN and your call sign will be registered together ("associated") with the FCC. Numerous services are now available to you for free, such as renewing your license or making an address change.

Getting to Exam Day

In the so-called Good Old Days, exam sessions for the higher-class licenses were conducted in federal office buildings by FCC employees. I vividly recall standing in line with dozens of other hams, waiting for my shot at a new license. Some of us drove for hours to reach the FCC office, nervously reviewing the material or listening to Morse code tapes between swallows of coffee. Inside, a steely-eyed examiner watched as we scratched out the answers.

These days, exams are certainly more convenient and the examiners are friendlier, but you'll still have some nervous anticipation as the day arrives. The best way to do well, of course, is to be prepared — for all aspects of the exams, not just the questions. The more you know, the less you have to worry about.

TIP

For some advice on getting ready for the big day, see the nearby sidebar "Taming the test tiger."

What to have with you

Be sure to have these items with you at the exam session, whether you're licensed or not:

>> Two forms of identification, including at least one photo ID, such as a driver's license or employer's identity card

>> Your FCC Federal Registration Number (FRN)

>> A couple of pencils

>> A calculator

>> (Optional) Scratch paper (but it must be completely blank)

TIP

If you don't have these forms of ID or an FRN, contact the exam session administrators for instructions.

If you already have a license and are taking an exam to upgrade to a higher class, you also need to have the following items. For a remote session, see the VEC website for instructions about these documents.

>> Your current original license and a photocopy

>> Any original Certificate of Successful Completion of Examination (CSCE) you have and a photocopy (see Chapter 7 for a sample CSCE form)

Note: The CSCE is your record of having passed an exam for one or more of the license elements. If you've just passed the Technician exam (Element 2), you have to wait for the FCC to grant you a call sign before you get on the air. For any other license changes, the CSCE allows you to operate immediately with your new privileges.

REMEMBER

If you have a disability and need to use an online device or software to assist you, coordinate their use with the VEC sponsoring the session.

What to expect

Each exam session involves three basic steps:

1. **Register for your exam.**

    Provide your name, address, and call sign if you already have one. The session administrators review your identification and documents. Finally, pay your exam fee. (As of mid-2020, VECs can charge up to $15.)

TAMING THE TEST TIGER

Follow these surefire pointers to turn that tiger of a test into a pussycat by keeping your thinker in top shape:

Do:

- Wear a couple of layers of clothing so you can be comfortable, whatever the room temperature.

- Visit the restroom before the session starts.

- Follow the directions for completing the identification part of your answer sheet or test software, even though you may be anxious to start right away.

- If you encounter a difficult question, study it and then move on to the next one. When you come back to the question later, it may be crystal-clear.

- If you are using a paper answer sheet, completely erase any wrong answer or indicate clearly that you made a change.

- If you really don't know the answer, guessing will not hurt your score. There is no extra penalty for a wrong answer.

- Double-check your answers before completing the exam to make sure that you marked or selected the answers you wanted.

Don't:

- Take the exam when you're hungry, sleepy, or thirsty.

- Drink extra coffee or tea.

- Change an answer unless you're quite sure.

- Leave a question unanswered (see the note above about guessing). If you have absolutely no idea about the answer, guess; you have a one-in-four chance of getting it right. If you leave the answer blank, you have no chance of getting the right answer.

- Rush. Remember to breathe; take a minute to stretch, roll your head, or flex your arms and legs. The exam isn't a race. Take your time.

2. **Take the exam.**

   When you start depends on how many people have signed up ahead of you and how many different classes of exam are being given. In a remote or small in-person session, you may start immediately; in a larger session, you may have to wait a while for your turn.

The exam questions are multiple-choice. For paper-based exams, you will receive a pamphlet containing the exam questions and an answer sheet for recording your choices. Follow the instructions for filling out the answer sheet and get ready to start. Remote sessions give exams using test administration software as described by the sponsoring VEC on their website.

Each exam takes 15 to 45 minutes. The session may be organized so that everyone starts and stops together, or exams may be taken continuously. The VEs will explain the process for your session.

3. **Complete your FCC and VEC forms (which I talk about in Chapter 7).**

What to do after the exam

When you're done with your exam, follow the administrator's instructions for completing the exam, sit back, and try to exhale! The VEs will now grade your paper answer sheet or the exam software will display your results. At least three VEs also verify the grades on all exams. Passing requires a score of 75 percent or better. (That's 26 questions on the Technician and General exams and 37 on the Extra.)

In all probability, because you studied hard and seriously, you'll get a big smile and a thumbs-up from the graders. Way to go! You can finally, truly relax and move on to the next stage.

If you didn't pass this time, don't be disheartened. Many sessions allow you to take a different version of the exam, if you want. Even if you don't take the second-chance exam again right away, at least you know the ropes of a session now, and you'll be more relaxed next time.

REMEMBER

Don't let a failure stop you. Many hams had to make more than one attempt to pass their license exam, but they're on the air today!

Chapter **7**

Obtaining Your License and Call Sign

After you pass your exam (see Chapter 6), only a small matter of paperwork separates you from your new license. The exam–session volunteers help you complete everything correctly and even send your application to the Federal Communications Commission (FCC). Some VECs can even access the FCC database directly!

You still need to understand what you're filling out, though; that's what I cover in this chapter. Fill your paperwork out correctly, and you won't delay the process of getting your call sign.

REMEMBER

When I refer in this chapter to "paperwork," I mean any necessary form, whether printed, online, or as a computer file. The FCC calls the forms you submit for a new license an "application."

Completing Your Licensing Paperwork

After you successfully complete the exam, you need to fill out two forms:

» **Certificate of Successful Completion of Examination (CSCE):** Figure 7-1 shows the ARRL-VEC CSCE. (As I discuss in Chapter 6, this certificate is issued by the American Radio Relay League's Volunteer Examiner Coordinator.) The VEC and FCC use the CSCE as a check against the exam session records. Your copy of the completed form documents your results. If you take the next level of exam before you receive your license or upgrade from the FCC, the CSCE enables the session sponsors to give you credit for the exam you passed.

REMEMBER

Keep your copy of the CSCE until the FCC sends you a new license or records the change in its database. You'll probably want to hang on to it as a record of your achievement.

» **NCVEC Form 605:** NCVEC Form 605 (shown in Figure 7-2) must be filed with the FCC to begin the process of granting your new license. You use this form to get a new license, upgrade to a higher class, renew your license, change your name or address, or pick a new call sign. You can also submit name, address, or call sign changes directly to the FCC by mail or online.

FIGURE 7-1: The ARRL VEC version of a CSCE form.

American Radio Relay League VEC
Certificate of Successful Completion of Examination ARRL *The national association for* AMATEUR RADIO®

NOTE TO VE TEAM: COMPLETELY CROSS OUT ALL BOXES BELOW THAT DO NOT APPLY TO THIS CANDIDATE.

Test Site (City/State): _____ Test Date: _____

The applicant named herein has presented valid proof for the exam element credit(s) indicated below.

CREDIT for ELEMENTS PASSED VALID FOR 365 DAYS
You have passed the written element(s) indicated at right. You will be given credit for the appropriate examination element(s), for up to 365 days from the date shown at the top of this certificate.

Element 3 credit
Element 4 credit

EXAM ELEMENTS EARNED	
Passed written Element 2	
Passed written Element 3	
Passed written Element 4	

LICENSE UPGRADE NOTICE
If you also hold a valid FCC-issued Amateur radio license grant, this Certificate validates temporary operation with the operating privileges of your new operator class (see Section 97.9[b] of the FCC's Rules) until you are granted the license for your new operator class, or for a period of 365 days from the test date stated above on this certificate, whichever comes first.

NEW LICENSE CLASS EARNED	
TECHNICIAN	
GENERAL	
EXTRA	
NONE	

LICENSE STATUS INQUIRIES
You can find out if a new license or upgrade has been "granted" by the FCC. For on-line inquiries see the FCC Web at http://wireless.fcc.gov/uls/ ("Click on Search Licenses" button), or see the ARRL Web at http://www.arrl.org/fcc/search; or by calling FCC toll free at 888-225-5322; or by calling the ARRL at 1-860-594-0300 during business hours. **Allow 15 days from the test date before calling.**

THIS CERTIFICATE IS NOT A LICENSE, PERMIT, OR ANY OTHER KIND OF OPERATING AUTHORITY IN AND OF ITSELF. THE ELEMENT CREDITS AND/OR OPERATING PRIVILEGES THAT MAY BE INDICATED IN THE LICENSE UPGRADE NOTICE ARE VALID FOR 365 DAYS FROM THE TEST DATE. THE HOLDER NAMED HEREON MUST ALSO HAVE BEEN GRANTED AN AMATEUR RADIO LICENSE ISSUED BY THE FCC TO OPERATE ON THE AIR.

Candidate's Signature _____ Call Sign _____ (If none, write none)

Candidate's Name _____

Address _____

City _____ State _____ ZIP _____

VE #1 _____
Signature   Call Sign

VE #2 _____
Signature   Call Sign

VE #3 _____
Signature   Call Sign

COPIES: WHITE–Candidate, YELLOW–VE Team, PINK–ARRL VEC
MVE 03/2020

TIP

As you may be asked in an exam question, you're required to maintain your current mailing address on file with the FCC. If your mailing address changes, keep the FCC database up to date using the ULS (see Chapter 6). Mail sent to the address in the FCC database should get to you in ten days or less. If the FCC is unable to reach you by mail, your license may be revoked.

NCVEC QUICK-FORM 605 APPLICATION
AMATEUR OPERATOR/PRIMARY STATION LICENSE

SECTION 1 - TO BE COMPLETED BY APPLICANT   PLEASE PRINT LEGIBLY!

PRINT LAST NAME	SUFFIX (Jr., Sr.)	FIRST NAME	M.I.	STATION CALL SIGN (IF ANY)

MAILING ADDRESS (Number and Street or P.O. Box)

FEDERAL REGISTRATION NUMBER (FRN)

CITY	STATE CODE	ZIP CODE (5 or 9 Numbers)

DAYTIME TELEPHONE NUMBER (Include Area Code)	E-MAIL ADDRESS (MANDATORY TO RECEIVE LICENSE NOTIFICATION EMAIL FROM FCC)

Basic Qualification Question: *ANSWER REQUIRED IN ORDER TO PROCESS YOUR APPLICATION*

Has the Applicant or any party to this application, or any party directly or indirectly controlling the Applicant, ever been convicted of a felony by any state or federal court? ☐ YES ☐ NO

If "YES", see "FCC BASIC QUALIFICATION QUESTION INSTRUCTIONS AND PROCEDURES" on the back of this form.

I HEREBY APPLY FOR (Make an X in the appropriate box(es)):

☐ **EXAMINATION** for a new license grant

☐ **EXAMINATION** for **upgrade** of my license class

☐ **CHANGE** my **name** on my license to my new name

Former Name: _____
    (Last name) (Suffix)   (First name)   (MI)

☐ **CHANGE** my mailing address to **above** address

☐ **CHANGE** my station **call sign** systematically

Applicant's Initials: To confirm _____

☐ **RENEWAL** of my license grant

Exp. Date: _____

Do you have another license application on file with the FCC which has not been acted upon?	PURPOSE OF OTHER APPLICATION	PENDING FILE NUMBER (FOR VEC USE ONLY)

I certify that:

- I waive any claim to the use of any particular frequency regardless of prior use by license or otherwise;
- All statements and attachments are true, complete and correct to the best of my knowledge and belief and are made in good faith;
- I am not a representative of a foreign government;
- I am not subject to a denial of Federal benefits pursuant to Section 5301 of the Anti-Drug Abuse Act of 1988, 21 U.S.C. § 862;
- The construction of my station will NOT be an action which is likely to have a significant environmental effect (See 47 CFR Sections 1.1301-1.1319 and Section 97.13(a));
- I have read and WILL COMPLY with Section 97.13(c) of the Commission's Rules regarding RADIOFREQUENCY (RF) RADIATION SAFETY and the amateur service section of OST/OET Bulletin Number 65.

Signature of Applicant: *By typing your name below, you are signing this form electronically. Understand that your electronic signature is legally binding, as if you had physically signed the document by hand.*

X_____   Date Signed: _____

SECTION 2 - TO BE COMPLETED BY ALL ADMINISTERING VEs

Applicant is qualified for operator license class:

☐ **NO NEW LICENSE OR UPGRADE WAS EARNED**

☐ **TECHNICIAN**   Element 2

☐ **GENERAL**   Elements 2 and 3

☐ **AMATEUR EXTRA**   Elements 2, 3 and 4

DATE OF EXAMINATION SESSION

EXAMINATION SESSION LOCATION

VEC ORGANIZATION

VEC RECEIPT DATE

I CERTIFY THAT I HAVE COMPLIED WITH THE ADMINISTERING VE REQUIRMENTS IN PART 97 OF THE COMMISSION'S RULES AND WITH THE INSTRUCTIONS PROVIDED BY THE COORDINATING VEC AND THE FCC.

By typing your name below, you are signing this form electronically. Understand that your electronic signature is legally binding, as if you had physically signed the document by hand.

1st VEs NAME (Print First, MI, Last, Suffix)	VEs STATION CALL SIGN	VEs SIGNATURE (Must match name)	DATE SIGNED
2nd VEs NAME (Print First, MI, Last, Suffix)	VEs STATION CALL SIGN	VEs SIGNATURE (Must match name)	DATE SIGNED
3rd VEs NAME (Print First, MI, Last, Suffix)	VEs STATION CALL SIGN	VEs SIGNATURE (Must match name)	DATE SIGNED

DO NOT SEND THIS FORM TO FCC – THIS IS NOT AN FCC FORM.
IF THIS FORM IS SENT TO FCC, FCC WILL RETURN IT TO YOU WITHOUT ACTION.

NCVEC FORM 605 - September 2017
(Updated March 2018)
FOR VE/VEC USE ONLY - Page 1

FIGURE 7-2:
The NCVEC Form 605 that is filed with the FCC. This version was released in September 2017.

The volunteer examiners (VEs) who administer the exam session send your completed CSCE and NCVEC Form 605 to their certifying VEC organization. Asking your examiners about the average wait before the FCC updates your information in its database is a good idea. It should take no more than two to five business days before your call sign is assigned. Applications with VECs that have direct FCC database access will be processed faster.

Finding Your Call Sign

If you are upgrading to a higher-class license, you already have a call sign, but you'll need to change it temporarily (see "Identifying with your new privileges," later in this chapter).

If you're a first-time licensee, you can begin watching the FCC database for your new call sign to appear after you complete the session and your paperwork is sent to the VEC. The next two sections walk you through the search process.

REMEMBER

As soon as your call sign appears in the FCC database with your name beside it, you can get on the air.

Searching the ULS database

You can search for your call sign in the FCC's Universal Licensing System (ULS) using the Federal Registration Number (FRN) that you created before taking the license exam. (I outline the process of registering for your own FRN in Chapter 6.)

REMEMBER

The FCC regularly updates its website, so the instructions in this section might not be up to date. Check the VEC or ARRL websites if you need help using the ULS.

Although you can search other databases (see the next section), the one maintained by the FCC is the one that really counts. Follow these steps to find your call sign online:

1. **Go to the ULS home page:** wireless.fcc.gov/uls.

2. **Scroll down to the Searching section and click the License Search link.**

   The License Search page loads.

3. **Click the Amateur link in the Service Specific Search column in the middle of the page.**

   You may have to scroll down to see it.

4. **In the Licensee section, enter your last name and Zip Code in the appropriate boxes.**

5. **Scroll to the bottom of the page and click the Search button in the bottom-right corner.**

   It may take a few seconds for your search results to appear.

6. **Browse the results page.**

   If the results take up more than one page, click the Query Download link to have the entire batch of results compiled into a single text file.

TIP

Feel free to browse through the database. By doing a little creative investigating using the License Search page, you can discover some interesting things about the ham population in your area. It's fun to see how many hams have the same last name as you, for example, or how many live in your Zip Code.

REMEMBER

Keep in mind that anyone else can do the same sort of search. When you submit your application, you identify yourself as a licensed ham and indicate where your *fixed station* (a station that doesn't move) is located. This information is made available to the public because you obtained a license to use the public airwaves; therefore, the public is entitled to know who's licensed, and with what call sign and FRN. Your personal information, however, is password-protected. (I discuss this in detail in Chapter 6.)

Searching other websites for call signs

You can also look for your call sign using these other websites:

>> **QRZ.com** (www.qrz.com): This is the best-known ham radio call-sign-lookup website. You need to register to gain access to call sign information. After you do, type the call sign in the Search box in the top-left corner of the home page and then click **Search.** This website accesses a number of non-US call sign databases so you can find information for DX stations. Some logging programs can be set up to access this website automatically so you can see more information about stations you contact.

   You can also search by name, address, county, or grid square by making a choice from the menu below the Search box. (I explain *grid square* in Chapter 11.)

>> **ARRL** (www.arrl.org/fcc/search): Enter your last name and zip code, and click the Search button for results.

WHAT IF YOU DON'T FIND YOUR CALL SIGN?

Patience is difficult while you're waiting, but be sure to wait at least one full calendar week before taking action. Here's what you can do:

- Contact the sponsor of the exam session, and ask whether the VEC accepted your paperwork and sent it to the FCC. Some problem may have caused a delay in the FCC's accepting the session results.

- If the paperwork went through okay, and it's been longer than the usual wait for the VEC that coordinated your session, ask the session leader to inquire about your paperwork. The VEC can trace all applications to the FCC.

Problems are rare. Paperwork is rarely lost or delayed without good reason, such as an error on a form.

Printing your license

You found your call sign! There is nothing in amateur radio quite like seeing your call sign for the very first time — savor and enjoy that feeling. After you come back to Earth (and maybe grab that radio and make a contact?) you can print out a copy of your license to carry with you or post in your station.

There are several methods of obtaining a printed license. The FCC no longer mails a printed copy of your license, but you can obtain a downloadable PDF of your official license that you can print yourself. Several methods to obtain an official copy are described at www.arrl.org/obtain-license-copy. The official copy will display the FCC logo and the watermark "Official Copy" will be printed across each page. You can print out and laminate the credit-card-sized license for carrying with you.

A "reference copy" can also be obtained from the ULS page that shows your license information. Look for the Reference Copy link at the top of the page. Clicking this link will create a PDF copy of your license with "Reference Copy" printed across the page diagonally. This would be handy as a backup or disposable copy.

TIP

Take or send your PDF to an office-supply store with professional printers and get a certificate-quality license for your station! They will also laminate the small version for you to carry with you.

Identifying with your new privileges

If you're upgrading an existing license, which means that you already have a call sign, you can go home after passing the exam and use your new privileges right away. You just have to add a temporary suffix to your call sign to let everyone know that you're qualified to use those privileges.

Here are the suffixes you must add to your call sign:

>> **Upgrade to General:** Add /AG to your call sign on Morse code or digital modes, and "slash AG" or "temporary AG" on voice.

>> **Upgrade to Amateur Extra:** Add /AE to your call sign on Morse code or digital modes, and "slash AE" or "temporary AE" on voice.

When your new license comes in the mail or your new license class is displayed in the FCC database, you can drop the temporary suffix.

Picking Your Own Call Sign

You can pick your own call sign (within certain limits, of course). If you're the sort of person who likes having a license plate that says IMABOZO or UTURKEY, you'll enjoy creating a so-called *vanity call sign.*

Short call signs and ones that seem to spell words are highly sought after. Many hams enjoy having calls made up of their initials. Whatever your preference, you'll likely find a vanity call sign that works for you.

TIP

Come up with at least two candidate call signs before you file an application (see "Applying for a vanity call sign," later in this chapter). That way, if your first choice is unavailable, you still have options. Remember that the FCC will soon begin to charge a $35 fee for each vanity call application so list all of the call signs you would want to use!

There are lots of options for selecting and searching for available calls, then applying for the ones you want. The ARRL vanity call pages at www.arrl.org/applying-for-a-vanity-call provide a lot of detail about the process.

Searching for available call signs

You can find available call signs by using the FCC's ULS search function (see "Searching the ULS database," earlier in this chapter), but that system can be quite cumbersome because it's designed to return information on only one call sign at a time. The following websites offer better and more flexible call-sign-search capabilities:

» **AE7Q's website** (www.ae7q.com) has lots of tools for searching the FCC database, including available calls.

» **RadioQTH** (www.radioqth.net/vanity) lets you search for call signs and offers practice exams, too.

» **WM7D call-sign database** (www.wm7d.net/fcc_uls) offers a good search function that allows wildcard characters, which speeds your search for that perfect call sign.

Depending on your license class (see Chapter 4), you can select any available call sign in the groups listed in Table 7-1. (*Note:* No new Novice or Advanced licenses are being issued, as I discuss in Chapter 4.)

TABLE 7-1   Call Signs Available by License Class

License Class	Types of Available Call Signs
Technician and General	2x3, with a prefix of KA–KG, KI–KK, KM–KO, or KR–KZ and a suffix of any three letters
	1x3, with a prefix of K, N, or W and a suffix of any three letters
Amateur Extra	2x3, with a prefix of KA–KG, KI–KK, KM–KO, or KR–KZ and a suffix of any three letters
	2x1, with a prefix beginning with A, K, N, or W and a suffix of any letter
	1x2, with a prefix of K, N, or W and a suffix of any two letters
	1x3, with a prefix of K, N, or W and a suffix of any three letters
Novice	2x3, with a prefix of KA–KG, KI–KK, KM–KO, or KR–KZ and a suffix of any three letters
Advanced	2x2, with a prefix of K or W and a suffix of any three letters

Call signs are referred to as *2-by-3* (2x3) or *1-by-2* (1x2), meaning the number of letters in the prefix (first) and the suffix (second). KDØPES is a 2 (KD) by 3 (PES) call, for example, and NØAX is a 1 (N) by 2 (AX) call. The FCC assigns certain types of call signs to the various license classes, with the higher-class licensees having access to the shorter calls. You can pick any type of call sign authorized for your license class.

Table 7-1 explains the structure of call signs, broken down by license class.

Note: This table doesn't cover call signs in Alaska, Hawaii, or the various U.S. possessions in the Caribbean and Pacific. Special rules apply to those locations.

TIP

Occasionally, you hear a *special event* call sign consisting of one letter, one numeral, and one number. Call signs of this type — called *1-by-1 (1x1)* call signs — are granted on a temporary basis to U.S. hams for expeditions, conventions, public events, and other noteworthy activities. The special event call sign program is administered by several VEC organizations for the FCC. For more information, visit www.arrl.org/special-event-call-signs.

Applying for a vanity call sign

When you've narrowed down your list of candidate vanity call signs, follow these steps to file your application:

1. **Go to** www.arrl.org/vanity-call-signs.

2. **List one or more call signs that you like.**

   All the call signs must be unassigned and available (which is why you need to search the vanity call websites first; see "Searching for available call signs," earlier in this chapter).

3. **Fill out the rest of the application.**

4. **Pay the $15 fee (as of mid-2020) by credit card or check.**

HAM RADIO LICENSE PLATES

You can also acquire a license plate with your call sign. The process is easy, and many states even have a special type of vanity plate just for hams. Contact your local department of motor vehicles and ask! For additional information, see www.arrl.org/amateur-license-plate-information.

One wrinkle for hams with call signs that contain the slashed-zero (Ø) character: In most states, you have to specifically request the slashed-zero. Talk to the clerk who handles your form and show examples of ham call signs with that character. Pictures of ham license plates are very effective in explaining what you're asking for. (The character code is ALT-0216 if that helps them find the character on their keyboard.)

Maintaining Your License

The FCC supports a set of common filing services at `wireless.fcc.gov/services`. These services are accessed using your FRN. Look for the list of services under "Personal Radio Services Resources." This page is the place to go if you want to do any of the following things:

Although free as of February 2021, these will soon have a $35 fee changed with each application:

>> Renew your license.

>> Apply for a vanity call sign (but it's a lot easier to use the ARRL VEC).

There are no fees for these items:

>> Change any of the information associated with the license, such as your name or address.

>> Get another copy of your license PDF.

>> Check on an application.

TIP

Check out this page when you first earn your license. That way, you'll be familiar with it when you need to take care of any licensing business later.

3

Hamming It Up

Chapter **8**

Receiving Signals

The chapters in Part 3 introduce you to the way hams operate. I've structured things this way so you'll know what you're trying to accomplish as you learn about setting up a station in Part 4. After you have your ham radio transceiver (or *rig*) set up and a license (*ticket*) clearing you for takeoff, you'll be ready to make your first contacts with other hams.

I start by showing you how to listen in and then explain the basics of how contacts are made. I also cover on-the-air manners and the "little things" that make contacts flow smoothly.

Learning by Listening

The ham bands are like a 24-hour-a-day party, with people coming and going all the time. Just as when you walk into any other big party, you need to size up the room before jumping in:

» **Tune:** Receive on different frequencies to assess activity

» **Monitor:** Listen to or watch an ongoing contact or conversation

By listening first, you discover who's out there and what they're doing, what the radio conditions are like, and the best way for you to make contact.

The most important part of learning how to make a contact, also known as a *QSO*, is listening — or, in the case of digital signals, watching what the computer displays. (*QSO* is one of many *Q-signals* — a type of ham radio abbreviation discussed in the next chapter's section "Q&A with Q-signals".) In fact, your ears (and eyes) are the most powerful parts of your station. It is often said that getting your ham radio license is really getting a license to learn, so let's begin!

TIP

Listening may seem like eavesdropping, but it is completely normal in ham radio. Everything is done and shared in public with whoever happens to be receiving. This is part of the fun and helps make ham radio tick! (Listening to contacts without joining the conversation is sometimes called *reading the mail*.)

Finding out where to listen

You can listen on the following groups of bands, which as Chapter 4 explained are just ranges of frequencies set aside for hams to use:

>> HF *(high frequency)* bands cover 3 MHz to 30 MHz and are usually thought of as the *shortwave* bands. (There are also three narrow bands below HF in the MF (*medium frequency*) and LF (*low frequency*) range.)

>> VHF *(very high frequency)* bands cover 30 MHz to 300 MHz.

>> UHF *(ultra high frequency)* bands cover 300 MHz to 3 GHz.

>> *Microwaves* are considered to start at about 1 GHz.

The shortwave or HF bands have a different flavor from the VHF and UHF bands and those are different than the microwave bands. On the HF bands, you can find stations on any frequency that offers a clear spot for a contact. Up on the VHF bands, most contacts take place by means of repeaters on specific frequencies or on channels, but there are random HF-style contacts, too. Up on the microwaves, you will find data networks and specialized communications such as *moonbounce* or *meteor scatter* (yes, really!) or *tropo*, which uses atmospheric conditions. How are you supposed to figure out where the other hams hang out?

REMEMBER

As a Technician licensee, you're likely to spend most of your time on the VHF and UHF bands at first, but take every opportunity to listen to the lower-frequency HF bands, which have a completely different flavor.

On both HF and VHF, hams engaged in certain activities tend to be found on or near specific frequencies. Digital operating fans who use the popular FT8 mode, for example, usually hang out near 14.074 MHz. Six-meter operators often listen at 50.125 MHz for voice signals to appear.

The word *mode* is often used to refer to a type of signal. There are voice modes, data and digital modes, image modes, and so on.

No rule says that they *must* operate on that frequency, but they gather there routinely anyway. That kind of consistency provides a convenient way for you to meet others who have similar interests and equipment. To continue with my party metaphor, it's like when a fellow partygoer tells you, "There's usually a group talking about jazz at that table in the corner."

When a frequency becomes known as a spot where you can find other hams using similar modes or operating styles, it becomes a *calling frequency.*

Understanding how bands are organized

In the United States, FCC regulations specify the segments in each band where the various types of signals may be transmitted and by which licensees. These are called *sub-bands*, and each license class has a different set of *privileges*. Figure 8-1 shows the sub-bands and privileges for the 80 meter band. The American Radio Relay League (ARRL) offers handy charts of U.S. sub-bands at www.arrl.org/graphical-frequency-allocations that are free to download and print.

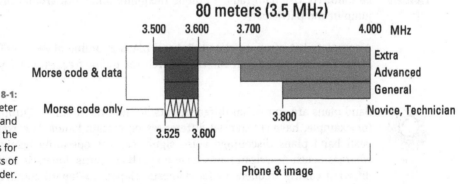

FIGURE 8-1:
The 80 meter band sub-band plan showing the privileges for each class of license holder.

Although it doesn't happen often, FCC rules sometimes change the organization of a band. As this book is being written (mid-2020), some proposals for changes are being considered. If they are adopted, the ARRL and other popular websites and social media should have the information. Don't worry about accidentally operating illegally because you'll get plenty of notice before the changes go into effect.

Outside the United States, regulations are much less restrictive. For example, you can often hear Canadian and overseas hams having voice contacts in a part of the 40 meter band where American hams aren't allowed to transmit *phone* signals.

(*Phone* is an abbreviation for *radiotelephone*, which includes all voice modes of transmission.) How unfair! Because of the number of American hams, the Federal Communications Commission (FCC) long ago decided that to maintain order, it was necessary to separate the wide-bandwidth phone signals from narrow-bandwidth Morse code and digital or data signals.

REMEMBER

Amateur radio has a very special status. Among all the communication services, it is by far the most flexible. Hams can transmit on any frequency and mode within our privileges, contact any other station, and build our own equipment. It's really great to have this much access to some very valuable spectrum space. Make use of that flexibility!

Within the bands, hams have organized the different operating styles on each band. Not all amateur users can coexist on the same frequency, so agreements about where different types of operations occur are necessary. These agreements are called *band plans*. Band plans are based on FCC regulations, but go beyond them to recognize popular calling frequencies and segments of bands on which you usually find certain operating styles or modes.

REMEMBER

A band plan isn't a regulation, and it only applies only during normal conditions. When a lot of activity is going on — such as during emergency operations, a contest, or even a big expedition to a rare country — don't expect the band plans to be followed. Use ham radio's unique flexibility and work around the activity, or jump in and participate.

A complete list of up-to-date U.S. band plans is online at www.arrl.org/band-plan. You can also do an Internet search for *calling frequency* and your favorite mode or activity.

TIP

Band plans are somewhat different outside the United States. Europe and Japan, for example, have substantial differences on certain bands. For example, European band plans discourage voice signal contest operating below 7.030 MHz whereas South American hams have no such concerns. Japanese hams also have different calling frequencies for domestic (Japan-to-Japan) contacts on the HF bands.

Using Your Receiver

Receiving or *tuning in* a signal means that you use the radio's main tuning control to change the radio's operating frequency. The control may change frequency smoothly across the band or in large steps, jumping from channel to channel. The type of control and "how you tune" depends on the technology used to build the radio and the types of signals you are looking for.

For signals you receive or *copy* by listening (called "copying by ear") — such as Morse code (CW) or voice transmitted as single sideband (SSB) or FM — your ears guide the tuning process. For digital or data signals — such as FT8, radioteletype (RTTY), or PSK31 — you use the messages and characters shown on the computer screen to set your receiver's frequency. Using other digital modes, such as packet radio, or accessing a radio "mailbox" to exchange email messages depend on the radio being set accurately to a published frequency or channel.

Software *tuning indicators* help you zero in on a digital signal. These indicators display a picture showing how well the receiver is able to process what it's hearing. You adjust the tuning until the display matches the "tuned in" pattern. Whether you tune in a signal from above or below its frequency doesn't matter, although you may develop a preference for one or the other.

Start by selecting the type of signal you want to receive. Set your radio to receive this type of signal as described in the operating manual (you *do* have your operating manual handy, right?) or the instructions in help screens. If you are using a computer to display the signals, you'll have to tell the software what type of signal you're trying to receive. There may be online videos that show you how to use your radio to tune in signals or select channels.

If you're using a stand-alone radio, the control that sets your operating frequency will be a rotating knob or dial. For a computer-based radio that uses software to receive the signal, the control may be a mouse-operated slider, cursor, or even specific keys. Some radios have both types of tuning controls — knobs and software control are both common. If you know what frequency you want, there will also be ways to enter that frequency directly with a keyboard or with front-panel buttons.

The control to set your receiving frequency is called a *VFO (variable frequency oscillator)*. In radios built with analog electronics, the VFO is an oscillator circuit. In most current models, the VFO is a special digital circuit or firmware. Whether the control changes the operating frequency of an actual oscillator circuit or sets a value used by software, hams refer to it as a "VFO." You'll probably encounter an exam question about VFOs and tuning.

Except for very simple radios, today's commercial models are controlled by a microprocessor that stores operating settings and control values. The VFO settings include a package of information: operating frequency along with type of modulation, filter bandwidth, fine tuning, noise reduction settings, antenna selection, and more. There are typically a pair of selectable sets of information, referred to as *VFO A* and *VFO B*. The radio has a front panel control that allows quick changes between A and B. If your radio has a *sub-receiver*, it may have a separate tuning knob, as well.

Along with the VFO tuning control, two other common tuning controls are the *bandswitch* and *incremental tuning*. The bandswitch may simply move the radio from band to band in order of higher or lower frequency — this is common on smaller portable or mobile radios. Larger radios with bigger front panels or software that controls a radio may have one band selection button for each band covered. Incremental tuning (RIT: *receiver incremental tuning*) is simply a "fine tuning" control that shifts operating frequency a little bit either up or down without changing the main VFO frequency.

Remember that no matter what type of radio you have — a vintage vacuum tube model, a hybrid of analog electronics and digital signal processing, or a full-blown software-defined radio (SDR) — they have a lot in common for tuning around the bands. In the next sections, I talk about how to use each type so you can get started listenin.

Tuning, and scanning with channels

As surplus FM equipment from the VHF and UHF business and public-safety bands became available, hams converted and used them on the amateur bands. Those services operate on fixed, specific frequencies called *channels*, and this type of operation is referred to as *channelized*. As a result, amateur FM operation is also channelized. Tuning FM radios consists of selecting different channels or changing frequency in fixed steps instead of making a continuous frequency adjustment.

TECHNICAL STUFF

Radios store your favorite frequencies and operating settings, creating *memory channels*, usually referred to just as *memories*. *Memory tuning* means adjusting the tuning control to jump from memory to memory. Remember this for the exam!

Figure 8-2 shows typical handheld and mobile radios used for FM operation. As a new ham, you'll probably have one of these as your first radio. The models shown in Figure 8-2 provide both analog and digital voice operation, but entry-level models are mostly analog-only or digital-only. The main tuning control is usually the largest knob on either type of radio. It rotates in fixed steps, with each step moving from channel to channel or from memory to memory.

If your radio is one of the popular *dual-band* models, it will also have a button to switch between VHF and UHF and a way to change frequency in channel-sized steps or in large steps of 1 MHz. (The larger step allows you to move quickly from one end of the band to the other.) The radio will also allow you to set the tuning rate (called *step size*) to suit the channel-to-channel spacing customary in your area.

Many FM radios can also *scan* multiple channels by listening for signals, then stopping or pausing on a channel if activity is detected. Instead of having to manually tune from channel to channel, the radio will do it for you! (There are also scanners designed to listen to public safety, commercial, aircraft, and government operation.) Scanning can cover a range of frequencies or a sequence of memory channels.

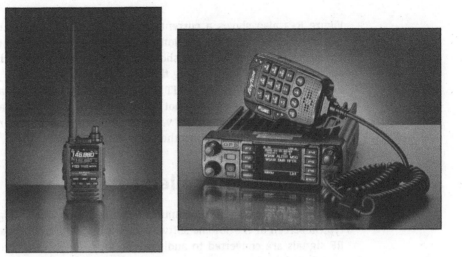

FIGURE 8-2: Handheld (left) and mobile (right) transceivers for the VHF/UHF bands.

Continuous tuning with a knob

On the HF bands and in the segments of the VHF/UHF bands where Morse (CW) and SSB operation takes place, *continuous tuning* is used. The operator (you!) spins the tuning knob to smoothly tune the radio across a range of frequencies, listening for signals.

The operating frequency is displayed by a physical dial in older radios or as a numeric display. Figure 8-3 shows an older analog receiver still popular among fans of vintage equipment. Rotating the main tuning knob also turns a plastic dial calibrated to show operating frequency. Some older radios have a *slide-rule dial* with frequencies laid out along a line and a sliding pointer that moves with the tuning knob.

FIGURE 8-3: Drake R-4C (left), top of the line 1970s analog receiver. The Kenwood TS-590S (right) is a modern microprocessor-controlled radio.

TECHNICAL STUFF

The dominant type of radio for many years, the *superheterodyne* or "superhet" is still in wide use. It uses multiple stages of frequency shifting circuits to convert a radio signal to the audio range where it is turned back into speech, Morse code, or digital waveforms. The VFO is used by one or more of these circuits to tune in a specific signal.

Figure 8-3 also shows a current model radio using knob tuning. The radio's microprocessor reads the motion of the tuning knob, which then controls the operating frequency which is shown on the display along with other operating status and control values. Note that the VFO selected, A or B, is also displayed. *Receiver Incremental Tuning* or RIT is used to tune in a voice signal for better fidelity, to accommodate a station not quite on the same frequency as you, or to avoid an interfering signal. If an RIT value of -0.15 is displayed, the receiver has been tuned 0.15 kHz (150 Hz) below the main VFO frequency.

Software-controlled tuning

Finally, some radios use a computer as their "front panel." Figure 8-4 shows a typical screen of the *SDRuno* software operating with an SDRplay RSPdx receiver. RF signals are converted to audio streams by the receiver and then input to a sound card that converts them to digital data.

Spectrum display

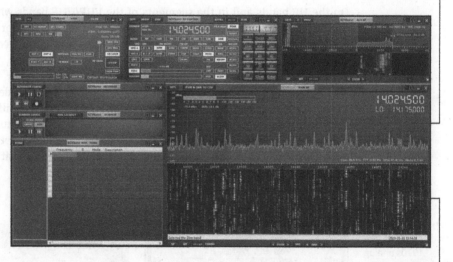

FIGURE 8-4:
The *SDRuno* receiving software simultaneously showing several 20 meter signals on a spectrum and waterfall display at the lower right.

Waterfall display

The software then processes the signals and shows them on a frequency scale from left (lower frequencies) to right (higher frequencies). Each signal is shown as a "spike" above the frequency scale. The taller the spike, the stronger the signal. You can zoom in on one signal or zoom out to see many signals — maybe an entire band or even multiple bands, depending on the type of receiver you have. This is a *spectrum* or *spectrum analyzer display* seen at the lower right of the figure.

To tune in a specific signal, a mouse or keyboard is used to move a cursor to the desired signal. After selecting the type of signal being received, it is then turned into audio to or data characters by the software. You can tune smoothly or skip from channel to channel. Different software packages may operate slightly differently, but the overall process and display are very similar.

The *waterfall display* shown just below the spectrum display is another way that software can show you what signals are present, the frequencies on which they reside, and how strong they are. The software draws a series of horizontal lines at the top of the display, with each new line moving the older ones farther down the screen. Signal strength is shown by brightness and by color. The result is something that looks like a slow waterfall. The waterfall and spectrum display are showing the same signals but in a different way.

TECHNICAL STUFF

Waterfall displays are sometimes referred to as *panadapters*. This was the name of a special piece of equipment required to create this style of display before SDRs became common.

In the waterfall display, frequency is still shown on the horizontal scale, low to high from left to right. However, instead of height to show signal strength, color is used with bright colors showing stronger signals. The spectrum analyzer display shows you the strength of signals at any given moment while the waterfall display shows you the history of where signals are present. Individual specific signals are received just like on the spectrum display, by moving a mouse or cursor to the desired signal for the software to turn into audio or data.

TECHNICAL STUFF

Switches, selector buttons, and rotary controls will all look about the same on all three types of radios. Slider controls, either vertical or horizontal, are much more common on software displays. They have the advantage of their setting, such as an audio level, being easy to read and adjust with a mouse or touch-screen.

Listening on VHF and UHF

Let's start on the VHF and UHF bands because new hams usually operate there as they get introduced to how ham radio works.

Most contacts on these bands are made using *repeaters*, which are radios that listen on one frequency and retransmit what they hear on another frequency at the same time. Repeaters are located in high spots such as hilltops or on tall towers so they can be used over a wide area. (You can find more information about repeaters in Chapter 9.)

The most common mode by far is FM voice, although more digital voice signals are appearing every day, as discussed below. *Frequency modulation*, or *FM*, is the preferred *analog* mode for VHF and UHF voice contacts because it rejects atmospheric and man-made noise for clearer reception. In an analog mode, the characteristics of a signal (such as frequency or amplitude or phase) vary smoothly to carry the information — speech, in this case. *Digital voice* is also a voice mode, even though the speech is converted to digital data. The transmitter and receiver process the speech as data instead of an analog signal. Digital voice systems use analog signals when interacting with a human listening or speaking.

TECHNICAL
STUFF

Contacts made via a repeater are called *duplex*, and those without the repeater's relay are *simplex*. When you hear one ham tell another, "Let's go simplex," it means to stop using the repeater and communicate station-to-station on another channel. There are channels set aside for simplex operation between the groups of repeater input and output channels.

Repeaters are used for local and regional communication because they enable individual hams to make contacts over a wide area with low-power handheld or mobile radios. The repeater input and output frequencies have to remain fixed, so the bands are organized into sets of channels. You can view a complete band plan for the 2 meter and 70 cm bands at www.arrl.org/band-plan. (Except for some at the upper end of the 10 meter band, repeaters aren't used on the HF bands.)

Many hams use repeaters as a kind of intercom to keep in touch with friends and family or club members as they go about their daily business. These contacts usually have an informal "conversational" style, and you're likely to hear contacts among the same groups of hams every day. Repeaters are where you find local hams and information about local events, traffic, upcoming meetings, hamfests, and so on.

TECHNICAL
STUFF

There are digital repeaters just like analog repeaters. Because the speech is already in digital form, digital repeaters are often linked together via the Internet to form networks over wide areas, even worldwide. To use digital repeaters, you need a radio that uses the same type of digital signal as the repeater.

Listening on HF

Most of the traditional shortwave bands between 1.8 MHz and 30 MHz are broadly organized into two segments. In the United States, Morse code (CW) and data signals occupy the lower segment, and voice signals occupy the higher segment. Voice contacts on HF are made using *single-sideband (SSB)*, which I explain in more detail later in this chapter. SSB comes in two modes: *upper sideband (USB)* and *lower sideband (LSB)*. Your radio must be set to the correct mode to receive the signals properly.

REMEMBER

Although the 160 meter band from 1.8–2.0 MHz is technically in the MF frequency range, it is very similar to the lower HF bands, so hams refer to it as part of "the HF bands."

Within each of these segments, the lower frequencies are where you tend to find the long-distance (*DX*) contacts, special-event stations, and contests operating. Casual conversations (*ragchews*) and scheduled on-the-air meetings (*nets*) generally take place on the higher frequencies within each band.

Organizing activity on HF bands

Table 8-1 provides some general guidelines on where you can find different types of activity. Depending on which activity holds your interest, start at one edge of the listed frequency ranges and start tuning as described in "Receiving Signals," later in this chapter.

TABLE 8-1   **Activity Map for the HF Bands**

Band	CW, RTTY, and Data Modes	Voice and Image Modes
160 meters* (1.800–2.000 MHz)	1.800–1.860 MHz (no fixed top limit)	1.840–2.000 MHz
80 meters (3.500–4.000 MHz)	3.500–3.600 MHz	3.600–4.000 MHz
60 meters (5 separate channels near 5.35 MHz)	Permitted, but the signal has to be centered in the channel	5330.5, 5346.5, 5357.0, 5371.5, and 5403.5 MHz (USB, CW, RTTY, and data only)
40 meters (7.000–7.300 MHz)	7.000–7.125 MHz	7.125–7.300 MHz
30 meters (10.100–10.150 MHz)	10.100–10.125 MHz CW 10.125–10.150 MHz RTTY and data	Not permitted
20 meters (14.000–14.350 MHz)	14.000–14.150 MHz	14.150–14.350 MHz
17 meters (18.068–18.168 MHz)	18.068–18.100 MHz	18.110–18.168 MHz
15 meters (21.000–21.450 MHz)	21.000–21.200 MHz	21.200–21.450 MHz
12 meters (24.890–24.990 MHz)	24.890–24.930 MHz	24.930–24.990 MHz
10 meters (28.000–29.700 MHz)	28.000–28.300 MHz	28.300–29.700 MHz (most activity below 28.600 MHz)

The new amateur bands at 176 kHz and 430 kHz are very narrow so they do not have segments as the higher frequency bands do. Their low frequency (and long wavelength) require some specialized antennas and operating techniques. Operating conventions for these bands are still being developed as this is being written in 2020. In general, they have characteristics that are similar to the 1.8 MHz (160 meter) band.

Because hams share the 60 meter band with government stations, there are special rules for operating on this band. Read the rules for 60 meter operation before getting on the air.

Adjusting for time of day

Because the ionosphere strongly affects signals on the HF bands as they travel from point A to point B, time of day makes a big difference. On the lower bands, the lower layers of the ionosphere absorb signals by day but disappear at night, allowing signals to reflect off the higher layers for long distances. Conversely, the higher bands require the Sun's illumination for the layers to reflect HF signals back to Earth, supporting long-distance *hops* or *skips*. (With the exception of sporadic effects, the ionosphere is much less a factor on the VHF and UHF bands at 50 MHz and above.)

Table 8-2 shows general guidelines on what you might hear on different HF bands at different times of day.

Using beacon networks and contact maps

It can be hard to tell if a band is *open* (meaning that signals can travel beyond line of sight), and in what direction. Propagation software is available to help with those decisions, but only as predictions. To help determine whether a band is actually open, hams use *beacons, beacon networks,* and *contact maps.*

Beacon stations

A *beacon* continuously sends a CW message at a modest speed on a published frequency. Amateurs who receive that beacon's signal know that the band is open to its location. Beacon transmitters are set up around the world and operate in special frequency ranges set aside for beacons. You can tell when you are receiving a beacon transmission because the station will append a "B" to the call sign, such as, for example, N1LF/B on 28.2685 MHz from Alabama.

TABLE 8-2   Day/Night HF Band Use

HF Band	Day	Night
160, 80, and 60 meters (1.8, 3.5, and 5 MHz)	Local and regional out to 100–200 miles.	Local to long distance, with DX best near sunset or sunrise at one end or both ends of the contact.
40 and 30 meters (7 and 10 MHz)	Local and regional out to 300–400 miles.	Short-range (20 or 30 miles) and medium distances (150 miles) to worldwide.
20 and 17 meters (14 and 18 MHz)	Regional to long distance; bands open at or near sunrise and close at night.	20 meters: Often open to the west at night and may be open 24 hours a day. 17 meters: Follows the same pattern but opens a little later and closes a little earlier.
15, 12, and 10 meters (21, 24, and 28 MHz)	Primarily long distance (1,000 miles or more); bands open to the east after sunrise and to the west in the afternoon. 15 meters: A good daytime band, especially to the Caribbean and South America. 12 meters and 10 meters: Usually have short openings in the morning and afternoon (unless there are lots of sunspots).	15 meters: Open to the west in very late afternoon but closes soon after sunset. 12 meters and 10 meters: Close immediately after sunset. 10 meters is often used for local communications 24 hours a day.

A network of 18 HF beacon stations is run by the Northern California DX Foundation (NCDXF) and the International Amateur Radio Union (IARU). These beacons have special permission to transmit on the 20 through 10 meter bands in a round-robin sequence. They also vary their transmitting power from 100 watts to 100 milliwatts so that hams receiving the beacon signal can judge the quality of propagation. A complete description of this useful network is available at www.ncdxf.org/beacons.

Other amateurs have set up beacons, too. A good example listing for all beacons on 6 meters is the excellent list at www.keele.ac.uk/depts/por/28.htm. Amateur VHF, UHF, and even microwave beacons are listed on several websites; just do an online search for *amateur beacon* and the band to locate beacon listings.

Beacon networks

Along with the beacons that transmit for you to hear, there are also stations that listen for signals and report them to a central server. This creates a network of stations listening 24 hours a day and reporting all of the signals that are heard.

Each station uses software that can automatically decode CW or digital mode signals. (There are no automated voice decoder stations on the amateur bands as of 2020.) The transmitting station's call sign and other information is reported. Ham operators access the reports from the server and determine "who's hearing whom" on any amateur frequency. The most active networks in operation today are discussed below: the Reverse Beacon Network, PSKreporter, and WSPRnet. The data collected by these networks is publicly available for research and propagation analysis.

PSKreporter (www.pskreporter.de) was the first network to automatically collect and store reception reports. Stations voluntarily relayed information online about stations they received or contacted. Originally, the network was set up to report on PSK31 signals, but it has expanded to many different modes including FT8, JT65, and others. At any time, there may be several thousand stations contributing reception reports.

The Reverse Beacon Network (RBN: www.reversebeacon.net) uses CW Skimmer and RTTY Skimmer decoding software that can decode CW and RTTY (radiotele-type) signals across nearly 100 kHz of a band. It listens for stations calling CQ, copies the call signs and relative strengths, then reports the information to the central storage server. Because the software can decode many signals at once, fewer stations are needed for worldwide coverage.

WSPRnet (www.wsprnet.org) is a special system set up to collect reception reports from stations using the ultra-low-power propagation test mode WSPR. This mode is part of the WSJT-X software package (physics.princeton.edu/pulsar/K1JT/wsjtx.html), which includes other digital modes such as FT8 and MSK144.

Contact maps

When reception reports are relayed to the beacon networks or spotting networks (discussed in Chapter 11) they are posted in a list that can be seen by other opera-tors or collected by software. Several websites show this information on a *contact map* by drawing lines between station locations. When the bands are open, contact after contact pops up. Sometimes, a band may be closed at your location, but by watching the online map, you can see propagation gradually moving in your direction.

DXMAPS is the best-known set of online contact maps (www.dxmaps.com). The website is run by Gabriel Sampol (EA6VQ), from the Balearic Islands, off the coast of Spain. You can watch maps of contacts on most amateur bands, send messages to other stations, check solar and ionospheric data, and do much more. Figure 8-5 shows 17 meter activity in late August of 2020. Click on the tab for the band you want and the map will show recent contacts.

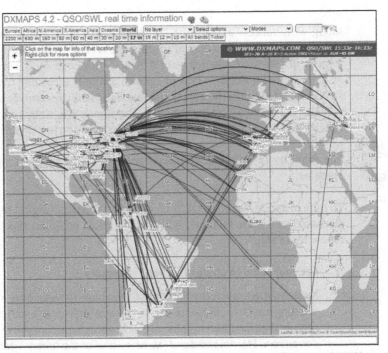

FIGURE 8-5:
The DXMAPS website displays contacts as soon as they are reported. You can see what bands are open and to what locations at a glance.

Map provided courtesy of DXMAPS.com

If you'd rather run a mapping application directly on your computer, some software hams use for making contacts and keeping records also has mapping features. Good examples are Ham Radio Deluxe (www.ham-radio-deluxe.com) and DXLab Suite (www.dxlabsuite.com). Other packages are listed at www.dxzone.com/best-qso-mapping-tools-to-display-contacts.

Receiving Signals

The next few sections give you an idea of how to tune in a few types of signals. Once you are used to using the tuning controls of your radio, you'll perform the same basic functions on each type. Tuning in different signals and learning to listen is the start of your ham radio adventure!

Receiving FM voice

Using a ham FM receiver is a lot like using your car radio to listen to a local FM broadcast station. You can manually tune from channel to channel or save your favorite stations and access them with a dedicated button. The signals used are very similar and so are the receivers.

Along with volume and tuning controls, ham FM receivers also have a feature called *squelch* which is a way of muting the receiver audio output when no signal is present. Without squelch, you'd hear continuous loud noise until someone transmitted! The simplest type of squelch operates according to signal strength: A strong signal exceeds the adjustable threshold and the receiver output is unmuted (squelch is "opened") so you can hear the signal. When the receiver is muted, the squelch is "closed." Other types of squelch operation are described in Chapter 9.

The exam has a couple of questions about different types of squelch and how it operates.

To tune in an FM voice signal, such as from a repeater (digital repeater signals are discussed in a later section), which is the most common type of FM operation, follow these steps:

1. **Set your radio to operate on FM if necessary.**

   Most VHF/UHF radios use only FM, so your radio may not have a control for selecting the mode. If you have a multi-mode radio or one with "extended receive" that can listen to commercial broadcasts, be sure you're using the *narrowband* FM, not *wideband* FM, which is what the broadcasters use for better audio quality.

2. **Set the squelch control so that you hear noise.**

   This procedure is called *opening the squelch.* The squelch control may be a rotatable knob or a slider-style on a computer screen, depending on the type of radio you're using. For now, if the *tone squelch* function is on, requiring the signal to have a specific sub-audible tone (I discuss these in Chapter 9), turn it off so you can listen to any signal.

3. **Reset the squelch so that it just stops the noise.**

   This step enables you to hear weak signals without having to listen to continuous noise. For very weak signals, you may have to reopen the squelch to receive them.

   Your radio may have a MONITOR or MON control that opens the squelch temporarily without changing the level. This is useful when checking to see if a weak signal is present.

4. **Determine which channel you'll listen to.**

   I assume you're going to listen to activity on a repeater. Most repeaters on the 2 meter band (the most popular) have outputs between 146.60 and 147.40 MHz. Not all repeater channels have an active repeater. If you are a member of a club with its own repeater, start there. Or ask club members what repeaters have good coverage.

To find other repeaters in your area or while traveling, check a repeater directory or website (see Figure 8-6). Repeater directories list the frequencies and locations of repeaters so you can tell which ones may be available in your area. Repeater directories also note special features, whether a repeater supports digital voice operation, and (for analog repeaters) any required access tones or codes.

Some of these directories are nationwide, such as *The ARRL Repeater Directory* (search for *repeater directory* at www.arrl.org/shop) or Repeater MapBook Directory (www.artscipub.com/mapbook); others focus on specific regions, such as the New England Repeater Directory at www.nerepeaters.com.

RFinder (www.arrl.org/shop/RFinder-The-World-Wide-Repeater-Directory) is an app that accesses a worldwide directory on a yearly subscription basis. Look for RFinder Ham on the home page. You can use the app while you are traveling or to find repeaters along a route. The information can be used to program your radio, as well.

5. **Set your radio's frequency.**

If you're using an FM-only radio such as a handheld or mobile unit, you can enter the frequency via a keypad, rotate a small tuning control until you have reached the desired channel, or select a memory channel. If hams are active on that channel, you hear the operator's voice.

TIP

Some VHF/UHF radios have an *auto repeater* feature that can tell when you are tuned to repeater channels and automatically set up the radio to transmit and receive on the different frequencies required. *Tone scan* is a similar feature that listens for and automatically sets your radio to use the right sub-audible control tone required for the repeater to retransmit your signal.

If you are using continuous tuning, you can tune in the signal by listening for the most natural-sounding voice with the least distortion. Some radios have a tuning indicator called a *discriminator,* sometimes labeled DISC. The discriminator shows whether you're above or below the FM signal's center frequency. You may have to tune back and forth to find the frequency where the voices sound best. If you're mistuned *(off frequency)*, the voices are muffled or distorted. When the signal is centered on the discriminator, you're tuned just right.

TIP

If your radio supports small tuning steps, tune an FM signal on a repeater and tune the receiver away from the designated frequency a kHz or two at a time while listening to the audio. As you get farther "off frequency" the signal audio will become distorted and finally unintelligible. Now you know what that sounds like!

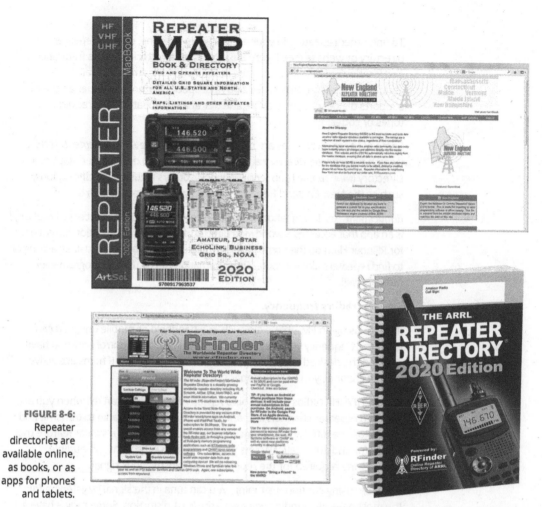

FIGURE 8-6:
Repeater
directories are
available online,
as books, or as
apps for phones
and tablets.

Courtesy American Radio Relay League, ArtSCI Publishing

Most handheld and mobile FM radios are sold with the tuning set for a standard spacing between channels. (This is called *channel spacing*, not surprisingly.) You can jump from channel to channel by turning the main frequency control knob or by using an UP/DOWN button on the front panel or microphone. The most common spacing is 20 kHz on the 2 meter band. Channel spacing can be adjusted by changing your radio settings.

TIP

You can also listen to stations transmitting to the repeater — this is called *listening on the input* — by using the "Reverse" feature usually activated by a button labeled REV or similar. This can be helpful if a station can't access the repeater on its own but is close enough to you that you can receive it more clearly. The exam has a question about doing this.

Receiving SSB voice

Single sideband (SSB) is the most popular mode of voice transmission on the HF bands. It is also used in the lower portions of the VHF and UHF bands. SSB is a type of *amplitude modulation*, or *AM*, in which the speech information is transmitted by variations in signal strength.

As shown in Figure 8-7, an AM signal has three parts: two identical copies of the voice information, called *sidebands*, plus a single-frequency signal called a *carrier*. Each sideband signal occupies a narrow range above and below the carrier frequency signal.

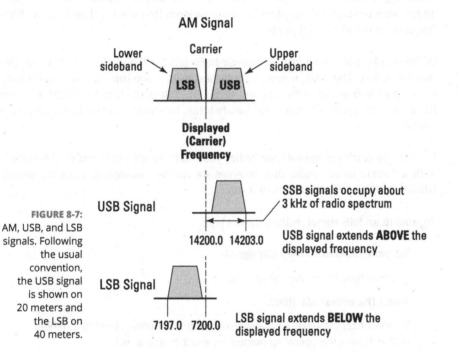

FIGURE 8-7: AM, USB, and LSB signals. Following the usual convention, the USB signal is shown on 20 meters and the LSB on 40 meters.

SSB starts with an AM signal then removes one of the sidebands and the carrier. All the power in an SSB signal is dedicated to that one sideband, so it is more efficient than AM and saves precious radio-spectrum space. (AM still has a dedicated following of hams who appreciate the mode's characteristic fidelity, relaxed style of operating, and equipment.)

On your radio you can choose to receive and transmit either the *upper sideband (USB)* or the *lower sideband (LSB)*. Even though it is not transmitted for an SSB signal, the carrier's frequency is the radio's *displayed frequency* when using SSB.

Above 10 MHz, amateurs use the upper sideband, or USB, and lower sideband or LSB below 10 MHz. The only exception is the 60 meter band near 5 MHz, where USB is required by the FCC.

For direct ham-to-ham contacts on VHF and UHF over distances at which an FM contact would be noisy and difficult, use the more efficient CW and SSB modes. These modes are referred to as *weak-signal modes* on VHF and UHF because you can make contacts with much lower signal levels than when using FM. (Weak-signal modes can produce very strong signals!) The lowest segments of the VHF and UHF bands are set aside for weak-signal operation.

Weak-signal contact are made in much the same way as SSB and CW operations on HF, with contacts taking place on semi-random frequencies close to the calling frequencies listed in band plans.

While tuning, use the widest *filters* your radio has for the mode (CW, SSB, or AM) that you select. That way, you won't miss a station if you tune quickly, and tuning in the station is easier. After you are tuned in, that is the time to "tighten" your filters by setting them to narrower bandwidths, limiting what you hear to just one contact.

Even if you don't yet have a ham radio, you might be able to listen to SSB signals with a "world band" radio that receives shortwave broadcasts. Look for switch labeled "SSB" or "BFO" and turn it on.

To tune in an SSB signal, follow these steps:

1. **Set your radio to receive SSB signals.**

   You may have to choose LSB or USB.

2. **Select the widest SSB filter.**

   To select filters, you use a Wide/Narrow control or buttons labeled with filter widths. (Check the operating manual for exact instructions.)

3. **Adjust the tuning dial until you hear the SSB signal.**

   As you approach an SSB signal's frequency, you hear either high-pitched crackling (like quacking) or low-pitched rumbling. You can tell from the rhythm that you're listening to a human voice, but the words are unintelligible. What you're hearing are the high- and low-frequency parts of the operator's voice.

A video I made showing you what an SSB signal sounds like and how to tune it in is available on the ARRL website. Read the article "About SSB" at www.arrl. org/chapters-5-and-6-signals-modes-and-equipment; the link to the video is embedded on the last page.

4. **Continue to tune until the voice sounds natural.**

   If the voice sounds too bass-y, your transmitted signal will sound too treble-y to the receiving operator, and vice versa.

REMEMBER

If every voice that you hear sounds scrambled, your radio is probably set to receive the wrong sideband. Change the sideband (USB or LSB) and try again.

TIP

If you tune across an AM signal (they are often found at the very upper edge of the 80 and 40 meter bands), you can tune in the signal using either USB or LSB. The whistling noise that gets lower and lower in frequency as you tune in the voice is the carrier signal between the two voice sidebands. When the pitch of the carrier signal becomes so low that you can't hear it anymore, the receiver is centered right on the signal. This is called *zero beat*, and you can now listen to either sideband by switching between USB and LSB.

Receiving digital voice

One of the biggest ongoing changes in amateur radio is the growing use of *digital voice* modes. There are now several different digital techniques for voice communication. Each turns the voice's audio into a stream of digital data by using a *codec* (short for code-decode) integrated circuit or software. The digitized voice is then packaged with additional data used to control the way the signal is exchanged with and processed by another station. The bundle of data is then fed to the transmitter, where it is added to the actual over-the-air signal. This process is reversed at the receiving station so that the digitized voice is turned back into analog speech.

FCC rules make no distinction between analog voice and digital voice signals, so they must use the same band segments and meet the same specifications for signal bandwidth and quality. Some of the digital voice protocols can also exchange graphic images, similar in quality and speed to *slow-scan television* (*SSTV*). When exchanging images, the FCC rules for image signals apply. I cover operating with these digital voice modes in Chapter 9.

AOR's digital voice and *FreeDV* are the most common digital voice signals found on HF as of mid-2020. If you tune them in with a radio set up for analog SSB, you'll hear a "roaring" noise. If you are tuning by using a spectrum display as discussed earlier that shows signal strengths across a band, digital voice signals appear "solid" in that they tend to occupy all of a channel without the peaks that follow speech patterns. The decoding equipment or software for these modes has tuning indicators to help you adjust the frequency for proper reception. When you're tuned in, the codec will sync up with the data, and a voice will suddenly be heard.

Several digital voice systems are in use on the VHF and UHF bands, with many repeaters for each system. They operate on the same channels that analog FM repeaters use. The most common as of mid-2020 are D-STAR, System Fusion, and DMR. Listening to these systems requires a radio designed to receive that type of signal. The RepeaterBook website makes it easy to search for repeaters of each type in your area:

>> **D-STAR:** www.repeaterbook.com/repeaters/niche/index.php?mode=DStar

>> **System Fusion:** www.repeaterbook.com/repeaters/niche/index.php?mode=YSF

>> **DMR:** www.repeaterbook.com/repeaters/niche/index.php?mode=DMR

Another option for listening in is to access the repeater system through the Internet and stream the audio to your computer speakers. Each system has a different method for accessing the audio and may require you to register before you can do so. A mentor who is familiar with the ins and outs of a particular digital mode is very helpful here!

TIP

Some digital system repeaters can relay both analog and digital voice signals. Typically, a repeater's users tend to be all-analog or all-digital, but some systems support both types.

One problem with all these digital systems is that they are not compatible with each other. Stations using AOR modems and *FreeDV* software can't talk to each other. The same is true for D-STAR, System Fusion, and DMR systems. The currently preferred solution is a *hotspot*: a small radio that understands one or more of the digital voice systems. A typical hotspot is shown in Figure 8-8 along with its web interface app and a pair of digital voice radios. Your radio communicates with the hotspot as if it was a separate station. The hotspot relays the digitized voice signals over the Internet to a digital voice network, even one that uses a different system. This is a convenient way to use a single radio with several different digital voice systems.

Receiving digital or data modes

Many types of digital signals exist, and all of them sound a little different on the air, from a collection of whistles (FT8) to buzzing (PSK31) to two-tone chatter (RTTY) to a sound like a rambling calliope (MFSK). Each type of signal requires a different technique to tune in, so in this section, I focus on an easy-to-use mode that you're likely to try right away: PSK31.

FIGURE 8-8:
The OpenSPOT2
hotspot
communicates
locally on the
70 cm band and
over the Internet
with several
popular digital
voice systems.

First, you need to be running some digital-mode decoding software. One very popular program that can decode a lot of modes is fldigi, by W1HKJ (www.w1hkj.com). You can download it for free and start using it immediately, as described in the software's help file.

If you start using the digital modes regularly, you'll want to use an audio interface gadget between the radio and your PC. But to try out listening to some digital signals, use a microphone connected to the computer's sound card input and turn up the radio's volume.

I'll start with one of the simpler modes, PSK31. If you have a desktop or laptop PC, install fldigi and tune in a signal as described in the following steps. There are more popular modes, but the radio and software setup are more complex, so I'll keep it simple at first. This video by Jeff McGrath (KG7HSN) called "A Beginners fldigi PSK31 High Level Introduction" may be helpful in getting fldigi set up and decoding PSK31 signals: www.youtube.com/watch?v=IhpX4Vo_Ng8&t=319s.

TIP

You can also use an app on your phone or tablet: DroidPSK works on Android and PSK31 on iOS devices. There are quite a few digital mode apps for both operating systems.

1. **Set your radio to USB, and tune to one of the PSK31 operating frequencies.**

   14.070 MHz, or 7.070 MHz will probably have some activity through the day or evening. You don't have to have a big antenna to listen in — a long piece of wire will do.

2. If several filters are available on your radio, select one that's suitable for voice, such as the standard 2.4-kHz filter.

(For details on selecting filters, see "Receiving SSB voice," earlier in this chapter.)

If PSK31 signals are present, you will hear whistling or buzzing. If you only hear a hiss or static for more than a minute or two, try a different band.

3. Run fldigi, and set it to receive PSK31.

(Refer to fldigi's instructions to get started.)

4. Turn on the waterfall display.

This display (see Figure 8-9) shows the signal as a yellow stripe against a blue-and-black background.

5. Adjust the receive volume until the background is mostly covered with blue speckles.

Assuming that signals are present, you'll see them as yellow stripes slowly making their way down the page. (The figure shows a signal as a white stripe near the left edge of the waterfall display area.)

6. Click the strongest signal.

The red PSK31 channel markers straddle the signal, and if it's a PSK31 signal, the decoded text is displayed in the yellow window.

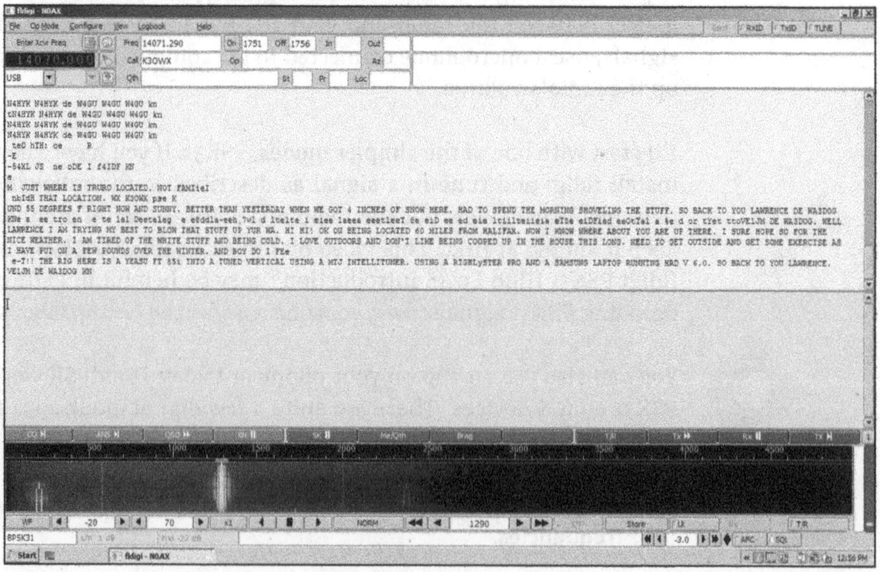

FIGURE 8-9:
The fldigi display with received text and a waterfall-style signal display.

That's really all there is to it: Run the software, connect the audio, and click a signal to start receiving. Practice this technique to receive other PSK31 signals. You may be surprised by what you can hear, because PSK31 is a very efficient mode.

Next, try tuning around the RTTY or MFSK calling frequencies, and see whether you can tune in a few of them. Don't forget to change the software's mode or it won't be able to decode the signal.

REMEMBER

There are lots of digital modes, and it's not uncommon for PSK, RTTY, MFSK, and other types of signals to be sharing the same section of a band. Try selecting a different mode in the software if a signal doesn't decode properly.

After you are comfortable with fldigi, you'll want to try one of ham radio's most popular digital modes: FT8. I don't recommend you start with that mode because you'll need a little more know-how to get its supporting software running and configured. FT8 is one of several modes supported by the *WSJT-X* software package. Since its introduction in 2017, it has enabled many hams to make simple QSOs with limited antenna systems and low power. You'll want to check out FT8 after you get set up and on the air!

Receiving Morse code

Morse code signals are often referred to as *CW*, which stands for *continuous wave*. Early radio signals died out quickly because they were generated by sparks. Soon, however, operators discovered how to make steady signals, or continuous waves, by turning the signals on and off with a telegraph key. Thus, *Morse code* and *CW* became synonymous.

IDENTIFYING BANDS AND MODES BY EAR OR EYE

With so many types of digital modes to choose from, and with more being invented all the time, how can you tell what you're hearing or seeing on the bands? Recordings of signals using many digital modes and pictures showing how they look on a waterfall display are available at www.sigidwiki.com/wiki/Category:Amateur_Radio to help you practice identifying them by ear or eye.

If you like to listen outside the ham bands, try www.sigidwiki.com/wiki/Signal_Identification_Guide for samples of commercial and military modes. Look under "Categories" for the type of signal you're interested in.

To tune in a Morse code signal, follow these steps:

1. **Set the radio to receive Morse code by selecting the CW mode and tuning to a frequency somewhere in the bottom 20 kHz to 50 kHz of an HF band.**

2. **If your radio has more than one filter, set it to use a wide filter.**

   A wide filter (500 Hz or more) allows you to find and tune in stations, whereas the narrower ones block out unwanted nearby signals. You select filters with a Wide/Narrow switch or with buttons or controls labeled with filter widths. Some radios also have adjustable-width filters — check your operating manual for instructions on how to adjust them.

3. **Adjust the tuning until you hear a Morse code signal.**

   The pitch will change as you change the receiver's frequency. Tune until the pitch is comfortable to your ear.

TIP

   A low tone (300–600 Hz) is most restful to the ear, but a higher tone (500–1200 Hz) often sounds crisper. Most radios are designed so that when you tune in a signal with a tone or pitch around 600 Hz, the transmitted signal is heard by the other station at a similar pitch. If you prefer to listen to a note more than 100 Hz higher or lower, check your radio's operating manual to find out how you can adjust the radio to accommodate your preferred pitch.

4. **When you tune in the signal at your preferred pitch, select a narrower filter (if one is available) to reduce noise and interference.**

   If the frequency isn't crowded or noisy, you can stay with the wider filter.

Chapter **9**

Basic Operating

Take another step into ham radio with this chapter's review of easygoing operating. The technical aspects of station configuration and operation are covered in Part 4. (Links to even more technical stuff are available on the *Ham Radio For Dummies* page at www.dummies.com.)

After you tune around the amateur bands for a while, you'll agree that the lion's share of the ham's life is making relaxed, casual contacts. Some contacts are just random "Hello, anybody out there?" encounters. You'll also hear contacts between hams who are obviously long-time friends or family members who meet on the air on a regular basis.

In this chapter, you find out about the different ways to conduct these casual contacts. As with most things in life, "There's kindy a knack to it," as my dear Aunt Lexie used to say. Learning the ways and means of ham radio will help you fit in quickly and enjoy ham radio more.

If the thought of actually transmitting makes you nervous, don't worry; all hams start out feeling just this way and they survived! You will, too. With a little preparation, you'll feel comfortable and confident, ready to get on the air and join the fun.

TIP

Before I start on this operating business, allow me to suggest that you get two books:

> » ***The FCC Rules and Regulations for the Amateur Radio Service:*** This book is available from the ARRL (www.arrl.org/shop) for only a few dollars. It conveniently includes not only the rules themselves, but also a clear

discussion of do's and don'ts, along with information on technical standards and the FCC Universal Licensing System. Hams really should have a copy in their shack, whether they're veterans or beginners.

>> *The ARRL Operating Manual:* The manual dedicates a separate chapter to all kinds of on-the-air operating; provides handy references, tables, and maps; and answers just about any operating question you can come up with. It's also available at www.arrl.org/shop. A series of how-to books is available for various operating specialties such as using digital modes on HF. Check the "Operating" products for the latest materials.

There are many other online resources that you can use if you have specific questions. YouTube and Instagram host videos on many ham radio topics. Just search for *ham radio* or *#hamradio* to find numerous examples. Other resources include

>> **The DX Zone** (www.dxzone.com): These sites offer links to many ham radio topics.

>> **QRZ.com** (qrz.com): The site started out as a U.S. call sign lookup service and has expanded to world-wide license databases, an active news area, and numerous forums on a variety of topics.

>> **eHam.net:** This ham radio portal includes numerous areas of interest to hams, including a handy "Elmers" page for mentoring (www.eham.net/forum/index). Click the Basic Operating link for information about repeater operating.

>> **ARRL Technical Information Service** (www.arrl.org/technical-information-service): This site has many public links and numerous in-depth articles for ARRL members.

You don't have to learn about ham radio all by yourself!

Understanding Contacts (QSOs)

You'll find that most contacts (called *QSOs* — see the section "Q&A with Q-signals" later in this chapter) are one of three types:

>> Casual conversations *(ragchews)*

>> Nets and talk groups

>> Contests and DXing

I discuss all three types in the following sections.

Common parts of contacts

Like most hobbies, ham radio has developed its own customs and ways of communicating. The following information will help you understand what's going on. These bits and pieces are involved in almost every contact.

Know your Alfa, Bravo, Charlies: Using phonetics

TIP

You'll hear hams using words called *phonetics* in place of letters when they make voice contacts. For example, *November, Alfa,* and *X-ray* represent the letters of my call sign. Phonetics are used because many letters sound alike (think *B, E, T,* and *P* in English), and the words help get the exact call sign across. You'll need to know what phonetics are for the exam but not the individual letters.

REMEMBER

When you talk with someone via radio, you can't see them — no lip reading, facial expressions, or hand-waving! Everything comes through a little speaker. You'll be surprised at how much different that is than chatting face-to-face. As a result, hams and others who communicate by voice over radios have developed techniques to get the message through reliably.

Table 9-1 lists the standard International Telecommunication Union (ITU) phonetics that hams use. Alternatives are also used, such as *Germany* instead of *Golf.* When in doubt, I respond or call with the same phonetics used by the station I want to contact. For example, if I hear my friend KX9X calling and using Kilo X-Ray Nine X-Ray, I will use those phonetics when calling him.

TABLE 9-1

ITU Standard Phonetics

Letter	Phonetic	Letter	Phonetic
A	Alfa	N	November
B	Bravo	O	Oscar
C	Charlie	P	Papa
D	Delta	Q	Quebec
E	Echo	R	Romeo
F	Foxtrot	S	Sierra
G	Golf	T	Tango
H	Hotel	U	Uniform
I	India	V	Victor
J	Juliet	W	Whiskey

(continued)

TABLE 9-1 *(continued)*

Letter	Phonetic	Letter	Phonetic
K	Kilo	X	X-ray
L	Lima	Y	Yankee
M	Mike	Z	Zulu

Q&A with Q-signals

Q-signals (also known as *Q-codes*) began in the early days of radio as a set of standard abbreviations that saved time and allowed radio operators who didn't speak a common language to communicate. Today, amateurs use Q-signals more for the speed than to break the language barrier. (Most amateurs use English well enough to make basic contacts.) The definitions have drifted a little over the past century of radio, but these are in wide use today. You already learned one Q-signal: QSO. Here are some more.

Table 9-2 lists many common Q-signals. Both QRM and QSY are included in the Technician Class exam.

TABLE 9-2 ## Common Q-Signals

Q-Signal	Meaning As a Query	Meaning As a Response
QRG	What is my exact frequency?	Your exact frequency is _____ kHz.
QRL	Is the frequency busy?	The frequency is busy. Please don't interfere.
QRM	Are you being interfered with?	I'm being interfered with (or just Interference).
QRN	Are you receiving static?	I am receiving static (or just Static).
QRO	Shall I increase power?	Increase power.
QRP	Shall I decrease power?	Decrease power.
QRQ	Shall I send faster?	Send faster (__words per minute [wpm]).
QRS	Shall I send more slowly?	Send more slowly (__wpm).
QRT	Shall I stop sending?	Stop sending.
QRU	Have you anything more for me?	I have nothing more for you.
QRV	Are you ready?	I'm ready.

Q-Signal	Meaning As a Query	Meaning As a Response
QRX	Do you want me to stand by?	Stand by.
QRZ	Who's calling me?	(Not used as a response)
QSB	Is my signal fading?	Your signal is fading.
QSL	Did you receive and understand my transmission?	Your transmission was received and understood.
QSP	Can you relay to ___?	I can relay to___.
QSX	Can you receive on ___ kHz?	I'm listening on ___ kHz.
QSY	Can you change to transmit on another frequency (or to ___ kHz)?	I can move to another frequency (or to ___ kHz).
QTC	Do you have messages for me?	I have messages for you.
QTH	What is your location?	My location is ___.

During contacts, Q-signals often take the form of questions. "QTH?" means "What is your location?" for example, and the reply "QTH New York" means "My location is New York." Sometimes, Q-signals are used as an abbreviation, such as QSO, which you already know means "contact." QST is "calling all amateurs."

Can you hear me now? — Signal reports

A *signal report* is an indication of your signal's strength and clarity at the receiving station. Given as two or three digits, the signal report tells the other station how well you're receiving their signal and whether their signal quality is good or not. The signal report is usually given during the first or second transmissions of the QSO. Table 9-3 provides the details.

>> **SSB:** A two-digit system communicates readability and strength, although sometimes you can use a single-digit quality report from Q1 to Q5.

>> **CW and RTTY:** The same two-digit system is used for readability and strength, but a third digit is added to indicate the purity of the signal's transmitted tones — rarely anything but 9 nowadays, because transmitting equipment is quite good. If you encounter a poor signal, however, don't hesitate to make an appropriate report to let the other operator know there is a problem.

>> **Digital modes:** Although the CW/RTTY system is the norm, it's becoming common to use a Readability, Strength, and Quality (RSQ) report. The final digit represents the quality of how a signal is being decoded. (For more

information on RSQ, see www.rsq-info.net.) If you are using one of the modes supported by the *WSJT-X* package, such as FT8 or MSK144, the signal report is built in to the messages that are exchanged.

>> **FM:** The signal report is the degree to which the noise is suppressed, called *quieting.*

TABLE 9-3   Reporting Signal Quality

Mode	System	Report Definitions
SSB	RS: Readability and Strength.	*R* is a value from 1 to 5. The value 5 means easy to understand, and 3 means difficult to understand; 1 and 2 are rarely used.
		S is a value from 1 to 9. This number generally corresponds to the radio's signal-strength-meter reading on voice peaks.
	Quality number(*number*): Indicates overall quality.	*Q* is a value from Q1 to Q5. Q5 indicates excellent readability; reports below Q3 are rare.
CW	RST: Readability, Strength, and Tone.	*R* is a value from 1 to 5; the values mean the same as for SSB.
		S is a value from 1 to 9; the values mean the same as for SSB.
		T is a value from 1 to 9. The value 9 is a pure tone, and 1 is raspy noise. The letter *C* is sometimes added to indicate a chirpy signal.
Digital (alternative)	RSQ: Readability, Strength, and Quality	*R* is a value from 1 to 5. The values mean the same as for SSB.
	FT8, MSK144, JT65, etc. — the signal report built into the messages is the signal-to-noise ratio (see Chapter 12 for information about decibels or B).	*S* is a value from 1 to 9. The values mean the same as for SSB.
		Q is a value from 1 to 9, reflecting the quality of your signal's modulation.
FM	Level of quieting (signal report is for the station calling, not the repeater's output signal strength).	*Full quieting* means that all noise is suppressed. *Scratchy* means that noise is present, possibly enough to disrupt understanding. *Flutter* means rapid variations in strength as a vehicle is moving. *Just making it* means that the signal is only strong enough to activate the repeater and not good enough for a contact.

Casual contacts

Keep these things in mind while you're making the contact:

>> **Start with your basic information:** Your call sign, signal report, operator name, and station location. Conversations may be between hams in the same town or a world apart so exchange the basics first.

Most digital mode software allows you to enter information about yourself and your station as stored messages. These are called *brag tapes* or *macros* — sent by pressing a single key — so you don't have to type everything in every time.

>> **After you exchange basic information, you may wander off in any direction.** Hams talk about family members, other hobbies, work, propagation, technical topics, operating, you name it. Just about anything may be discussed.

In general, hams avoid talking about politics or religious topics, and they don't use profanity. There are plenty of online forums for contentious topics — please use one of them. Those restrictions still leave a lot to talk about, however, and hams seem to cover most of it.

>> **Wrap up the contact when you run out of things to talk about, when conditions change, or when maintaining contact gets difficult.** Thank the other station for the contact, exchange call signs once more, and tune away.

>> **Keeping it short:** It's okay to just make a short contact without a long conversation. Just say (or send) "Thanks for the short contact" and wrap it up. (See the sidebar "The long goodbye" later in this chapter.) Some hams just like seeing where their signal can travel.

Nets and talk groups — On-the-air meetings

Nets (short for *networks*) meet at a specified time and frequency. Each net has a theme — emergency-communications training, maritime or mobile service, or specific topics such as antique radios or technical Q&A.

Most nets are *directed*, meaning there is a *net control station* or *NCS* directing operation and activities. *Nets* generally meet for a short time on a preannounced frequency and time. Ongoing disaster relief nets, though, may be on the air for extended periods. Service nets like the Maritime Mobile Service Network are active every day for several hours. The Technician exam has several questions about net operating.

Talk groups (also *talkgroups*) or *rooms* are similar: These are like repeater channels but are created by digital voice systems to allow multiple groups to share one repeater without interfering with each other. They may be accessed via a specific repeater channel but are *virtual* groups created by the digital system. They can be focused on a specific topic, a geographic region, or perhaps a type of service activity. In between nets, talk groups generally host open discussions and contacts. I discuss them in more detail later in the chapter.

TIP

You can find nets for a specific topic or frequency online. The ARRL Net Directory, for example, at www.arrl.org/arrl-net-directory lists nets on HF and VHF/UHF frequencies. Digital voice mode conversations can often be accessed via the Internet, as well. Take a look at the Digital Ham Nets page on Facebook to get an idea of what's out there on the different digital voice systems: www.facebook.com/groups/DigitalHamNets.

If you are going to join or *check in* to a directed net, follow these tips:

> » **Tune in to the listed net frequency, channel or talk group. Listen for a request to check in with the net control station (NCS), give your call, and when recognized, state your business according to the NCS instructions.** The NCS orchestrates all exchanges of information and formally terminates the net when business is concluded. (The net may meet until all business is taken care of or for a fixed amount of time.)

> » **If you're a visitor, find out when you can check in.** Nets often have specific times when visitor stations can participate, usually at the end of the check-in or just before the net concludes. Guests and visitors will be invited to call.

TIP

Listen to the way other stations with business for the net let the NCS know. Each net operates slightly differently, and you'll have a better check-in experience if you have listened to their procedures a couple of times first. There is often announcement at the beginning of the net with some instructions on how to check in and let the NCS know if you have any business or messages (*traffic*) for the net.

Nets and net operation are discussed in more detail in Chapter 10.

TECHNICAL STUFF

Roundtables are similar to nets. They are shared QSOs in which a group of hams on one frequency exchange information informally. Each ham transmits in turn, and everyone gets a chance to transmit in sequence. Some roundtables have moderators.

Contests and DXing — Radiosport

Many hams like to participate in radio contests, which are competitions in which call signs and short messages are exchanged as quickly and accurately as possible. Similar to the short-term contests is *chasing DX* or *DXing* — pursuing contacts with distant stations. (I discuss both in detail in Chapter 11.)

In a contest, whether you're competing or just "giving out some points," here are some guidelines to follow:

>> **Keep contest contacts short.** In a contest, the object generally is to make the largest number of contacts, so a lot of conversation isn't desirable.

>> **Pass along just the minimum amount of information, called the** *exchange.* Then tune in search of more contacts.

REMEMBER

Good contest practices work for DXing, too. If many stations are calling, just exchange call signs and signal reports. This allows other hams to make those sought-after contacts, too.

Ragchews with DX stations can work if conditions support good signals in both directions. If conditions don't support a conversation or if the DX station is trying to work stations quickly, keep your contact short. Try to judge conditions and tailor your contact appropriately.

>> **If you encounter stations making contest QSOs, listen until you figure out what information is being exchanged before calling.** By far the most common types of information exchanged are signal reports, locations (often expressed as a numbered geographical *zone* or abbreviation defined by the contest sponsor), and *serial numbers*. Serial numbers count each contact made in the contest. If you're making your fifth QSO in a contest, for example, your serial number is 5.

If the rate of making contacts is relatively relaxed, just ask "What do you need?" Contesters are happy to explain what information they want to get from you. Usually, you can find the complete rules for contests on the sponsor's website or in online contest calendars such as www.contestcalendar.com and www.arrl.org/contest-calendar. If contestants are making contacts lickety-split, you may want to wait, or figure out what they need on your own and then make a quick contact.

TIP

Remember that not everyone speaks English. Most hams who don't speak English still know enough words of English to communicate a name, location, and signal report. Otherwise, an international set of Q-signals allows you to exchange a lot of useful information with people who speak a different language. Websites like www.infoplease.com/common-english-phrases-translated-other-languages

and the Google Translate service can help, too. DXing is a great way to exercise that rusty high-school Spanish or German. The DX operator will appreciate your efforts, too!

How contacts get started

To make contacts, you have to start contacts! In this section, I give you an example of how hams start contacts, using my call sign (NØAX) and either of my two sons' (KD7FYX and KG7FUT) as examples. (Don't forget that Ø is the way hams write a zero.) Replace my call sign with your own.

TIP

In ham radio, the custom for calling a station is to give their call sign first, then yours. For example, I would call my son this way: "KD7FYX this is NØAX." Calls are given in the same order of a tag on a present — To followed by From. This alerts the station you're calling, then tells them who is calling.

Starting repeater contacts

After you get a Technician license, you will most likely begin operating by making contacts on a repeater. So we'll start there! This section describes analog FM repeaters — still the most common type — but be aware that digital systems like D-STAR, DMR, or System Fusion have different "radio manners."

Hams generally use repeaters as regional intercoms and the signals are strong in the repeater's coverage area. That means you don't have to make extended calls to start a QSO. (You won't hear "CQ CQ CQ" on a repeater, for example.) Hams will listen to or monitor a repeater channel to contact friends or to be present in the event someone needs a contact. It is common for someone interested in making a QSO to simply say, "This is KG7FUT. Is anyone around?" This is an invitation for anyone to call.

It's also an invitation to call if someone has announced their presence by simply giving their call sign. You will also hear stations saying something like "This is KD7FYX monitoring," or making some other general "I'm here and listening" announcement. In response, a caller makes a 1-by-1 call ("KD7FYX this is NØAX") to see whether the other ham responds or "comes back" to them. Calling a station immediately after a contact is finished is another good way to start a contact with them.

Starting an SSB voice contact

Single-sideband (SSB) is used on HF and VHF/UHF but is most common on HF. Although the rest of this section assumes you're listening on HF, you will hear similar exchanges on the calling frequencies for VHF (50.125 MHz on 6 meters and 144.200 MHz on 2 meters are the most used), particularly if you live in a populated area.

Because there is no formal structure of channels on the HF bands, one way of making a contact is tuning to a calling frequency or net frequency where certain types of activity are conducted. Or you can tune around to find a station to contact or a clear frequency where you can throw your call sign out there to see who hears you!

TIP

When you come across a ham who's *calling CQ* (see "Calling CQ," later in this chapter and it's on the exam, too), they are making a general "Come in, anybody" call. This situation is the easiest way for you to make a contact. You'll hear something like this: "CQ CQ CQ, this is Kilo Delta Seven Foxtrot Yankee X-ray, standing by . . ." (See the section "Know your Alfa, Bravo, Charlies: Using phonetics," earlier in the chapter to understand their call sign.)

TIP

Paper and pencil can be very handy when you're just starting out. After all, there's so much to remember! When you're getting ready to start a contact, make a note of the other station's call sign and any other information so you won't forget. This is good practice when you're listening because it helps you remember how contacts are made.

An answering station will give the CQing station's call sign once (there's no need to repeat it; they already know it!) and then give theirs three times — this is a *1-by-3 call*. If the calling station is strong, just give your call twice instead of three times — a *1-by-2 call*.

As the contact gets going, you may hear something like this: "NØAX [possibly in phonetics] from KG7FUT, thanks for the call. Your signal report is . . ." When you do, the QSO has begun!

Starting a CW or digital mode contact

With CW and digital modes like RTTY, Olivia, or PSK31, the process is much the same as it is for voice contacts.

REMEMBER

If a digital mode allows each operator to send characters back and forth as they are typed (or at the end of each line when the Enter key is pressed) that is called *keyboard-to-keyboard* communication. This is *chat* operating with back-and-forth conversation, just like online messaging.

TECHNICAL STUFF

Some digital modes, called *structured modes*, like FT8, packet radio, or PACTOR/WINMOR have special procedures for starting a contact. These modes are discussed in Chapter 11.

TECHNICAL STUFF

If you are using software or a code reader to receive or *copy* signals off the air, be sure to use "monitor" mode which displays every character that is decoded.

Starting a contact proceeds like an SSB contact, so you will receive something like this: "CQ CQ CQ DE KD7FYX KD7FYX KD7FYX K."

TIP

DE is telegrapher's shorthand for *from*. *K* means "end of transmission, go ahead." (*Note*: Morse code doesn't use uppercase or lowercase characters, so *de* is equivalent to *DE*.)

Like on SSB, the responding station will send or type something like this: "KD7FYX DE NØAX NØAX NØAX K." (This is a 1-by-3 response as I discussed earlier.)

The response will be something like this: "KD7FYX DE NØAX TKS FOR THE CALL MY NAME IS. . . ." (*TKS* and *TNX* are shorthand for *thanks*.) The contact proceeds similarly to SSB contacts from there.

TIP

Telegraphers and typists tend to use all sorts of abbreviations to shorten text. A table of abbreviations is available at en.wikipedia.org/wiki/Morse_code_abbreviations.

Joining a contact

Sometimes, you need to call a station before they complete a contact. Joining another contact is called *breaking in* (or *breaking*). The proper procedure is to wait for a pause in the contact and quickly say "Break" (or send *BK*) followed by your call sign. Or just send your call sign.

Why would you want to do this? Perhaps you have an emergency and need to make contact right away. More frequently, you tune in to a contact, and the participants are talking about a topic with which you're familiar; if you wait for the contact to end, you may not be able to contribute or help. Maybe one of the stations is a friend and you want to say hello!

Breaking in follows these steps:

1. **Listen for a good opportunity to make your presence known.**

When the stations switch transmitting and receiving roles, that's usually a good time to break in. You may hear something like this: "So, Sharon, back to you. AE7SD from KG7FUT."

2. **Quickly make a short transmission.**

   Don't be shy and wait for the other station to begin transmitting. Say your call sign right away, just once. You can say "break," but it's not necessary since you're just trying to get the QSO-er's attention.

3. **Wait to see whether either station heard your transmission.**

   If a station hears you, the operator may say something like this: "This is AE7SD. Who's the breaking station?" (or maybe just "Who's calling?")

   If no one hears your transmission, start over with Step 1.

4. **Respond as though you're answering a CQ.**

   Say: "AE7SD, this is NØAX NØAX. Over."

5. **Depending on the circumstances, give your name and location before proceeding to explain why you broke in.**

   At that point, the stations may engage you in further conversation, and you'll be in a *three-way QSO*. Sometimes, however, they won't want to have a third party in the contact, in which case you should just courteously sign off and go on to the next contact.

Failing to make contact

What if you try to make a contact and your call doesn't get a response? Your signal may be too weak, or the station may have strong noise or interference. In such a case, just find another station to call and keep trying. The most important thing is to keep from getting discouraged!

Assuming that your signal is strong enough for other stations to hear, however, several other things may have happened:

>> **Other hams are calling at the same time.** You can either wait around until the station you intend to contact is free and then try again, or you can tune around for another contact opportunity.

>> **The calling station can hear you but can't make out your call sign.** The ham may either ask you to call again or respond to you, but doesn't have your call sign correct. The station may say or send "Station calling, please come again" or "QRZed?" or "Who is the station calling?" At this point, just repeat your call two or three times, using standard phonetics, and say "Over" when you finish.

TIP

QRZed? is the international Q-signal for "Who's calling me?" (refer to Table 9-2, earlier in this chapter). Hams often use *zed*, the British pronunciation of the letter *Z*. On the digital modes or Morse code, just *QRZ?* is used.

>> **The station gets your call wrong by a letter or two.** First, stand by for a few seconds to make sure that another station with a similar call sign isn't on the same frequency. (I'm often on the air in contests at the same time as NØXA, for example, and we're always getting confused.) If a few seconds go by and you don't hear another station responding, respond as follows: "KD7FYX, this is NØAX NØAX. Do you have my call correct? Over."

Don't be bashful about correcting your call sign. After all, it's your radio name. The other station will want to have it correct, too! After your call is received correctly by the other station, proceed with the rest of the contact.

If repeated attempts at making contacts aren't producing results, check out your equipment. The easiest way is to have a friend make a contact with you. That way, you'll know whether your transmitter is working and your signal is understandable.

You can also run through the following checklist to make sure you're transmitting what, when, and where you think you should be:

>> **Are you transmitting on the right frequency?** Key (turn on) the transmitter and watch the radio's display very carefully. If you are operating on SSB or CW, the indicators for frequency and sideband should stay exactly the same. If not, you're transmitting on a different sideband or frequency from the one you think you're using. On a repeater, the frequency should change from the repeater's output channel (where you're listening) to the repeater's input channel (where the repeater is listening). If it doesn't you may not have your radio configured properly.

>> **Is your transmitter producing power?** Watch the radio's power output meter while you transmit to see that you are actually producing a signal. On SSB, the power output should follow the loudness of your voice. On CW, it should follow your keying, and on the digital modes and FM, the power output will remain fairly constant while transmitting.

>> **Is the antenna connected properly?** You should be receiving most signals as moderate to strong, with an indication of 4 to 9 on the radio's *S meter*, which displays signal strength. If the signals are very weak, you may have an antenna or cable problem. This problem also shows up as a reading of more than 5:1 on your radio's *SWR meter*, which measures how well your transmitter power is getting to the antenna. (A reading of 2:1 or less is the best case with higher values indicating that you may have an antenna or feed line problem.)

During a contact

Now that you know how contacts get going, what do hams talk about, anyway? As in most casual contacts with people you don't know, warming up to a new contact takes a little time.

During the initial phase of a contact, both parties exchange information about the quality of the signals, their names, and their locations. This phase is a friendly way of judging whether conditions permit you to have an extended contact. Follow with information about your station and probably the local weather conditions. This information gives the other station an idea of your capabilities and indicates whether static or noise is likely to be a problem.

Common information to exchange

Along with a signal report, here are a few other items often exchanged when making a contact:

>> **QTH (location):** On HF, where contacts take place over long distances, you generally give your town and state or province. You can give an actual address if you're asked for it, but if you aren't comfortable doing so, you don't have to. On VHF/UHF, you report the actual physical location, particularly if you're using a mobile radio, such as a *grid square* (www.arrl.org/grid-squares). Online call sign look-up services and databases like qrz.com can tell you the other station's location, too.

>> **Radio:** You can just report the power output shown on your transmitter's power meter (such as 25 watts) or give the model number and let your contact assume that the transmitter is running at full output power.

>> **Antenna:** Typically, you just report the style and number of elements, such as a dipole, 3-element beam, or 5⁄8-wave whip. You can also report the specific model if it's a commercial product. If you made the antenna yourself, you can proudly tell the other station it's *homebrew*!

>> **Weather:** Remember that stations outside the United States report the temperature in degrees Celsius. Standard weather abbreviations that you can use for CW and digital modes include SNY, CLDY, OVRCST, RNY, and SNW. A Russian ham in Siberia once sent me his weather report as "VY SNW" (very snow)!

TIP

Identifying yourself on the air is the subject of several exam questions. You are required to identify with your call sign every ten minutes during a contact and when you stop communicating, such as at the end of a contact. Just give your call sign once. There is no need to add "for ID" because that is why we have call signs in the first place!

After you go through the first stages of the QSO, if the other ham wants to continue, you can try discussing some other personal information. The possible topics are almost endless — your age, your other hobbies, what you do for a living, your family members, any special interests, and ham-radio topics such as propagation conditions or particularly good contacts that you made recently.

TIP

The exam expects you to remember that the FCC forbids obscene speech (which is pretty rare on the air). The three topics that seem to lead to elevated blood pressures are politics, religion, and sex — hardly surprising. So hams tend to find other things to talk about. Oh, sure, you'll find some arguments on the air from time to time, just as you do in any other group of people. Don't be drawn into arguments yourself, however; no one benefits. Just "spin the big knob" and tune on by.

Try to keep your transmissions short enough that the other station has a chance to respond or that someone else can break in (see "Joining a contact," earlier in this chapter). If propagation is changing, or if the band is crowded or noisy, short transmissions allow you to ask for missed information. But you can have a QSO just as long and detailed as both parties want it to be.

TIP

At the conclusion of the contact, invite the other station to call you again. Lifelong friendships are forged on the ham bands!

Getting used to repeater style

Because VHF and UHF repeater voice contacts are usually local or regional, they tend to be used for personal contacts rather than to make random acquaintances. Most hams use a few favorite repeaters or simplex frequencies as a sort of regional intercom. They turn on a radio at home or in the car and monitor a channel or two to keep an ear out for club members or friends. Even though several people may be monitoring a repeater, they mostly just listen unless someone calls them specifically or they hear a request for information or help.

This style can be a little off-putting to new hams and can even seem unfriendly at times. Rest assured that the hams aren't being unfriendly; they're just not in "meet and greet" mode on the repeater or a favorite simplex channel. Imagine the difference between meeting someone at a party versus at a grocery store. At the party, everyone expects to make new acquaintances and has conversations. At the store, people aren't there as a social exercise and may even seem to be a little brusque.

As you monitor the different channels, you will learn which repeaters encourage conversations (ragchews) and which don't. If you find a repeater that's ragchew-friendly, you'll have a fairly easy time making a few casual contacts.

Joining a group

The best way to become acquainted with a group is to participate in its activities, such as a net. In areas with good repeater coverage — such as cities and suburban areas — nets are a very common group use of FM repeaters that Technicians can use.

The most common local nets are for emergency-services groups, weather and traffic (the automobile kind), and training. Technical assistance or question-and-answer nets are common in the evenings and on weekends. For-sale or "swap-n-sell" nets are usually held on weekends. Your club or group will be able to tell you what nets are available and how to participate.

TIP

Many general-service clubs have "new ham" nets intended to help newcomers learn the ropes and get answers to questions. All are welcome, even if you're not a member. These nets sometimes meet on HF frequencies available to Technicians. Check the websites of local clubs to see if they sponsor this type of net and become a regular check-in. You might even be able to answer a few questions yourself!

If you aren't part of a group that has regular nets, you can use the ARRL net-search page at `www.arrl.org/arrl-net-directory-search` to find local nets on the VHF and UHF bands. Check the websites of local clubs, too. Your mentor (or Elmer; see Chapter 3) may be able to help you with times and frequencies, and so will other radio club members.

Almost all nets include a special opportunity to call in for visitors, generally at the end of the net check-in period and again before the net closes. That's your chance! When you check in (by giving your call sign and maybe your name; follow the given directions), ask for an after-net contact with the net control station or a station that said something of interest to you. After-net contacts are initiated on the net frequency when the net is completed. Sometimes, they're held on the net frequency; at other times, the stations establish contact there and then move to a different frequency.

During the after-net contact, you can introduce yourself and ask for help finding other nets in the area. If you have specific interests, ask whether the station knows about other nets on similar topics. Ideally, you'll get a referral and maybe even a couple of call signs to contact for information.

After you check in to a few nets, your call sign starts to become familiar, and you have a new set of friends. If you can contribute to a weather or traffic net, by all means do so. Contributing your time and talents helps you become part of the on-the-air community in no time, and it's good practice for your skills, too.

Calling CQ

After you make a few non-repeater contacts, the lure of fame and fortune may become too strong to resist. It's time to call CQ yourself!

TIP

Learning how to call CQ is a valuable skill. It's the way you find out "who's out there" and where your signal can be heard. An effective CQ lets everyone know what your call sign is and that you are actively wanting to be called by them. Think of each CQ as a short advertisement for your station that makes the other operator think, "Sure, I'll talk to them!"

There is a knack for knowing the right way to CQ for band conditions and signal strength. You can develop this knack by listening to other stations calling CQ. Did their CQ make you want to call them? Or keep tuning! Was their signal clean and easy to understand or was it distorted or garbled? Strive to make your CQ-ing signal a good one.

For casual CQs, you can start with the time-tested "three by three" consisting of "CQ CQ CQ" followed by "this is" and your call sign given three times. Then do that two more times. Pause for five to ten seconds and listen for someone calling you back. It's not uncommon to do this several times before someone responds so be patient.

During a contest or if you are operating a special event station or from a sought-after location, you might only have to send the CQ sequence once, perhaps followed by your location if that matters to the calling stations. Keep your CQs short and don't wait a long time between CQs. Think of CQing as fishing, with your CQ as the lure you cast into the water!

Anatomy of a CQ

A CQ consists of two basic parts, repeated in a cycle:

>> **CQ itself:** The first part is the CQ itself. For a general-purpose "Hello, World!" message, just say "CQ." If you're looking for a specific area or type of caller, you must add that information, as in "CQ DX" or "CQ New England."

>> **Your call sign:** The second part of the CQ is your call sign. You must speak or send clearly and correctly. Many stations mumble or rush through their call signs or send them differently each time, running the letters together. You've probably tuned past CQs like those.

A few CQs, followed by "from" or "DE," and a couple of call signs make up the CQ cycle. If you say "CQ" three times, followed by your call sign twice, that's a *3-by-2* call. If you repeat that pattern four times, it's a *3-by-2-by-4 call*. At the end of the cycle, you say "Standing by for a call" (or "Over" or just send K), to let everybody listening know that it's time to call.

Here's an example of a 3-by-3-by-3 on CW or a digital mode: "CQ CQ CQ DE NØAX NØAX NØAX CQ CQ CQ DE NØAX NØAX NØAX CQ CQ CQ DE NØAX NØAX NØAX K."

On RTTY, PSK31, and other keyboard-to-keyboard modes, it common to end CQs with "PSE K" (please respond now, go ahead) instead of just plain "K."

Depending on conditions, repeat the cycle (CQ and your call sign) two to five times, keeping it consistent throughout. If the band is busy, keep it short. If you're calling on a quiet band or for a specific target area, four or five cycles may be required. When you're done, listen for at least a few seconds before starting a new cycle to give anyone time to start transmitting.

CQ tips

Here are a few CQ do's and don'ts.

Do:

>> Keep your two-part cycle short to hold the caller's interest.

>> On voice modes, use standard phonetics for your call sign (refer to Table 9-1) at least once per cycle.

>> Send CW at a speed you feel comfortable receiving.

>> Make an effort to sound friendly and enthusiastic.

>> Wait long enough between CQs for callers to answer.

>> Speak in a normal, clear voice.

Don't:

>> Mumble, rush, or slur your words.

>> Send erratically or run letters together.

>> Drag the CQ out. A 3-by-3-by-3 call is good for most conditions.

>> Turn up your microphone gain or speech processing too much. Clean audio gets the best results.

Treat each CQ like a short advertisement for you and your station. It should make the listener think, "Yeah, I'd like to give this station a call!"

TIP

If your radio has the capability to listen between CW dits and dahs (which your operating manual calls *break-in* or *QSK*), use it to listen for a station sending dits while you are CQing. That means "I hear you, so stop CQing and let me call!" Then you can finish with "DE [*your call*] K," and the other station can call instead of waiting until you end the CQ. If you don't get a response, try again after the next CQ. They might not have heard your signal for some reason. If you can't get through, turn to "Failing to make contact," earlier in this chapter.

Casual Conversation — Ragchewing

In this section, I go into deeper detail about the etiquette of the ragchew, ham radio's conversational art. Ragchewing is an excellent way to build your operating skills, too.

"Chewing the rag" is probably the oldest activity in ham radio. It is a very old expression, as I explain in Chapter 1. If you like to chat, you're a *ragchewer*, and you will be following the footsteps of many hams. The ragchewers space themselves out around the nets, roundtables, calling frequencies, and data signals, taking advantage of ham radio's unique frequency agility to engage in a preferred style of operating with a minimum of interference. We all have to share!

Hams come from all walks of life and have all kinds of personalities, of course, so you'll come across both garrulous types, for whom a ragchew that doesn't last an hour is too short, and mike-shy hams, who consider more than a signal report to be a ragchew. Relax and enjoy the different people you meet.

Knowing where to chew

The parts of the HF phone bands between DXing and nets are full of ragchew contacts. Ragchewing on CW is found above the DXers and below the digital signals. Digital mode ragchews mix right in with the DXers. You'll find most ragchew-style contacts around the higher portions of the band where Advanced and General Class hams operate. This is a little above where DX and contest operators tend to be. This is oversimplified, of course, but gives you an idea of how to start.

Sometimes, if a super-rare DX station comes on the air, the sheer numbers of DXers calling (a *pileup*) crowd out sustained contacts, so the ragchewers move up the band. Because DXing tends to attract a crowd, this is somewhat incompatible with the more ordered style of nets. Therefore, nets tend to gather at the other end of the bands.

On the VHF and UHF bands, you usually find ragchew contacts on the repeater and simplex channels, but there are also stations active in the so-called weak signal segments around the calling frequencies listed in Table 9-2. Starting near the calling frequencies, scan the wide-open spaces and you may come across a local group using a frequency as their "watering hole." Join in and say hello!

TIP

The bottom portion of the VHF/UHF bands is referred to as *weak signal,* although that's really a misnomer. The reason for the name is that contacts via CW and SSB can be made with considerably weaker signals than on FM. Most of the weak-signal signals you hear are sufficiently strong for excellent contacts, thank you!

Table 9-4 lists the calling frequencies and portions of the VHF/UHF bands. The operating style in this portion of the bands is similar to HF as far as calling CQ and making random contacts go, but the bands are far less crowded because propagation generally limits activity to regional contacts.

TABLE 9-4   **VHF/UHF CW and SSB Calling Frequencies**

Band	Frequencies (in MHz)	Use
6 meters	50.0–50.3	CW and SSB
	50.070, 50.090	CW calling frequencies
	50.125 and 50.200	SSB calling frequency; use upper sideband (USB)
2 meters	144.0–144.3	CW and SSB
	144.200	CW and SSB calling frequency; use USB
222 MHz (1¼ meters)	222.0–222.15	CW and SSB
	222.100	CW and SSB calling frequency; use USB
440 MHz (70 cm)	432.07–433.0	CW and SSB
	432.100	CW and SSB calling frequency; use USB

You may think that ragchewers are buffeted from all sides, but that's not really the case. Ragchew contacts take place all the time, so they tend to occupy just about any spare bit of band. To be sure, in the case of disasters (when a lot of nets are active), major operations from rare places, or weekends of big contests, the bands may seem to be too full for you to get a word in edgewise. In those cases, try a different band or mode; you'll probably find plenty of room.

TIP

Why do the activities tend to separate? DXing tends to take place at the low end of the HF bands because it's easier for the DX stations and those who want to contact them to find each other. That's why the low-frequency segments are considered an incentive to upgrade to the higher classes of license. It's prime territory for juicy DX contacts!

REMEMBER

The FCC can declare a communications emergency and designate certain frequencies for emergency traffic and other communications. Keeping those frequencies clear is every amateur's responsibility. The ARRL transmits special bulletins over the air on W1AW, by email, and on its website if the FCC does make such a declaration. The restrictions are in place until the FCC lifts them.

Identifying a ragchewer

If you're in the mood for a ragchew, and you're tuning the bands, how can you tell whether a station wants to ragchew? The easiest way is to find an ongoing ragchew and join it. You can break in or wait until one station is signing off then call the remaining station.

Look for a station that has a solid signal — not necessarily a needle-pinning strong station, but one that's easy to copy and has steady signal strength. The best ragchews are contacts that last long enough for you to get past the opening pleasantries, so find a signal that you think will hold up.

TIP

Roundtables — contacts among three or more hams on a single frequency — are also great ways to have a ragchew. Imagine getting together with your friends for lunch. If only one of you could talk at a time, that would be a roundtable. Roundtables aren't formal, like nets; they generally just go around the circle, with each station talking in turn. Stations can sign off and join in at any time.

One cue that the station isn't looking for a ragchew is a targeted CQ. You may tune in a PSK signal and see "CQ New York, CQ New York de W7VMI." W7VMI likely has some kind of errand or message and is interested in getting the job done. Perhaps the station on voice is calling "CQ DX" or "CQ mobiles." In that case, if you're not one of the target audience, keep on tuning.

Another not-a-ragchew cue is a hurried call or a call that has lots of stations responding. This station may be in a rare spot, in a contest, or at a special event. Keep tuning if you're really looking for a ragchew.

Calling CQ for a ragchew

Although responding to someone else's CQ is a good way to get started, it's also fun to go fishing — to call CQ and see what the bands bring.

The best CQ is one that's long enough to attract the attention of a station that's tuning by but not so long that that station loses interest and tunes away again. If the band is quiet, you may want to send long CQs; a busy band may require only a short CQ. As with fishing, try different lures until you get a feel for what works.

On voice modes, the key is in the tone of the CQ. Use a relaxed tone of voice and an easy tempo. Remember that the other station hears only your voice, so speak clearly, and be sure to use phonetics when signing your call. Sometimes, a little extra information — such as "from the Windy City" — helps attract attention. Don't overdo it, but don't be afraid to have a little fun.

THE LONG GOODBYE

If hams do one thing well, it's saying goodbye. Hams use abbreviations, friendly names, phrases, and colloquialisms to pad their contacts before actually signing off. You rarely hear anyone say "Well, I don't have anything more to say. W1XYZ signing off." Sometimes, signing off takes as long as signing on. This is an endearing feature of hams and ham radio in general!

Toward the end of the contact, let the other station know that you're out of gas. Following are some good endings:

- **I AM QRU:** QRU means I am out of things to talk about.

- **See you down the log:** Encourage another contact at a later time.

- **BCNU:** Abbreviation for be seein' you.

- **CUL:** Abbreviation for see you later.

- **73:** Don't forget to send your best regards. (It's just "73" and not "73s.")

- **Pulling the big switch** or **Going QRT:** If you're leaving the airwaves, be sure to say so after your call sign on the last transmission. On voice, say "clear." On digital or Morse, send "NØAX SK CL." Anyone receiving these transmissions will know that you're vacating the frequency.

Making Repeater and Simplex Contacts

Most new hams begin operating as Technician class licensees, with access to the entire amateur VHF and UHF bands. By far the most common means of communicating on those bands is FM voice through a repeater. In this chapter, I begin with the basics of regular FM voice repeaters then move to the more sophisticated digital systems.

Understanding repeater basics

TIP

Figure 9-1 explains the general idea behind a repeater. A repeater receives FM signals on one frequency and simultaneously retransmits (or *repeats*) them on another frequency. The received signals aren't stored and played back; they're retransmitted on a different frequency at the same time they're received, in a process called *duplex operation*. (Talking directly from one station to another without relays is called *simplex operation*.)

REMEMBER

Instead of "channel," hams refer to repeaters by frequency. There is no standard list of repeater channels like there is on marine VHF or Citizen's Band.

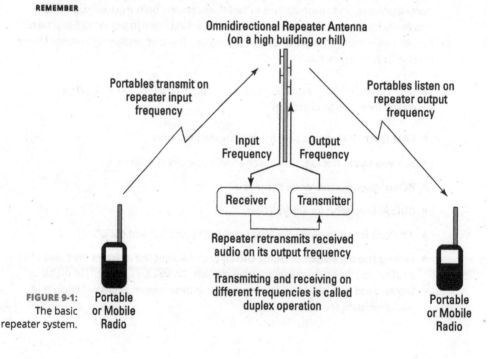

FIGURE 9-1: The basic repeater system.

TECHNICAL
STUFF

Most voice repeaters use analog FM, although there are more digital systems every day. FM is used instead of single sideband (SSB) because of the relative simplicity of the transmitters and receivers — an important consideration for equipment that's operating all the time and needs to be reliable. FM is also relatively immune from static if signal strength is good, so it makes for a more pleasant contact.

TIP

Except for a small segment of the 10 meter band, FM is rarely found on HF due to restrictions on signal bandwidth and to FM's relatively poor quality for weak signals when compared with SSB. FM's qualities are ideal for local and regional coverage on VHF and UHF.

If the repeater is located on a high building, tower, or hill, its sensitive receiver picks up signals clearly from even tiny handheld radios. Then it uses a powerful transmitter to relay that input signal over a wide area. Stations can be separated by tens of miles yet communicate with a watt or two of power by using a repeater. It is common for repeaters to be *linked* together using the Internet or *control links* to extend the range of a single repeater. Digital systems (described later in this chapter) can often be accessed over the Internet from anywhere in the world!

TIP

Ham radio repeaters are constructed and maintained by radio clubs or individual hams as a service to the community. Installing equipment on towers and high buildings often involves rental or access fees, even for not-for-profit amateur groups. If a repeater users' group or club exists in your area, consider joining or donating to it to help defray the cost of keeping the repeater on the air.

Understanding repeater frequencies

To communicate through a repeater you have to know the frequency on which it's listening and the frequency on which it's transmitting. The listening frequency (the one that listens for your signal) is called the repeater's *input frequency,* and the frequency that you listen to is called the repeater's *output frequency.* The difference between the two frequencies is called the repeater's *offset.* The combination of a repeater's input and output frequencies is called a *repeater pair.*

TIP

As Figure 9-2 shows, repeater pairs are organized in groups, with their inputs in one part of the band and their outputs in another, all of them having a common offset. Each pair leapfrogs its neighbor, the channels all spaced equally in frequency; this is the *channel spacing.* The input channel may be at a lower frequency than the output, or vice versa.

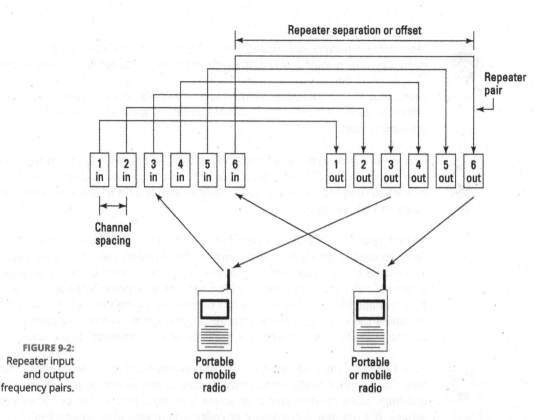

FIGURE 9-2:
Repeater input
and output
frequency pairs.

Figure 9-3 shows the locations of repeater segments on the five primary VHF/UHF bands. The 6 meter band has three groups of repeaters: 51.12 to 51.98 MHz, 52 to 53 MHz, and 53 to 54 MHz. The 2 meter band also has three groups: 144.6 to 145.5 MHz, 146.01 to 147 MHz, and 147 to 147.99 MHz. The 70 cm band hosts one segment with a large simplex segment in the middle: from 440 to 449.99 MHz. The 222 and 1296 MHz bands have a single group. Repeaters are allowed on the 902 MHz and 2304 MHz bands but aren't common. If you have a license with HF privileges, you may want to give the 10 meter FM repeaters a try. They have output frequencies between 29.610 and 29.700 MHz and an offset of –100 kHz.

Finding repeater frequencies

Not all channels are occupied in every area. Also, around the country, some local variations exist in channel spacing and, in rare cases, offset. To find out where the repeater inputs and outputs are for a specific area, you need a repeater directory (see Chapter 8). You'll also find a growing number of repeaters in the 902–928 MHz (33 cm) band.

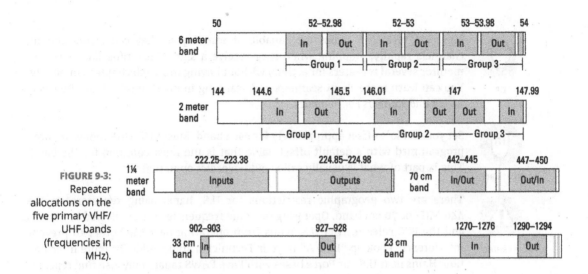

FIGURE 9-3: Repeater allocations on the five primary VHF/UHF bands (frequencies in MHz).

If you don't have a repeater directory and are just tuning across the band, try using Table 9-5, which lists the most common output frequencies and repeater offsets to try. Tune to different output frequencies, and listen for activity.

TABLE 9-5   Repeater Channel Spacings and Offsets

Band	Output Frequencies of Each Group (In MHz)	Offset from Output to Input Frequency
6 meters	51.62–51.98	–500 kHz
	52.5–52.98	
	53.5–53.98	
2 meters (a mix of 20 and 15 kHz channel spacing)	145.2–145.5	–600 kHz
	146.61–147.00	–600 kHz
	147.00–147.39	+600 kHz
220 MHz or 1-1/4 meters	223.85–224.98	–1.6 MHz
440 MHz or 70 cm (local options determine whether inputs are above or below outputs)	442–445 (California repeaters start at 440 MHz)	+5 MHz
	447–450	–5 MHz
902 MHz or 33 cm	902-903 and 927-928	+25 MHz
1296 MHz or 23 cm	1282–1288	–12 MHz

Most FM VHF/UHF radios are capable of *scanning* so they can switch among channels rapidly, stopping when they receive a signal. Scanning allows you to monitor several repeaters for activity without having to switch channels manually. You can learn more about scanners and scanning in my Dummies book *Two-Way Radios and Scanners For Dummies.*

Set your radio's offset appropriately for each band. Most VHF/UHF radios are pre-programmed with a default offset value that is the most common for the band. This is part of what is usually called *auto-repeater* operation.

There are two geographic restrictions for U.S. hams using repeaters on the 440 MHz or 70 cm band. Operating on some frequencies is not allowed in Canada and the FCC rules restrict U.S. hams from operating near the border to prevent interference. Look up "Line A" in your Technician study guide for more information. Hams near U.S. Air Force bases with Pave PAWS radars may also find repeater operation curtailed. The 70 cm band is shared with the primary radiolocation and military users so hams have to avoid interference with these important installations.

Making a repeater contact

Begin by picking a repeater with a good, strong signal at your location. Set your radio for the right repeater offset and maximum power output (for now). Turn on *tone encoding* if required by the repeater. (The radio's operating manual can tell you how to select and activate the tones.) Listen for a minute to be sure no one else is using the repeater or that there isn't a net going on. Review the section "Starting repeater contacts," earlier in this chapter.

For your very first contact, it's a good idea to prearrange the contact with a friend so you'll be less nervous. This is called *making a schedule* (or just *sked*). They will help you do things just right.

The next step is to see if your signal is strong enough to activate the repeater. To do so, you'll need to make a short transmission. Press the PTT switch and say something like, "This is NØAX. Can someone give me a signal report, please?" (You can also ask for a "radio check.") A station may respond after a few seconds. If you did activate the repeater, when you release the PTT switch after a couple of seconds the repeater transmitter turns off and a short burst of noise known as a *squelch tail* is heard.

Because unidentified transmissions are not allowed, the practice known as *kerc-hunking* is discouraged — pressing the microphone PTT switch for a second or two without saying anything. The kerchunker then unkeys the microphone and listens to see if the repeater transmitter is on, indicating the repeater heard the transmitted signal. To other stations, the repeater turning on and off sounds a little like *kerchunk* and it's aggravating. All you have to do is to say your call sign and you'll get the same result without breaking a rule.

Listen for a station accessing the repeater, which sounds something like this: "N7WA monitoring" or just "N7WA." When you hear that transmission, respond with a quick 1-by-2 call using phonetics, such as "N7WA, this is NØAX, November Zero Alpha X-ray, over." If they respond, you can begin a contact as I discussed earlier.

TIP

If you haven't made many contacts, let the other station know, and he or she will walk you through the basic steps.

TECHNICAL STUFF

Digital voice systems have different procedures for accessing a repeater than an analog FM repeaters do. For example, in a DMR system, you have to press the PTT switch and wait for a *confirmation tone* before your signal will be accepted by the DMR system.

REMEMBER

Repeater contacts can be extended ragchews, but remember that the contact is occupying the channel over a wide area. Pause frequently in case someone else needs to use the repeater.

Repeaters have a *time-out* function that prevents their transmitters from being on continuously. Each time the transmitter is turned on the time-out timer starts. If the timer expires (usually three minutes), the transmitter is turned off, even if a station is still talking! After some time to let the transmitter reset, the transmitter can be turned on again by a received signal. There may be some announcement like "Time out!" as the transmitter is turned off or is able to be turned on again. There's no need to be embarrassed about timing out a repeater (unless you keep doing it), because it's a rare ham who hasn't done it.

TIP

If you are monitoring a repeater and hear a station with an unfamiliar call ask for a contact, a radio check, or just announce their presence — give that station a call! You probably appreciated someone responding to your call when you were "the new ham in town," so why not return the favor?

To fit in with FM's strong-signal intercom style, keep your transmissions short. This also prevents activating a repeater's time-out system.

Using access control

To minimize interference from other repeaters and strong signals from nearby transmitters, extra audio tones are used as described in the following paragraphs. Sometimes, individual hams use these control tones when talking to each other in crowded areas, such as at hamfests or conventions. Find out what system is used and how to configure your radio to use it. Repeater directories provide the information you need.

TIP

Most repeaters now use a system called *tone access* — also known as *sub-audible*, *PL*, or *CTCSS* (*Continuous Tone Coded Squelch System*). There are several exam questions involving this system. You may have used tone access on the popular Family Radio Service (FRS) and General Mobile Radio Service (GMRS) radios, where the tones are known as *privacy codes*. Tone access keeps a repeater or radio output quiet (or *squelched*) for all signals except those that carry the proper tone. You won't hear the low-frequency tones, but if you don't transmit the required tone, the repeater won't retransmit your signal, and you won't be heard.

Using tone access

Regardless of what it's called, tone access works this way: When you transmit to a repeater, a low-frequency tone (between 67 Hz and 254.1 Hz) is added to your voice. (You can find a list of these tones and how to select them in your radio's operating manual.) When the repeater receives your transmission, it checks your transmitted signal for the correct tone. If it detects the correct tone, the repeater forwards your voice to the repeater output. This system prevents interfering signals from activating the repeater transmitter; those signals won't carry the correct tone signal and aren't retransmitted.

TECHNICAL STUFF

For a more detailed discussion of tone access, see *The ARRL Handbook* or the article "Decoding the Secrets of CTCSS," by Ken Collier (KO6UX), in the Clallam County ARC library at olyham.net/library.htm.

Most radios also offer *tone squelch*, which uses the same set of tones to control the squelch circuit. It works just like the repeater receiver's tone access, in that only signals with the right tone are output as audio. Most of the time, the same tone is used for both transmitting and receiving.

TIP

Many recent radio models also have a tone-decoder function that detects which tone a repeater is using. If your radio doesn't have this function, and you don't know the correct tone, you can't use the repeater. How can you find out what the proper tone is? Check a repeater directory, which lists the tone and other vital statistics about the repeater. When you determine the correct tone, either via the tone-decoder function or the repeater directory, you can program your radio to send the correct tone and activate the repeater.

Using Digital Coded Squelch (DCS)

DCS is another method of reducing interference. It allows you to hear only audio transmitted by selected stations. DCS consists of a continuous sequence of low-frequency tones that accompanies the transmitted voice. If your receiver is set to the same code sequence, it passes the audio to the radio's speaker. If the transmission uses a different code, your radio remains silent.

Most people use DCS to keep from having to listen to all the chatter on a repeater, hearing only the audio of others who use the same DCS sequence, such as friends or club members.

Not all repeaters pass the tone access or DCS tones through to the transmitter and may filter them out.

Miscellaneous repeater features

You can find an amazing set of features in repeater land. Many repeaters have voice synthesizers that identify the repeater and announce the time and temperature. Hams who use the repeater can activate and deactivate some functions, such as the ability to make phone calls on the air and automated announcements of time or temperature, by using the keypads commonly used on microphones. Repeaters are linked to provide wide coverage even across bands or may have connections via the Internet, as described in "Digital repeater networks," later in this chapter.

Autopatch

Some repeaters have a feature called *autopatch*, which allows a repeater user to make a telephone call through the repeater. You may think that mobile phones make autopatch unnecessary and mostly, you're right! Autopatch is not widely used today. However, even in a world full of mobile phones, autopatch can still come in handy if your mobile phone's battery goes out or the network malfunctions or overloads in a disaster or emergency.

All autopatch calls take place "in the clear" over the air so anyone can listen; they're not private.

Remote receivers

Within a local or regional area, many repeater systems may use *remote receivers* that relay weak signals from outlying areas back to the main repeater transmitter. The relayed signal is transmitted over an *auxiliary station* operating on a VHF or UHF band. It's not unusual for a repeater to have three or more remote receivers.

Open and closed repeaters

In a repeater directory, you may see some repeaters that are marked as *closed*, meaning that they're not open to the ham-radio public. Some repeaters are closed because they're dedicated to a specific purpose, such as emergency communications. Other repeaters are intended to be used only by members of their supporting group.

Rest assured that you can use a closed repeater in case of emergency, but respect the wishes of its owners and don't use it for casual operation. If you aren't sure whether a repeater is closed or not, transmit to it and say something like this: "This is NØAX. Is a control operator monitoring?" If you get a response, that person is the one to ask.

Maximizing your signal

When you are getting started with repeaters, it can be hard to understand what to do when you are having trouble hearing another station or being heard. The right choice depends on what the problem is.

If you are not being received clearly by the repeater, other stations will receive a strong signal from the repeater but your signal is scratchy and noisy. You can make your signal stronger by using more power, a better antenna, or getting to a location with a better path to the repeater's receiver.

If you are not receiving the repeater clearly, everyone's signal through the repeater will be scratchy and noisy. Use a better antenna or get to a better location. Using more power will not help you receive the repeater better.

You may also experience what is called a *dead spot* where your signal fades suddenly as you move. This occurs if a building or hill blocks the path to the repeater receiver. If the problem is an obstruction blocking the path, you need to get out from behind or under it. Sometimes, more power or a better antenna will help, as well.

Signals can also take two or more paths to the repeater receiver (or from the repeater's transmitter) and cancel out — this is called *multi-path*. If the problem is multi-path, you may be able to move a short distance to where the signals no longer cancel. The distance can be as short as one-half of a wavelength (1 meter or 35 cm on the 2 meter or 70 cm band, respectively)! If you are in a vehicle, try moving a short distance. If you are at home, try moving your antenna. If you have a beam antenna, try pointing it in different directions to see whether one path is better with less fading.

Setting up your radio

Now that you know about offsets, tones, and repeater frequencies, it's time to set up your radio to use them. While you're in there, take a few more minutes to check out the rest of your radio's operating manual, too!

TIP

The new digital systems discussed later in this chapter require their own type of setup. This section deals with the most common type of VHF/UHF radio used for analog FM.

To use your radio effectively, you need to know how to do each of the following things. Find the pages in your manual showing you how to do each one and practice a little bit.

TIP

The exam has several questions about these settings on a radio.

>> **Set the radio's receive frequency and transmit offset.** Know how to switch to simplex (no offset) or to listen on the repeater's input frequency. Some radios have a **REV** (reverse) button just for this purpose.

>> **Switch between VFO and Memory modes.** In VFO mode, a radio can be tuned to any frequency; this is usually how you select a frequency to be stored in a memory channel. In Memory mode, tuning the radio changes from channel to channel.

>> **Turn subaudible tones on or off and change the tone.** If your radio can detect and display the tone frequency being used (a feature called *tone scanning)*, learn how to use it.

>> **Control the Digital Squelch System (DCS) function.** You'll need to know to activate and deactivate DCS in your radio. To use DCS, you also need to know how to pick the tones used to make the DCS code.

>> **Store the radio settings in a memory channel, and select different memory channels.** Storing information in a memory channel (called *programming a channel)* usually requires you to first use the VFO mode to configure the radio's operating settings. Then you press a "memory write" button, select the channel number you want, and press the memory write button again to program the radio's memory for that channel. Some radios have hundreds of programmable memories.

TIP

To avoid carrying an operating manual around, make a cheat sheet for the most common functions or get a condensed operating guide like the *Nifty Mini-Manuals* from Nifty Ham Accessories (www.niftyaccessories.com). Nifty offers condensed operating guides as small as wallet cards and on weatherproof laminated paper. I have one for each of my radios.

Using radio programming software

The manufacturers of portable radios recognize that having so much flexibility and so many features can make setting up a radio a long process. Most of them offer a simple software package to help you configure your radio. You can do it through the radio's keypad or front panel, but this is often a cumbersome process for all but the basic steps, particularly creating easy-to-read labels for channels and storing messages.

Software for programming radios is also available from independent groups. Two of the most popular are from RT Systems (www.rtsystemsinc.com) and Chirp Software (chirp.danplanet.com). Figure 9-4 shows a programming table for a handheld radio. The RT Systems software programs each memory channel with frequencies and all the necessary control or access code information. This is far easier than entering each channel through the radio's small keypad! Chirp is a free open-source programming package. RT Systems is not free, but it is a more complete programming system with cables and USB adapters.

FIGURE 9-4:
An RT Systems setup table.

The typical programming package is downloadable, requiring a programming cable that connects to a PC. The interface to most radios is a serial port (RS-232 or COM). Few portable or mobile radios have USB interfaces as of 2020. You will need a USB-to-COM port adapter if your PC doesn't have a COM port. The cable usually plugs into some kind of audio jack that does double-duty for programming. Make sure you have the right cable as the required connections are usually different than standard audio cables and the software won't work with the wrong cable.

TIP

Check for sales and package deals, particularly during holidays and during bigger hamfests, when vendors may include programming software and cables in a single package at a discount.

The process begins with installing the software and following the instructions about when to connect your radio and turn it on. The software then "detects" your radio (it's not plug-and-play, but similar). If necessary, you can "read" your radio and transfer all its internal channels and other setup information to the software. At this point, save your radio's data (see the software help function) in the event there is a mistake or other problem and you want to return the radio to its previous configuration.

Now go through the process of configuring all the options and features your radio offers. This is a good time to enter all the local repeaters and simplex channels with a label so you don't have to remember all the frequencies. For example, I labeled my local St. Charles Radio Club primary repeater "SCARC1." You can also program in call signs if that's easier to remember. Be sure to get all the access control tones and codes in there, as well. Then save a copy of all the information so you can reload it later, if necessary. I use the radio model number and date as the file name, such as VX8- -7-SEP-2020.

TIP

Most 2 meter (VHF) radios can also receive frequencies close to the ham bands, such as 136 to 174 MHz. When setting up your radio, program a few channels to receive the NOAA National Weather Service broadcasts of weather information and severe weather alerts (www.weather.gov/nwr). There are seven channels just above 162 MHz, WX1 through WX7, and at least one will be receivable almost anywhere in the U.S. If your radio has "extended receive," it may also be able to receive commercial wideband FM broadcasts (88–108 MHz), aircraft Civil Aviation Band channels (wiki.radioreference.com/index.php/Aircraft), and even AM broadcasts (550–1700 kHz, AM). If you live near water, you may also want to program in some of the Marine VHF channels from 156.05 to 162.025 MHz (wiki.radioreference.com/index.php/Marine), especially the national distress frequency, Channel 16, on 156.80 MHz.

Cloning radio configurations

Most radios can also be "cloned," that is, the configuration of one radio can be transferred to another radio of the same model without having to do the full programming operation. This time-saving feature can be very useful in the field where a PC with the host software isn't available. The ability to clone radios is a good reason for public service or emergency response teams to use radios of the same type.

Your manual shows the exact the steps for your radio. The usual procedure involves turning both radios off and holding down a switch or key when turning the power back on. The user's manual provides the necessary instructions and the order in which each step is to be performed. The cloning process then transfers the necessary information from the "master" radio. The radios notify you when the process is finished. You then cycle power on each radio, and the job is done!

WARNING

You *must* use the right cable for cloning. Cloning cables or adapters are often available with the software programming package. The order of the required steps in the manual must be followed exactly to start the process and to be sure you clone the right radio! Don't interrupt the cloning process or the radio being reprogrammed may require a complete reset operation.

Making a simplex contact

TIP

When one station calls another without using a repeater, both stations listen and talk on the same frequency. This is called *simplex* operation. Hams often use FM simplex when they're just making a local contact over a few miles and don't need to use a repeater. Interspersed with the repeater frequency bands shown in Figure 9-1, earlier in this chapter, a few channels designated for simplex operation.

TIP

If the station you're contacting via a repeater is nearby, it's good manners to switch to a simplex frequency instead of tying up the repeater. This is a good idea at hamfests and conventions. Public service teams may also use simplex channels to keep a main repeater channel clear. Practice switching to simplex quickly and have a few of the simplex channels programmed into your radio.

Having a common simplex channel is a good way for a local group of hams to keep in touch. Simplex frequencies are usually less busy than repeater frequencies and have a smaller coverage area, which makes them useful as local or town intercoms. Clubs and informal groups often decide to keep their radios tuned to a certain simplex frequency just for this purpose. If these groups aren't having a meeting or conducting some other business, feel free to make a short call (such as "NØAX monitoring" or "KDØYJN this is NØAX") and make a friend.

On bands that have a lot of space, such as the VHF and UHF bands, making contacts outside the repeater channels is easier if you know approximately where the other hams are. That's the purpose of calling frequencies: to get contacts started. On a calling frequency, you may hear hams call CQ (the general "Come in, anybody" call) or just announce they are present and listening. After establishing contact, they then move to a nearby frequency to complete it. If I call CQ on the 6 meter FM simplex calling frequency (52.525 MHz), and AE7SD answers me, I say, "AE7SD from NØAX. Hi, Sharon. Let's move to five-four, OK?" This transaction means that I am receiving AE7SD's signal and am changing frequency to 52.54 MHz, a nearby simplex channel.

REMEMBER

Making a couple of complete contacts on calling frequencies is okay if the band isn't busy. Otherwise, move to a nearby frequency.

A national FM simplex calling frequency is set aside on each band just for general "Anybody want to chat?" calls. These frequencies are 52.525, 146.52, 223.50, 446.00, and 1294.5 MHz. When driving between cities, I often tune my radio to one of these channels in case I hear a call from other travelers on the highways. With in-vehicle GPS now common and scanning functions built in to most radios, the simplex channels are not used as much. Nevertheless, having them stored in a memory channel of your radio is a good idea.

If you're traveling and want to make a contact on the simplex calling frequencies, the best way to do so is to just announce that you're monitoring, as you would on a repeater channel, or make a transmission similar to this one:

> This is NØAX November Zero Alfa X-ray mobile, headed southwest on Interstate 44 near Leasburg. Anybody around?

Repeat this transmission a few times, spacing the repeats a few seconds apart. If you're moving, try making a call once every five minutes or so.

TIP

Because simplex communications don't take advantage of a repeater's lofty position and powerful signal, you may have to listen harder than usual on these frequencies. Keep your squelch setting just above the noise level. After making a call, you may want to open the squelch completely (use the monitor or MON function) so you can hear a weak station responding.

Solid simplex communications usually require more power and better antennas than typical handheld radios have — at least, on one end of the contact. To get better results on simplex with just a few watts, try replacing a handheld radio's flexible antenna with a full-size ground plane or a small beam. See the section "Maximizing your signal" earlier in this chapter for more about these antennas in Chapter 12.

Digital Voice Systems

The traditional analog voice modes of SSB and FM have some competition! In recent years, individual hams and manufacturers have developed a variety of methods for sending voice signals over the air as digital signals. More are added to the amateur's toolbox every year. Being digital, most of the new modes offer ways to connect beyond station-to-station contacts over the air. At a minimum, they all employ multiple layers of technology to get the job done. As such, they are all rightfully considered to be *digital voice systems* and not just a signal mode.

This section describes the most common digital voice systems as of mid-2020. You can find out more about each of them in publications like *The ARRL Handbook*, magazine and web articles about the different modes, and the websites of manufacturers that produce equipment compatible with those modes. The best compilation of technical information about all the digital voice modes are *The ARRL Handbook* chapters on digital protocols and modes and digital communications.

Except for using a digital signal, the procedures and customs for making digital voice QSOs are about the same as for the analog voice modes of FM and SSB. A little more technology is involved, but you can listen and respond and call in the same basic ways. (The DMR system described below has a very different setup and more tuning selections.) The websites shown for each of the systems explain the details of configuring your radio and making a contact.

HF digital voice

The following two digital voice systems are active on the HF phone bands. If you tune them in with an analog SSB receiver, you'll hear a "roaring" noise. The decoding equipment or software that turns the "noises" into "voices" has tuning indicators to help you adjust the frequency for proper reception. When the frequency is adjusted properly, the associated software can decode the digital system and regenerate the original voice.

TECHNICAL STUFF

At their core, digital voice systems depend on a *codec* (short for "code-decode") to convert between the analog voice and digital data. Some codecs are proprietary and can only be used with permission or licensing. Each type of codec, whether it is software that runs on a PC or firmware in a special IC, does the job a little differently and so "colors" the voice a little bit. Different codecs are selected or developed for the type of signals used to carry the voices around. A codec designed for telephone audio won't work as well for audio transmitted over the amateur HF bands.

>> **AOR Digital Voice** (www.aorusa.com) — This is the oldest of the digital voice modes used on HF. It requires modems produced by the AOR company which are connected between a microphone and the microphone input of an SSB transceiver to create the digital voice signal. The modem changes the headphone or speaker output of the radio back to analog speech.

>> *FreeDV* (freedv.org) — *FreeDV* is an open-source software system including software package that runs on a PC and controls tuning, receiving, and transmitting. The codec used by *FreeDV* is called *codec2*. Both *FreeDV* and *codec2* are used on both HF and VHF/UHF bands. Using *FreeDV* requires a PC sound card to be connected to an SSB transceiver's microphone input and audio output. The microphone and speakers are connected to the PC.

VHF/UHF digital voice

Several digital voice systems are in use on the VHF and UHF bands. They are organized on the same channels that analog FM repeaters use.

Regular analog FM voice radios can't receive digital voice signals, and vice versa. An analog FM radio hears digital voice signals as white noise. A digital voice radio simply rejects the analog FM signal. (Some radios can operate in either analog FM or digital voice mode.)

A digital voice repeater receives the digital stream of data and retransmits it just like an analog FM repeater does, but it is still digital data. Because the voice signal is already digital when it's transmitted, it can be shared or relayed over digital networks easily. Some digital voice repeaters can receive and retransmit either analog FM or digital voice signals automatically.

Some of the digital system repeaters support both analog and digital voice signals. Typically, a repeater's users tend to be all-analog or all-digital. Some systems support both types of users although sharing is somewhat difficult. You can check a repeater directory or just search online for repeaters supporting a specific system in your area. For example, searching for *d-star repeater missouri* turned up several lists of local repeaters and links to national repeater databases.

TECHNICAL STUFF

The data rates available to amateur radio may seem pretty slow and compared to WiFi or household broadband service they are. This results from the relatively narrow bandwidths available to hams. Commercial wireless data occupies many more megahertz of spectrum than hams. Amateur data rates are faster on the shared microwave bands where reprogrammed commercial wireless routers and access points can be operated on the ham bands at full speed and with much higher power.

TECHNICAL STUFF

All information transmitted by amateur radio should be assumed to be public communications. Encryption and other security mechanisms are not allowed. Your voice and what you say are not private like over a mobile phone.

The most popular digital voice systems in use today include

>> **D-STAR** (www.dstarinfo.com): This is a combination voice-and-data mode that was designed under an open standard. Both Icom and Kenwood make D-STAR-compatible equipment. D-STAR is used on the VHF and UHF bands and has three modes: DV (digital voice and 1200 bits/sec data), Fast DV (4800 bits/sec data), and DD (high-speed data, approximately 100k bits/sec). D-STAR also links to a world-wide system of repeaters via microwave and Internet links. Several applications are available to exchange data over D-STAR links.

>> **System Fusion** (systemfusion.yaesu.com/what-is-system-fusion): This is a voice-only system created by Yaesu for use on the VHF and UHF bands. It is also referred to as C4FM, which is the type of modulation used to

send the digital data over the air. System Fusion radios and repeaters can operate as either analog or digital voice. System Fusion equipment can also use the Yaesu WIRES-X system to link stations via the Internet (systemfusion.yaesu.com/wires-x).

>> **DMR (Digital Mobile Radio;** www.trbo.org/docs/Amateur_Radio_Guide_to_DMR.pdf**):** Created by Motorola for commercial and government users under the trade name MOTOTRBO, amateurs have adapted DMR to ham radio and administer world-wide networks that link DMR equipment together via the Internet. DMR is very similar to "trunking" systems that allow many users to share a few radio channels through the use of "talk groups" and routing protocols via the Internet.

Data-only network systems such as Broadband-Hamnet (also known as HSMM-MESH) create networks using Internet protocols and reprogrammed wireless data router equipment. This technology has been extended to an amateur network known as Amateur Radio Emergency Data Network (www.arednmesh.org). These systems operate on the same bands that are shared between hams and WiFi equipment — 2.4 GHz and 5.6 GHz. Digitized voice can be sent across these networks using standard Internet protocols like VoIP (Voice over IP) just as from a home computer.

TECHNICAL STUFF

Because the various digital voice modes are incompatible over the air, amateurs have developed "translators" known as *hotspots*. A hotspot contains a low-power VHF/UHF digital transceiver and connects through a host PC to the Internet. The hotspot transceiver communicates with the users low-power handheld or mobile transceiver over any supported digital voice mode. (Most hotspots support all three current digital voice modes.) The hotspot also communicates with a digital voice repeater on the Internet, acting as a bridge between the user's transceiver and the network. Hotspots are available for D-STAR, System Fusion, and DMR systems.

Digital repeater networks

A full discussion of the way repeater networks operate is beyond the intentions of this book. However, an overview of the way the systems of repeaters and stations are built is worth including.

If you decide to dive deeply into the networks, there is a lot of information online at the manufacturer website or through Internet searches. Building and using repeaters is a fast-moving area of amateur radio and hams are creating interesting apps and features every day.

The systems described below are active and more repeaters are linked to the systems every day. The IRLP system includes about 3,000 stations around the world. EchoLink currently lists more than 2,000 repeaters and more than 200 conference servers. The D-STAR system lists about 1,000 gateways, many serving more than one repeater. To find other hams who use these systems, you'll want to join an online user group. The following are just the most visible groups — there are lots of others:

>> **EchoLink:** groups.io/g/EchoLink

>> **D-STAR:** www.dstarusers.org/

>> **IRLP:** irlp.groups.io/g/IRLP/

>> **System Fusion:** groups.io/g/YaesuSystemFusion

>> **Digital Mobile Radio (DMR):** www.facebook.com/groups/DMRTrack

Complete information about how the repeater networks are connected and used is available on the home pages for each system, along with extensive "getting started" instructions, FAQs, and resources for finding and even installing your own system repeater or station.

Voice over Internet Protocol (VoIP) systems

Two ham radio systems link repeaters via the Voice over Internet Protocol (VoIP): EchoLink (www.echolink.org) and the Internet Relay Linking Project (IRLP; www.irlp.net). Voice signals and control information is exchanged in much the same way that the Skype web communication application does. Figure 9-5 illustrates the basic IRLP system.

TIP

The links between repeaters and individual stations in the IRLP and EchoLink systems are controlled manually by the system users using DTMF (phone dialing) tones. When you're connecting to either system beyond your local repeater, you must enter an access code manually using the radio's keypad. The code identifies the repeater system you want to use; then the system sets up the connection and routes the audio for you. When you finish, another code or a disconnect message ends the sharing. This overview is very simplified, of course, and both systems offer useful features beyond simple voice links. Linked repeater systems are the subject of a couple of Technician exam questions.

EchoLink systems can also be accessed directly from a computer via the Internet, so users don't have to have a radio at all, making EchoLink a popular way to communicate if you don't have a radio handy or are traveling away from your home repeater but still want to have contacts with local stations. The best guide for learning about and using EchoLink is the *Nifty E-Z Guide to EchoLink Operation* by Bernie LaFreniere, N6FN (www.niftyaccessories.com/EchoLink_book.php).

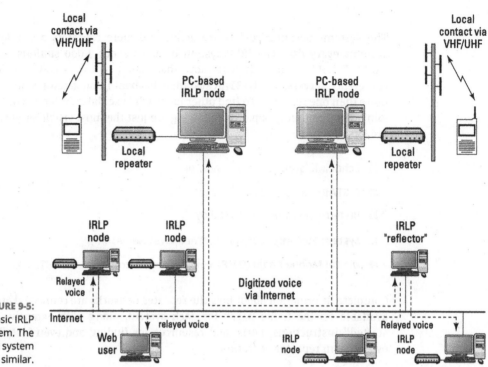

FIGURE 9-5:
The basic IRLP
system. The
EchoLink system
is very similar.

An IRLP *node* is a regular FM repeater with an Internet link for relaying digitized voice. A user or control operator can direct an IRLP node to connect to any other IRLP node. When the node-to-node connection is made, the audio on the two repeaters is exchanged, just as though both users were talking on the same repeater. It's common for a ham in Europe to communicate with a ham in New Zealand, for example, with both parties using handheld radios that put out just a watt or two.

You can also connect several nodes by using an IRLP reflector. The reflector exchanges digitized audio data from any node with several other nodes in real time. Even a user who doesn't have a radio can join in by logging on to an IRLP reflector or node. (All users who create radio transmissions have to be licensed, however.)

To use the IRLP system you don't need anything more than your radio, the IRLP system's control-tone sequences for your repeater, and a list of the four-digit node on-codes that form the IRLP address of an active IRLP repeater. Using the IRLP system is described at www.irlp.net/guidelines.html.

The D-STAR system

D-STAR is not only a type of repeater system, but also a complete set of digital communication protocols for individual radios. D-STAR's digital voice protocol must be used by everyone on the system, requiring the use of radios that are compatible with D-STAR. D-STAR's fully digital system in Figure 9-6 also enables data-only operation, and many software applications have been developed to use that capability.

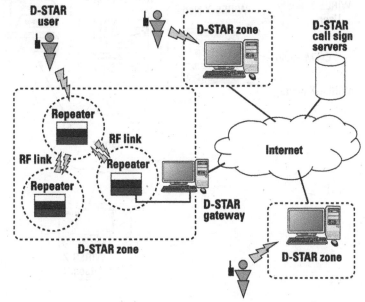

D-STAR repeaters are connected to the D-STAR system through *gateways:* computers that are connected to the repeater and to the Internet. Hams establish links from one D-STAR repeater to another by entering call signs into their D-STAR radios. The D-STAR system's servers use these call signs to look up the low-level Internet network addresses of individual repeaters. Then the system directs each repeater to make the connection and share the voice data.

Call-sign lookup is an interesting feature because hams don't have to know where another ham is to contact that person; the D-STAR system remembers where each ham last used a D-STAR repeater. Calls to a specific ham are routed to his or her last-known D-STAR repeater. This feature is a lot like the mobile phone system, which keeps track of where your phone is connected to the network so it can route your calls to that point.

The WIRES-X system

This system shown in Figure 9-7 is similar to the IRLP and EchoLink systems. Individual stations are connected to the Internet through WIRES-X routers, creating a WIRES-X *node*. The node has a unique ID that can be accessed by other WIRES-X nodes through the Internet.

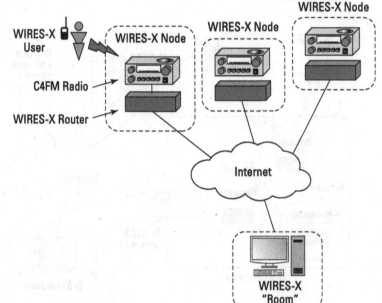

FIGURE 9-7:
An overview of the WIRES-X system.

The radio connected to the router can be used directly by a user so that talking into the microphone communicates through the Internet without transmitting an RF signal. Another radio, such as a nearby handheld or mobile radio, can communicate over the air to the node-connected host radio and be connected to other WIRES-X nodes through the Internet as well. This creates user-to-user communication.

On the Internet, computers can host a WIRES-X *room,* like an Internet chat room. When a node is connected to a room, the node user hears all the audio from every node connected to that room. This supports roundtable contacts and nets. Rooms can be semipermanent and have a theme, such as "DXing" or "Local Traffic" or they can be temporary, created for a specific need.

The DMR system

The DMR or Digital Mobile Radio system was originally designed by Motorola for shared communications systems used by public safety, government, and commercial users. It operates similarly to the earlier "trunked" systems that were intended to support fleet and department-level operations with many users. Hams adapted this technology as they had done with analog FM systems decades before. Because DMR was designed for commercial use, much low-cost equipment is available, although the user interface and programming are not as user-friendly as equipment designed for ham radio. DMR's popularity is increasing rapidly.

In the DMR system (see Figure 9-8), each radio connects to the local repeater and is assigned one of two alternating time slots during which it can transmit and listen to the digital voice stream of the repeater. (DMR digital voice transmissions consist of two alternating streams of data. This is referred to as *TDMA — Time-Division, Multiple Access*.) Much like a mobile phone and its host system, the radio must request permission to transmit from the repeater.

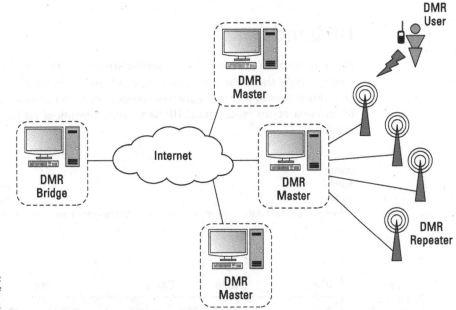

FIGURE 9-8:
An overview of
the DMR system.

TIP

The repeater is in constant communication with a local or regional "master" that manages all the data streams from the repeaters connected to it. This allows the DMR system to create talk groups so that groups of users can share the same repeater without interference. In addition, masters are also in communication with central "bridge" systems, creating networks that route communications between thousands of users. There are two primary bridge systems today (2020), the MARC (Motorola Amateur Radio Club) and the Bridge-Com systems.

There are numerous other features that make DMR the most flexible amateur communications network currently in use. Talk groups can be local, regional, or even international in scope. Various levels of authorization and authentication can be created. If you are interested in the technical details of DMR, the *Amateur Radio Guide to Digital Mobile Radio (DMR)* by John Burningham (W2XAB) is an excellent introduction (`www.trbo.org/docs/Amateur_Radio_Guide_to_DMR.pdf`).

Casual Operating on HF

When you start operating on the HF bands, you'll recognize many similarities to the VHF and UHF bands you may be used to. This will help you get comfortable as you adjust to the different modes and styles of operating that I review in the next few sections. Just like on the VHF and UHF bands, the best way to get started is to jump in and begin learning!

HF bands

Below 30 MHz, all the bands have a similar structure. CW (Morse code) and digital modes occupy the lower third (more or less), and voice modes occupy the upper two-thirds (less or more). Figure 9-9 shows where you can expect to find contacts of different styles on a typical HF band. (Ragchewing is covered earlier in this chapter.)

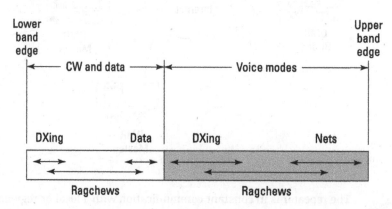

FIGURE 9-9: The general operating conventions on the HF bands.

You can always find CW at the low frequencies of the HF bands. The faster operators tend to be at the bottom of the bands, with average code speed dropping slowly as you tune higher.

WARNING

Because hams must keep all of their signals within their allocated bands, you need to remember where your signal is actually transmitted. Most voice signals occupy about 3 kHz of bandwidth. If the radio is set to USB, your on-the-air signal extends 3 kHz *above* the displayed frequency. Similarly, on LSB, the signal extends 3 kHz *below* the displayed frequency. When you are operating close to the band edges, make sure that your signal will stay "in the band." For example, because the 20 meter band's upper edge is 14.350 MHz, you should never transmit an USB signal with the display showing a higher frequency than 14.347 MHz.

Digital signals tend to cluster near the published calling frequencies listed in the band plans for that particular mode unless a major contest or some other event is going on. Stations spread out higher in frequency from there.

TIP

When the bands sound too full to use, try reducing your receiver's sensitivity. Receivers are very sensitive and can easily be overloaded by strong signals, causing distortion that sounds like interference. It could be overloading or just making it too easy for you to hear noise. Turn off the preamp, turn down the RF Gain, or turn on the attenuator . . . or all three! You will probably be able to hear just fine, and the noise and "crud" will be a lot less objectionable. While you're at it, turn off the Noise Blanker, which can generate distortion all by itself if strong signals are present. Use the minimum amount of sensitivity required to make the contact, and you'll enjoy listening a lot more.

Picking good times to operate

You may want to revisit Chapter 8 when considering what band is best to use. If you like to talk regionally, try one of the low-frequency HF bands. For a coast-to-coast talk, one of the high-frequency bands is your best bet. The better your antenna system, the more options you have.

REMEMBER

I keep coming back to this point, but listening is the best way to learn operating know-how in ham radio. The most important part of any amateur's station is between the ears. If you want to operate successfully, spend some time listening to more experienced hams do it.

Lots of hams do their operating on weekends, but that's also when special events and contests are held. Be prepared for a full band every weekend of the year. The silver lining of this cloud is that plenty of hams will be on the air for you to contact. If you know that one mode or band is hosting some major event, you can almost always find a quiet spot on another mode or band.

TIP

The WARC bands (see the nearby sidebar "What's a WARC?") never have contests and usually are wide open for ragchews and casual operating.

Because weekends are busy times, you should check the contest and special-event calendars (see Chapter 11). A little warning keeps you from being surprised when you get on the air and allows you to be flexible in your operating.

WHAT'S A WARC?

Hams refer to 30, 17, and 12 meters as the WARC bands. *WARC* refers to an international World Administrative Radio Conference. At the 1979 conference, amateurs worldwide were granted access to three new bands: 30, 17, and 12 meters. These new bands were immediately nicknamed the WARC bands to distinguish them from the older ham bands at 160, 80, 40, 20, 15, and 10 meters. The 60 meter band was opened to amateurs in 2002 on a limited basis while the 630 and 2200 meter bands were added in 2017. Although hams didn't gain access to them through the WARC process, they are still lumped in with the true WARC bands as not being one of the traditional bands.

By convention, contest operation on HF and below doesn't take place on the WARC bands and 60, 630, and 2200 meters. If you prefer not to deal with contest QRM (interference), these bands are good choices.

REMEMBER

It can be a very different time of day at the other end of a DX contact.

When the bands seem to be frustratingly full, here are some helpful strategies that keep you doing your thing:

>> **Try a nontraditional band.** Most nets and all contests are run on the traditional bands. The WARC bands almost always have sufficient space for a QSO.

>> **Try a different mode.** Very few big contests have activity on more than one mode. You can change modes and find the band much less crowded.

>> **Be sure that you know how to operate your receiver.** Cut back on the RF gain, use narrower filter settings, know how to use controls such as the IF Shift and Passband Tuning controls on superhet radios, try optimizing DSP settings on an SDR, and generally be a sharp operator. You can remove much interference and noise just by using all the adjustments that your receiver provides.

>> **Always have a backup plan.** There's no guarantee that any particular frequency will be clear on any given day. Hams have frequency freedom second to none, so use that big knob on the front of your radio.

TIP

Be sure your transmitted audio is not distorted and that your RF signal is not splattering. Have a friend listen in and make sure you have a clear, clean signal. Pay attention to the ALC (automatic level control) meter indication and set your microphone gain and any speech processing as shown in your radio's user manual. Take note of your transmitter settings and how the meters respond when you're speaking.

REMEMBER

Evaluate on-the-air technique as you tune across the bands. Consider what you like and dislike about the various styles. Adopt the practices you like, and try to make them better; that's the amateur way.

Contacts on CW and digital modes

You'll find that contacts using a keyboard or CW has many similarities to voice contacts. The general structure is the same and much of the same information is exchanged. There are some minor differences you'll quickly get used to. Like learning voice operating, just monitor QSOs and you'll quickly learn how other operators do things.

TIP

The basic style of most CW and digital QSOs is the same on RTTY as it is for PSK31 and other keyboard-to-keyboard modes. Message-oriented modes like FT8 follow a fixed format that doesn't (yet) support long conversations. I introduce these modes in Chapter 11.

Figure 9-10 shows some of a PSK31 QSO between Gil, F2SI (in Marseilles, France) and Skip, KH6TY (in Charleston, SC). Monitored by a third station, this QSO is too long to fit entirely in the decoded information window, but I'll do a play-by-play, since this contact has some important elements:

	(Before the figure starts)
F2SI Calls CQ:	CQ CQ de F2SI F2SI F2SI K
KH6TY Responds:	F2SI DE KH6TY KH6TY KH6TY/4 K
F2SI starts the QSO:	KH6TY KH6TY de F2SI
	You are 589 589 — Good copy.
	(The figure starts here)
	QTH MARoeEILE MARSEILLE . . . sohern FRANCE

This is why even digital modes need to sometimes repeat information — some noise or fading or interference caused the "oe" and misspelling of "southern" to appear on our screen.

	How copy?
	KH6TY KH6TY de F2SI F2 | (noise or fading here)
KH6TY calls again:	F2SI F2SI F2SI DE KH6TY KH6TY/4 K

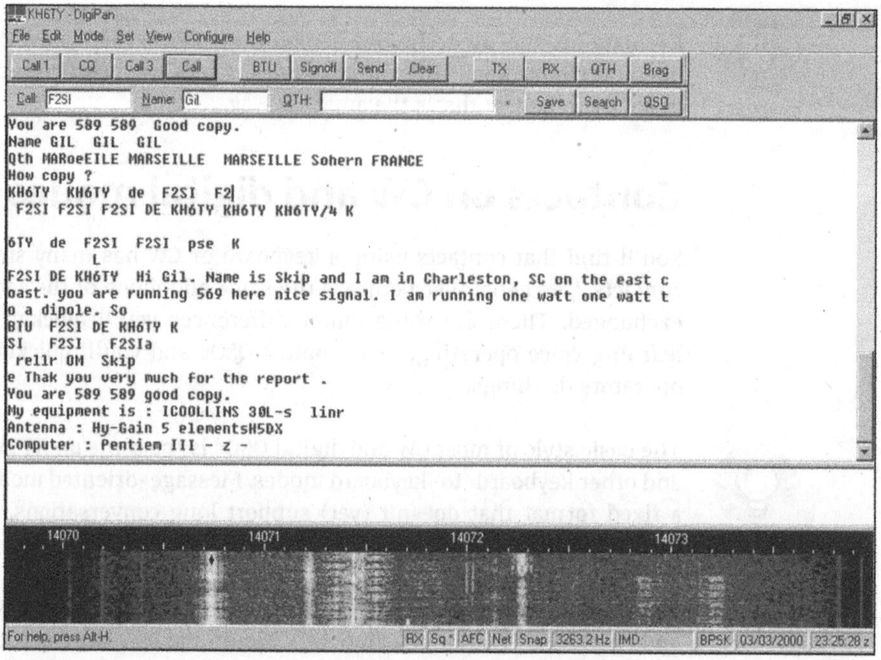

FIGURE 9-10:
A PSK31 QSO
using DigiPan
software.

Skip is not sure that Gil was talking to him and gave him another call. The fading or noise goes away and Gil responds to confirm he is contacting Skip.

Gil confirms: 6TY de F2SI F2SI pse K (go ahead, please, KH6TY!)

L (a noise burst decoded as spaces and "L")

Skip gives his info: F2SI DE KH6TY Hi Gil, Name is Skip and I am in Charleston, SC on the east coast. You are running 569 here nice signal. I am running one watt one watt to a dipole. So BTU (back to you) F2SI DE KH6TY K

There is some fading or noise and we only see the last part of Gil responding to Skip's information:

F2SI responding: . . . SI F2SI F2SIa

rellr OM Skip

e thak you very much for the report.

You are 589 589 good copy.

Now, Gil sends a preprogrammed set of information called a *brag tape* (see below):

My equipment is : ICOOLLINS 30L-s linr

Antenna : Hy-Gain 5 elementsH5DX

Computer : Pentium III – z –

And so on. An experienced ham would read right through the noise causing the extra or missing characters to understand that Gil is using a Collins 30L-1 S-line linear amplifier, a TH-5DX antenna, and a P3 computer. The noise "takes out" a few characters here and there but like smudges on a printout of an article, missing a few characters doesn't render the information unreadable.

TIP

Why show a noisy contact? It's unreasonable to give you the idea that ham radio contacts are all perfect! You'll get the hang of dealing with the various impediments — that's part of what makes it fun and not just another mobile phone exercise. Your abilities to interpret what is received play an important part, too, even on digital modes!

REMEMBER

Most digital software has a default window width of about 80 characters. If you don't break up your lines of text, the characters may scroll out of the receive window of the receiving station! Keep your lines of characters reasonably short. For CW, you send each character manually at an appropriate rate. Digital QSOs are another matter. If you are a quick typist, you may feel comfortable typing characters while the transmitter is sending them. This makes many new hams uncomfortable and feel like they are under pressure to keep the characters coming! You can get around this with macros as explained below. It's really okay, too, to just let the transmitter *idle* (send null characters) while you think of something to type. This is the digital mode equivalent of "Uhhhh . . ."

If you find yourself at a loss for words, just enter "BTU" (back to you) as Skip did, both call signs, a "K" to indicate you're done transmitting, and follow it with a new-line character (Enter). The other station will wait to see his or her call and K before assuming your transmission is complete. We've all been flustered at one point but don't be embarrassed; just "turn it around" and let the other station type!

Finally, the end of a digital mode QSO looks a lot like the end of a CW QSO, which is a lot like the end of a phone QSO. Refer to the sidebar "The long goodbye" earlier in this chapter for that discussion.

Brag tapes

Rather than type everything each time, common information such as operator or station information is saved in a brag tape file. A typical brag tape (also known as a *macro*) has information about the operator or the station or something else that is sent frequently enough to have recorded and not have to type every time.

For example, here's a typical brag tape with station details:

Rig: IC-7300 and MFJ-939 Auto-tuner

Power: 25 watts

Ant: HF-6V Vertical with 32 radials

Software: FLDIGI 4.1.15

OS: Windows 10

You can have brag tapes for all sorts of things and have them ready to play back with a single key press. If you find yourself having to type in the same information over and over, perhaps putting it in a brag tape is a good idea.

WARNING

It's easy to overdo it with brag tapes. Keep them short and to the point because it can take a while to send this information, which might not be all that interesting to the other station! It's best to have several short macros that you can send as needed instead of one big long one.

REMEMBER

Brag tapes come from the old days of electromechanical teleprinters when "tape" literally meant "paper tape"! RTTY operators usually had several rolled up and ready to go with one loaded into the tape reader for CQs and that sort of thing. Today, brag tapes are typically computer files sent with a key press.

Chapter **10**

Public Service Operating

As your experience with ham radio grows, you'll find more and more practical uses for your communications skills. Your ham radio skills can also benefit others, which is where the *service* part of *amateur radio service* comes in. These services are important to you for two reasons: You can use them yourself and you can provide them to others. In this chapter, you find out what those services are and how to get started with the groups that provide them.

WARNING

You may have obtained your ham radio license to provide personal communications for emergencies or disasters. You're part of a long tradition in ham radio — welcome! However, ham radio is not like a flashlight that can be dropped in a drawer with some batteries. It's more like a first-aid kit that you have to train with to use effectively. If you don't practice regularly, you won't know how to use the radio, whom to contact, or how, and if you do contact someone you won't know what to say or do. Why waste your time and effort? Practice and train for when you need ham radio. You will probably find yourself having a lot of unanticipated fun along the way!

Note: This chapter, written primarily for American and Canadian hams, describes the U.S. emergency communications organizations. Elsewhere around the world, you can find similar organizations. Contact your national amateur radio society for information about them.

Joining a Public Service Group

An important element of the Federal Communications Commission's (FCC's) Basis and Purpose of amateur radio (Part 97.1) is providing emergency communications. Just during the writing of this edition, hams provided critical services during hurricanes along the Gulf Coast and in the Caribbean, and are supporting firefighting teams confronting wildfires across the West. You never know when the need will present itself, so start preparing as soon as you're licensed.

In return for privileges that go with the license — access to a broad range of frequencies, protection from many forms of interference, maintenance of technical standards, and enforcement of operating rules — amateurs "give back" as trained operators and communication systems. *Emergency communications* (known in the radio biz as *emcomm*) is loosely defined as any communication with the purpose of reducing an immediate threat of injury or property damage. This includes everything from reporting car accidents to supporting large-scale disaster relief. Public service communications are more frequently conducted in support of events or organizations and provide great training for emergencies and disasters.

REMEMBER

Emergency communications are what happens *during* an life- or property-threatening event. Disaster relief or recovery communication takes place *after* a disaster as systems and services are reestablished.

In this section, I introduce the elements of public service communication and show you how to get started. The ARRL provides many additional training resources and tutorials on its website for Public Service Communications at www.arrl.org/public-service.

TIP

The process of becoming an effective communicator is a lifelong learning process, not only for public service but for all of ham radio. Remember that your license is a "license to learn" and that no one knows everything. We all have lots to teach each other and that is particularly true for public service. If you take the attitude that every event and drill and meeting is an opportunity to learn something new, you will have a great time!

Finding a public service group

Whether your interests are to support yourself and your family or to participate in organized public service, you need to know how amateurs are organized. Otherwise, how will you know where to find or interact with them? This section gives you several good places to start. These are a few of the U.S.-wide organizations you will encounter as a communications volunteer.

REMEMBER

The different levels of emergencies and disasters, with varying degrees of resource needs, require different responses by government agencies. As a result, a one-size-fits-all plan isn't appropriate to handle all events. You should expect to adapt to whatever your *served agencies* (the organizations you support) need to handle the situation in your area.

ARES

ARRL's Amateur Radio Emergency Service (ARES) is the largest nationwide ham radio emergency communications and public service organization and it is managed by the ARRL's Field Organization. ARES is administered at the ARRL section level, which may be as large as a state or as small as a few counties, depending on population.

ARES does not manage or control the way you, the individual volunteer, respond to emergencies and disasters. That is up to the leadership of your public service team or group. The goal of ARES is to train operators who can provide communications services and interact effectively with responding organizations. Training materials, resources, and guidelines are available through ARES. You can find out more at www.arrl.org/ares, including planning and training documents.

RACES

Radio Amateur Civil Emergency Service (RACES), organized and managed by the Federal Emergency Management Agency (FEMA), is a national emergency communications organization governed by special FCC rules. Its mission, like that of ARES, is to provide volunteer communications assistance during a civil emergency or disaster. RACES teams are organized and managed by local, county, or state civil-defense agencies. Dual membership in ARES and RACES is fairly common. The ARES-RACES FAQ page at www.arrl.org/ares-races-faq helps explain how the two services relate to each other.

MARS

The Military Auxiliary Radio System (MARS) provides an interface between the worldwide military communications systems and ham radio. MARS is sponsored by the U.S. Department of Defense. If you are willing to meet the time commitment and learn the necessary procedures and techniques, MARS members receive special licenses and call signs that allow them to operate on specific frequencies just outside the ham bands. You can learn more about the MARS programs through Wikipedia articles and links at en.wikipedia.org/wiki/Military_Auxiliary_Radio_System.

SATERN and Red Cross

The largest faith-based emcomm group, the Salvation Army's Team Emergency Radio Network (SATERN: satern.org) is a system of communications volunteers who support the Salvation Army's Emergency Disaster Services ministry. The Red Cross also uses ham volunteers for communications. Each of the Red Cross chapters organizes volunteers according to its own needs and plans. If you volunteer for these organizations, they will provide training and certification so you can work effectively and safely.

TIP

CERT (Community Emergency Response Team) training goes hand-in-hand with ham radio public service training. You may have even learned about ham radio from a CERT class since many teams incorporate it into their operational training. CERT and ham radio complement each other very well.

Volunteering for ARES

Unless you are already associated with a public service group, you can start by participating with an ARES-based group or team. This section shows you how to volunteer. After you have gained some experience, you may want to investigate some of the other groups serving your area.

NATIONAL INCIDENT MANAGEMENT SYSTEM (NIMS)

Emergency response organizations at all levels of management have come together to develop a common plan for responding to emergencies and disasters. This plan is called the National Incident Management System (NIMS) and it is mandatory for all public-safety and other government agencies in the United States. Having a common plan greatly improves the capability of different organizations to work together efficiently and effectively.

Because these agencies are primary customers for amateur radio emergency communications, ARES teams have adopted NIMS and made it a central part of their training, as have many other nongovernmental organizations. You can find out more about NIMS and the Incident Command System (ICS) at training.fema.gov/nims. Your emergency response team will show you more as it applies to your community.

FEMA also provides free training courses to help you understand the ICS. Your ARES or other public service team will help you get the right training and certification for your area. Many teams and agencies require that you to have completed a specific set of courses in order to be certified and participate as a member.

REGISTERING AND GETTING CHECKED

Be aware before you volunteer that you might be asked to approve a law enforcement background check and maybe even a credit check. You should ask about this when evaluating the organization. If this level of checking is okay with you, by all means go ahead!

You can volunteer for ARES by filling out the ARES application form FSD-98 (www. arrl.org/files/file/Public%20Service/fsd98.pdf) and sending it to the ARRL or your local Emergency Coordinator (EC). The easiest way to find out about the ARES organization in your area is to contact your ARRL Section Manager (www. arrl.org/sections). You can also search www.arrl.org/arrl-net-directory-search for ARES nets in your area; check in to the net as a visitor, and ask for information about ARES.

If you want to help administer and manage public service activities in your ARRL Section after you have some ham radio experience, consider applying for an ARRL Field Organization appointment (www.arrl.org/field-organization). There are positions available for hams of all backgrounds and skill levels, such as public information officer (PIO), assistant section manager (ASM), or technical specialist (TS).

Preparing for Emergencies and Disasters

Getting acquainted with emergency organizations is fine, but it's only a start. You need to take the necessary steps to prepare yourself so that when the time comes, you're ready to contribute. Preparation means making sure that you know four things:

>> Whom to work with

>> Where to find your group on the air

>> What gear to have on hand

>> How to respond and work with your group

Knowing who

Earlier in this chapter, I discuss the organizations that provide emergency communications. First, become familiar with the leaders in your ARRL Section; then get acquainted with the local team leaders and members.

The call signs of the local public service teams and stations operating from an *emergency operations center* (*EOC*) are valuable to have at your fingertips in times of emergencies. The best way to get familiar with these call signs (and make your call sign familiar to them) is to be a regular participant in nets, training exercises, and public service events. Checking in to weekly nets takes little time and reinforces your awareness of who else in your area is active. If you have the time, attending meetings and other functions such as open houses and work parties also helps members put faces with the call signs. Building personal relationships pays off when a real emergency comes along.

TIP

A *call directory* or *call list* is very useful. Your group may have a list of call signs that are used by EOCs, shelters, churches, and clubs. Be sure you have a copy in your go kit!

Knowing where

When an emergency occurs, you don't want to be tuning around the bands hoping to find your local teams. Keep a detailed list of the emergency net frequencies, along with the names of the leaders in your area. (I provide a link to a downloadable chart for you to fill in at this book's page at www.dummies.com.)

You may want to reduce this list with a photocopier or scanner and laminate it for a long-lasting reference the size of a credit card that you can carry in your wallet or purse or vehicle. (Office supply and copy centers can do the reduction and lamination for you.)

Knowing what

If the need occurs and your equipment isn't ready, you can be under pressure to get your gear together and respond. In your haste, you might omit some crucial item or won't be able to find it on the spur of the moment. I recommend assembling a *go kit* (similar to a first-aid kit) as an antidote to adrenaline-induced confusion and mistakes.

Assembling a go kit

Your go kit is a group of items necessary during an emergency; collect them in advance and place them in a handy carrying case or bag. Having a go kit allows you to spend your time responding to the emergency instead of racing to get your gear together. Preparing the kit in advance also makes you less likely to forget important elements.

Figure 10-1 shows a portable go kit that can sustain a person for 24 hours. With more food and water, it could support two people.

FIGURE 10-1:
A typical
personal go kit.

Courtesy Ralph Javins (N7KGA)

Before making up your go kit, consider what mission(s) you may undertake. A personal checklist is a good starting point. You can find a generic checklist in the ARRL's ARES Field Resources Manual at www.arrl.org/shop/ARES-Field-Resources-Manual.

What goes into a go kit varies from ham to ham, but every kit should contain the following essentials:

>> **Nonperishable food:** Remove the uncertainty of not knowing when or if food will be available by having your own food that doesn't require refrigeration. If you bring canned food, don't forget the can opener!

>> **Appropriate clothing:** You can get cold even on summer nights or in an exposed location, so include a jacket; if you get too hot, you'll want to exchange your current clothing for something more lightweight. Preparation allows flexibility. Having separate winter and summer go kits can be useful and keeps weight and volume down.

>> **Radios and equipment:** Bring all the equipment you may need: radios, antennas, power supplies, and batteries. Make sure everything is lightweight, flexible, and easy to set up. Practice setting up to be sure all of the pieces are present and functional. Rotate or recharge batteries on a regular basis.

>> **References:** You need lists of operating frequencies, any special access codes or digital network IDs, as well as a personal phone list and a list of emergency-related telephone numbers. Don't forget the email addresses!

TIP

For a complete list of go kits for every occasion and need, check out Dan O'Connor's Personal Go Kit for Emergency Communications presentation at `www.soara.org/classes/technician/personal_go_kit_ke7hlr_2011.pdf`. Rather than make one all-inclusive go kit, you might find it more reasonable to have two or three go kits, each with a specific purpose.

Preparing your home

You may not need a formal go kit if you operate from home, but you still need to prepare for emergencies such as an extended power outage or the failure of your main antenna.

TIP

An important rule of emcomm is take care of your family first! The website `www.ready.gov` provides a lot of good information about preparing your home and family for emergencies of all sorts.

Your primary concern is emergency power. Many radios draw more than 1 amp even when they're just receiving. You'll need a generator or vehicle to power them during any extended power outage. If you have a home generator, make sure that you can connect to it safely and that it can adequately power the AC circuits in your station.

Most radios with a DC power supply can run from an automobile battery which is another backup power source. Getting power from your car to your radio isn't always easy, however. Decide which radios you want to operate from your car, and investigate how you can power and connect an antenna to each of them.

TIP

Overall, the most important step is to simply attempt to answer this question: "How would I get on the air if I'm unable to use my regular station?" or "How quickly can I get on the air on HF or VHF/UHF from my car?" Just by thinking things through and making plans, you'll be well on the road to being prepared for any emergency.

Knowing how

Knowing the procedures to follow is the most important part of personal preparedness. Whatever your experience and background are, you have to know the specific details of working with your emergency organizations. If you don't, you won't be prepared to contribute when you show up on the air from home or at a disaster site.

Do everyone a favor — including yourself — by getting trained in the necessary procedures and techniques. Your public service team has plenty of training opportunities and training nets for practice. Participating in public-service activities, such as acting as a race-course checkpoint in a fun run or as a parade coordinator, is very good practice, and it exercises your radio equipment as well. (By the way, you'll make good friends at these exercises who can teach you a lot.)

TIP

Don't miss out on Field Day (www.arrl.org/field-day) as a valuable training and education activity. Many emcomm teams use Field Day as part of their operator training or as an opportunity to try out new tools and equipment. VHF contests are great training for Technicians, as well. (I discuss contests in Chapter 11.)

TIP

The ARRL offers online emergency communications training courses. Check out these courses at www.arrl.org/emergency-communications-training. FEMA also offers many courses that support general emergency operations and specifics of the IMS (training.fema.gov/is/crslist.aspx). CERT, the Red Cross and similar organizations may have their own training courses, as well, usually for free or at very low cost. Your public service team or served agency may require you to take one or more of these courses to be a full member. Check with your team to see what you need. Most courses only take a few hours to complete online or in a local classroom.

TIP

After you start training in emergency communications, you'll find that training is available for many other useful skills, such as cardiopulmonary resuscitation (CPR), first aid, orienteering, and search and rescue.

Operating in Emergencies and Disasters

All situations are different, of course, so no single step-by-step procedure is always going to be useful. But here are some solid general principles, based on the *ARES Field Resources Manual*, to follow when disaster strikes:

1. **Make sure that you, your family, and your property are safe and secure before you respond as a volunteer.**

2. **Monitor your primary emergency frequencies.**

3. **Follow the instructions you receive from the net control. Check in if and when check-ins are requested.**

4. **Contact your team leader or designee for further instructions.**

TIP

Everyone is likely to be fairly excited and tense. Keep calm and follow your training so that you can help rather than hinder in an emergency situation.

REMEMBER

"Response" is the actions taken immediately after an incident occurs. This includes damage assessment, for example. "Recovery" is the longer process of what happens to return things to their normal status.

Reporting an accident or other incident

Accident reports are more common than you may think. I've personally used ham radio to report accidents, stalled cars, and fires. Know how to report an incident quickly and clearly — don't assume that people with mobile phones are already doing it, particularly in rural areas where service might be spotty.

Follow these steps to report an incident via a ham repeater:

1. **Set your radio's output power to maximum, and clearly say "Break" or "Break emergency" at the first opportunity.**

   A strong signal can get the attention of listening stations. Don't hesitate to interrupt an ongoing conversation.

2. **After you have control of the repeater or the frequency is clear, state that you have an emergency to report and ask for a relay to 911.**

   Report all the necessary material and then stand by on frequency until the relaying station reports to you that the information has been relayed and the call is complete. (If the repeater has an *autopatch* function, the responding station may elect to make the call that way.)

3. **If you are the relaying operation, dial 911 and when the operator responds, state your name and state that you're reporting an emergency via amateur radio.**

4. **Follow the directions of the 911 operator.**

   If the operator asks you to stay on the line, do so, and ask the other repeater users to please stand by.

5. **When the operator finishes, announce that you are finished with the emergency traffic and that the repeater is available for normal use.**

To report the need effectively, you must be able to generate clear, concise information. To report an automobile accident, for example, you should know the following details:

>> The street name or highway number

>> The street address or approximate highway mile marker

>> The direction or lanes in which the accident occurred

>> Whether the accident is blocking traffic

>> Whether injuries are apparent

>> Whether the vehicles are on fire, are smoking, or have spilled fuel

WARNING

Don't guess if you don't know something for sure! Report what you know, but don't embellish the facts. Don't be afraid to say, "I don't know." Bad information is worse than no information.

TECHNICAL STUFF

Autopatch is a link between the phone line and the repeater audio and control circuits. At one time, autopatch facilities were common, but fewer repeaters offer them now that mobile phones have reduced the need. You can find out more about autopatch at this book's page at www.dummies.com.

Making and responding to distress calls

Before an emergency occurs, be sure that you know how to make a distress call. You should also know the frequencies where hams are likely to be listening, such as a wide-coverage repeater in your area. Boaters know about the Maritime Mobile Service Net on 14.300 MHz (www.mmsn.org). Store a few such frequencies in your radio's memory channels.

REMEMBER

Anyone, licensed or not, can use your radio equipment in an emergency (defined as "an immediate threat to life or property") to call for help on any frequency. Write down the emergency frequencies or channels and post them near the radio. You won't have time to be looking up frequencies in an emergency.

Making a distress call

Do the following things when you make a distress call:

1. **For immediate emergency assistance, say "Mayday" or send the Morse code signal SOS (yes, just like in the movies).**

   Maydays sound something like "Mayday, mayday, mayday, this is NØAX," followed by

   - The location (latitude/longitude) or address of the emergency

   - The nature of the emergency

   - The type of assistance needed (such as medical or transportation aid)

2. **Repeat your distress signal and your call sign for several minutes or until you get an answer.**

   Even if you don't hear an answer, someone may hear you.

3. **Try different frequencies if you don't get an answer.**

   If you decide to change frequencies, announce the frequency to which you're moving so that anyone who hears you can follow.

Responding to a distress call

Here's what to do if you hear a distress signal on the air:

1. **Immediately record the time and frequency of the call.**

   If possible, start keeping a log of what is said and the time of each communication. You may not remember clearly later if the information becomes important.

2. **Respond to the call.**

   Say something like this: "[The station's call sign], this is [your call sign]. I hear your distress call. What is your situation?"

3. **Collect and record the following information:**

   - The location (latitude/longitude) or address of the emergency
   - The nature of the problem
   - The type of assistance needed (such as medical or transportation aid)
   - Any other information that might help emergency responders

4. **Ask the station in distress to remain in contact with you.**

5. **Call the appropriate public agency or public emergency number, such as 911.**

   Explain that you're an amateur radio operator and that you've received a distress call. The dispatcher will either ask you for information or transfer you to a more appropriate agency.

6. **Follow the dispatcher's instructions to the letter.**

   The dispatcher may ask you to act as a relay to the station in distress.

7. **As soon as possible, report back to the station in distress.**

   Tell the operator whom you contacted and any information you've been asked to relay.

8. **Stay in contact as long as the station in distress or the authorities need your assistance.**

TIP

If the disaster or emergency situation occurs outside your immediate vicinity, make yourself available to the on-site communication team — but only if you're called upon to do so. If you aren't in the disaster area, you can use the Internet to find the frequencies or channels that are in use. Bulletins may be posted or broadcast by the ARRL or responding organizations. Pay attention to instructions or restrictions. In the meantime, listen.

REMEMBER

The most important information from a disaster flows *out*, not in! You don't want to get in the way of that information. Listen, listen, listen.

Providing Public Service

Between emergencies, hams perform many other valuable public services. After you become associated with a local emergency communications group, you can use your ham radio skills for the public's benefit.

Weather monitoring and SKYWARN

One of the most widespread public-service functions is amateur weather watching. In many areas, particularly those that have frequent severe weather conditions, nets devoted to reporting local weather conditions meet regularly on HF or VHF/UHF. Some nets meet once or twice every day; others meet only when a threat of severe weather exists. Ask around to see if one operates in your area. These nets usually are active at commuting drive times. (For more information on these groups, see "Participating in Nets," later in this chapter.)

Many weather nets are associated with the SKYWARN program (www.skywarn. org), operated by the National Weather Service (NWS). Groups reporting weather conditions under the SKYWARN program relay information to the NWS, which uses the reports in forecasting and severe weather management. In some areas, a net control station may operate a station from the NWS itself. For information on whether a SKYWARN net is active in your area, click the Local Groups link on the SKYWARN home page or enter *skywarn net* in an Internet search engine.

TIP

Your local National Weather Service office may provide free weather-spotting training classes (training.weather.gov). The ARRL also publishes *Storm Spotting and Amateur Radio*.

REMEMBER

If you live in the southeastern United States, it is very useful to monitor the Hurricane Watch Net on 14.325 MHz (www.hwn.org) and the Caribbean Emergency and Weather Net on 3.815 MHz with a secondary frequency of 7.188 MHz depending on condition (cewn.org), even if you are not licensed to transmit on those frequencies.

Parades and charity events

Amateurs often assist parades and charity events like "fun runs" by providing communications. This gives the event managers timely information and helps with coordination. In return, amateurs get training with procedures and operations that simulate real-life emergencies. You can think of a parade, for example, as being similar to a slow-speed evacuation. A lost-child booth at a parade is similar to a small SAR operation. Helping keep track of race entrants in a marathon or bikeathon is good practice for handling health-and-welfare messages.

A leader of the amateur team usually coordinates plans with the event managers; then the group deploys as the plan requires. Depending on the size of the event, all communications may take place on one frequency or several channels may be required. Information may be restricted to simple status or actual logistics information may be relayed. Communications support includes a wide variety of needs. Be flexible! If you want to participate in these events, start by contacting your ARRL section manager who will let you know about upcoming events and what club or team to contact.

TIP

Here are a few tips for supporting an event:

>> Check to see if any pre-event registration or personal background information is required by the organizers or public service team.

>> Get the appropriate identification and any required insignia. Dress similarly to the rest of the group members.

>> Take along a copy of your amateur license and a photo ID.

>> Take water and some food with you in case you're stationed somewhere without support.

>> Don't assume that you'll be out of the weather. Protect yourself against the elements.

>> Have your identification permanently engraved on or attached to your radio equipment so the radio can be returned to you if borrowed, lost, or misplaced.

WARNING

Remember the prohibition against hams being paid for communicating on behalf of third parties! You and your team may be reimbursed for direct expenses (such as mileage or materials) but not compensated for your services. It's up to the leaders of your team to ensure that your support of any event or organization complies with this important rule. The exam asks questions about getting compensated for operating.

Participating in Nets

Short for "network," *nets* are regularly scheduled on-the-air meetings of hams with a common interest or purpose. Nets are among the oldest ham radio activities. Because the range of early ham stations was limited, nets helped relay *traffic* (messages) over longer distances. Sometimes, nets are strictly for enjoyment, discussing topics such as collecting equipment or pursuing awards. Other nets are more functional, such as those for training, public service, or weather reporting.

TIP

You can find net schedules online. First, your club may list one or more nets on its website or social media page. ARRL section managers (www.arrl.org/sections) will list a number of service and emergency nets. And the ARRL's online Net Directory (www.arrl.org/arrl-net-directory) lists nets of all sorts.

If a net follows standard operating procedures, it's called a *directed net*. Nearly all directed nets have a similar basic structure. A *net control station (NCS)* initiates the net operations, maintains order, directs the net activities, and then terminates net operations according to a standard procedure. Stations that want to participate in the net check in at the direction of the NCS. A *net manager* defines net policy and focus, and works with the NCS stations to keep the net meeting on a regular basis.

REMEMBER

Nets are run in many ways. Some nets are formal; others are more like extended roundtable QSOs. The key is to listen, identify the NCS, and follow directions. The behavior of other net members is your guide.

TECHNICAL STUFF

Digital voice networks (discussed in in more detail in the section "Digital Message Networks," later in this chapter) each have their own procedures to join a net. There may be special codes or tone sequences, you might be able to connect through an online website, or you may have to enter addressing information in your radio.

Checking in and out

To participate in a net, *check in* (register) with your call sign and location or status by calling the NCS when check-ins are being taken. Be sure that you can hear the NCS clearly and that you can understand his or her instructions. If you're not a regular net member, wait until the NCS calls for visitors.

REMEMBER

If you have an emergency or priority message, you should call any time the NCS is standing by. He or she will give instructions as to what to do next.

When you check in, give your call sign once (phonetically if you're using voice). If the NCS doesn't copy your call sign the first time, repeat your call sign, or the NCS can ask one of the listening stations to relay your call sign.

TIP

The two most common ways for an NCS to accept check-ins are random (anyone can call) or *roll call* (the NCS calls each net member in turn, usually by alphabetical order of the call sign's suffix).

You can check in with business (such as an announcement) or traffic (messages) for the net in a couple of ways; listen to the net to find out which method is appropriate. The most common method is to say something like "NØAX with one item for the net." The NCS acknowledges your item, and you wait for further instructions. Alternatively, you can check in with your call sign, then when the NCS acknowledges you and asks whether you have any business for the net, reply, "One item." Follow the examples of other net members checking in — when in Rome, check in as the Romans do.

If you want to contact one of the other stations checking in, say so when checking in or wait until the check-in process is complete and the NCS calls for net business. Either way, the NCS will ask the other station to acknowledge you. You will then be directed to make contact according to net procedures. The contact may be on the net frequency or the NCS may direct you to another frequency or channel.

To check out, contact the NCS and request to be excused. The NCS may release you immediately or may ask if any other station has business or traffic for you. If you will need to be released before the regular net session concludes, let the NCS know that when you check in.

Exchanging information

Formal net traffic consists of messages in a standard format. If your served agency operates according to the ICS (Incident Command System) they will probably use those message formats, such as the ICS-213 General Message Form shown in Figure 10-2. Another common format is the ARRL Radiogram, which is similar to a telegram. Your public service team will determine what form is appropriate for the need. The Wireless Society of Southern Maine has published a nice overview of both forms at www.ws1sm.com/Message-Handling.html.

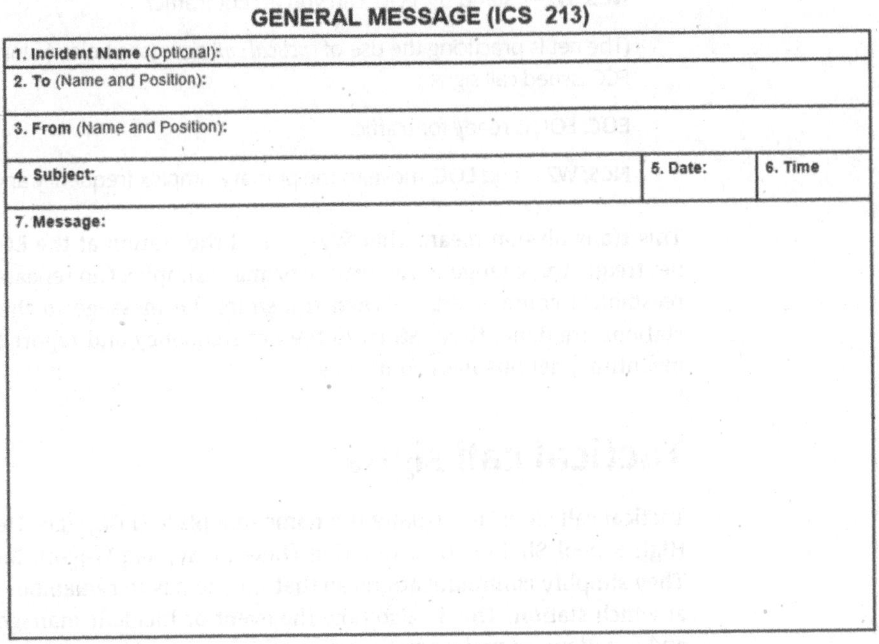

GENERAL MESSAGE (ICS 213)

1. Incident Name (Optional):		
2. To (Name and Position):		
3. From (Name and Position):		
4. Subject:	5. Date:	6. Time
7. Message:		

FIGURE 10-2:
An ICS-213
General Message
form.

TIP

Have blank forms printed out and handy whenever you check into a public service or traffic-oriented net. Part of your go kit should be a set of forms. If they are available in electronic format, load the files onto a USB thumb drive for printing when needed.

Informal verbal messages may be exchanged on the net frequency or on a nearby frequency or channel at the direction of the NCS. Nets for selling and trading equipment, for example, keep all their transmissions on one frequency so that everyone can hear them. This is too inefficient for a net that's intended to route traffic and formal messages. The NCS of these nets will send stations off frequency to exchange the information and then return to the net frequency.

TIP

If you are going to participate in nets on a regular basis, practice changing your radio's channels and frequencies so you can do it quickly, without looking at the manual.

Here's an example of an NCS directing an off-frequency message exchange during a net that's using a repeater.

W2---: I have one piece of traffic for the EOC.

(W2--- is either relaying the message from another ham or is the originating station.)

NCS: W2---, stand by. EOC, can you accept traffic?

(The net is practicing the use of *tactical call signs* (see below) along with the FCC-issued call signs.)

EOC: EOC is ready for traffic.

NCS: W2--- and EOC, move to the primary simplex frequency and pass the traffic.

This transmission means that W2--- and the station at the EOC are to leave the net frequency, change to the team's primary simplex (no repeater) frequency, and reestablish contact. W2--- then transmits the message to the EOC. When both stations are done, they return to the net frequency and report to the NCS. In the meantime, net business continues.

Tactical call signs

Tactical call signs are usually the name of a place (EOC, Race Headquarters, West High School Shelter) or a function (Bike Relay, Sag Wagon, Medical Response). They simplify communications so that no one has to remember which operator is at which station. This is also how the event or incident managers refer to places and functions so no translation is required.

As a control operator, you still have to give your FCC call sign along with the tactical call sign whenever you begin operation, every ten minutes during operation, and whenever the operator changes. All you have to say is something like "Lead Car, NØAX" with no extra words or phrases. Your public service team will provide training to help you satisfy these simple rules.

Radio discipline

You may hear that term in meetings or training exercises. What does it mean? Mostly it means to be efficient and keep the net operating smoothly. Here are some pointers:

» Do not transmit out of turn or if not authorized to transmit.

» Be ready to respond promptly. If you have to step away from the radio for a significant period, let the NCS know you are temporarily unavailable.

>> Use a minimum of words and don't repeat information unless requested. Practice not saying, "Uhhh. . ." and unnecessary phrases like "Please copy." Even "Please" and "Thank you" eat up a lot of air time.

>> Use plain language. Say "Yes" and "No" and not the unnecessary "Affirmative" or "Negative" no matter what you see in the movies.

>> Follow procedures. Follow instructions.

Your team leaders and experienced net control operators will help you become a top-notch, smooth operator with practice!

Digital Message Networks

Hams have adapted a number of commercial networking technologies to amateur radio. New digital modes and services are always being evaluated and adopted! This section covers systems that are widely used as of mid-2020 — Winlink, AREDN (Amateur Radio Emergency Data Network), and NBEMS (Narrow Band Emergency Messaging System). Work with your served agencies to determine what networks meet their needs.

WARNING

Even though you are using digital techniques, the FCC regulations on amateur communication apply. You can't encrypt messages, send business traffic or obscene content, or use radio links on behalf of third parties in countries where such use is prohibited. Ham radio networks aren't a web browsing service, but web content is often added to email or other online messages. That content shouldn't be accessed or distributed via ham radio.

TIP

Keep in mind that although the network stations may be connected via the Internet, your station is connected to them by a radio link that is pretty slow compared to broadband speeds. If your data rate is limited due to the radio link, don't try to send big attachments.

WARNING

Don't forget — to fulfill your responsibilities as a control operator you have to be sure you won't interfere with ongoing communications and you must monitor your transmissions. This is particularly important on HF where the network stations share channels with many other users. Your software or modem's "busy" light may not recognize their signals and you will create interference when you start transmitting. No one has priority on any particular frequency so unless you have an emergency need, if the frequency is busy you'll have to use a different channel or be patient and wait until the channel is available. Try to always have a Plan B!

Winlink — email by radio

The dominant ham radio email system is Winlink (www.winlink.org). Winlink is a worldwide network of RMS (Radio Message Server) stations operating 24 hours a day on the HF bands as well as on VHF and UHF. It has grown from a network used by boaters to a sophisticated, hardened network used for disaster relief and emergencies of all kinds. Stations communicate with its mailbox stations using packet radio, PACTOR, ARDOP, or VALA digital modes. (I talk about digital modes in Chapter 11.)

WARNING

There are four levels of PACTOR, 1 through 4. As of mid-2020, PACTOR 4 is not legal for U.S. amateurs because its symbol rate is higher than FCC regulations allow. Rule-making proposals under consideration by the FCC may increase the maximum rate but have not yet been released. Be sure your PACTOR modem is configured to comply with FCC rules.

The ability to send and receive email has proven to be extremely useful, particularly during regional emergencies. For example, after hurricane Maria severely damaged commercial phone and Internet service on Puerto Rico in 2017, email sent by ham radio carried thousands of health-and-welfare and logistics messages. Email over ham radio has become an important part of disaster recovery and relief efforts!

WARNING

Although email over ham radio might sound attractive, it's not the easiest thing to do successfully or correctly. If this is something you want to do, get a club, public service team, or mentor to show you how to do it right. You'll get much better results, and you won't create interference or delays by not knowing how to operate properly.

To find the frequencies and locations of Winlink stations, visit winlink.org/RMSChannels. This extensive and growing network covers much of the world. (See Figure 10-3.) These stations are linked to a system of hardened servers via the Internet, creating a global home for Winlink users. The Winlink system is also connected to the Automatic Packet Reporting System (APRS; www.aprs.org) so that position, weather, and other information can be exchanged or viewed via the Internet. (I discuss APRS in Chapter 11.)

To use the Winlink system, you must register as a user on the Winlink network so that the system recognizes you when you connect. When you're a recognized user, your messages are available from anywhere on Earth, via whichever Winlink station you use to connect. You must also download and install a Winlink-compatible email *client* program, such as *Airmail* or *Paclink*, which are available on the Winlink website.

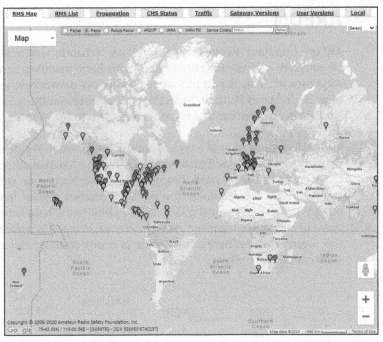

FIGURE 10-3: Winlink mailbox stations using WINMOR as of September 2020.

Along with a computer that runs the email client software, you need a way to generate the signals for the digital mode you choose to access the RMS stations. On HF, you need a sound card and software to send and receive ARDOP or VARA mode signals (www.winlink.org/content/ardop_overview and www.winlink.org/tags/vara). The SCS PTC or P4 stand-alone modem (www.scs-ptc.com/pactor) supports the PACTOR modes. On VHF and UHF, you can connect to a local RMS station with a standard packet radio Terminal Node Controller (TNC) or a D-STAR radio.

Once you have all the necessary software and hardware to connect to the Winlink stations, download the Winlink "Frequently Asked Questions" (FAQ) document from winlink.org/content/winlink_faq_frequently_asked_questions_answers. There is also an excellent set of "Quick Start Links" (winlink.org/content/quick_start_links_mariners_and_everyone) that will help get you started.

If you are using the Winlink system to support your public service team or activities, you'll need to follow procedures for generating and exchanging email. Your team may have a set of forms all ready to go, for example. Winlink is very powerful but you still have to use it according to the needs and requirements of the "customer" (the agencies or organizations you support).

AREDN

AREDN (Amateur Radio Emergency Data Network; www.arednmesh.org) uses commercial wireless Internet routers with ham-modified firmware to create a *mesh network*. This is part of ham radio wireless networking technology referred to as High-Speed Multimedia (HSMM). It operates on the 2.4 GHz, 3.4 GHz, and 5.6 GHz microwave bands with data rates of up to 144 Mbps.

The basic organization of a typical network is shown in Figure 10-4. The wireless network routers are configured to form network *nodes* that automatically link with other nodes in the network through a *backbone*. Each node is then connected to standard WiFi access points. You can then connect regular laptops, tablets, and other WiFi-enabled gear through the WiFi link to the ham network. The network may not connect to the commercial Internet and is not intended to be an alternative Internet access network. (See the previous warning about using ham radio and commercial content.)

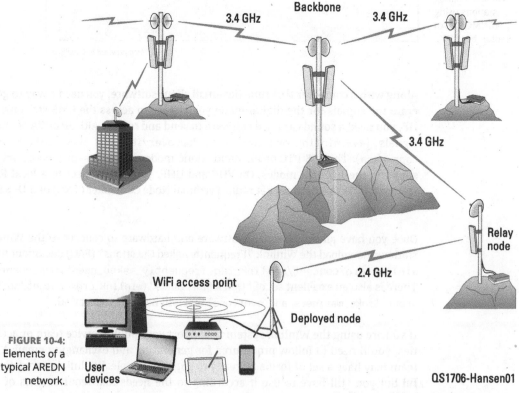

FIGURE 10-4: Elements of a typical AREDN network.

QS1706-Hansen01

Graphic courtesy of the American Radio Relay League.

AREDN network nodes can be permanent or temporary. A populated area may have an established AREDN network that runs all the time. An AREDN network can be set up just for one event or emergency need. The temporary AREDN network can be completely stand-alone or it may connect to a permanent network if a link can be established.

Once the network is established, software on the computing equipment can be used as if it is connected to the Internet. (Internet-connected services will not be available, of course.) Because of the high data rate compared to HF/VHF/UHF networks, AREDN can be used to transfer images and larger files that are normally used by the served agencies.

NBEMS

NBEMS (Narrow Band Emergency Message System) is a service for passing digital messages that does not require a network or Internet access. It is part of the *fldigi* digital mode software package. NBEMS is a suite of applications that work together to pass messages in standard formats, using a variety of digital modes on HF and VHF/UHF. A thorough explanation of NBEMS in provided by SATERN at satern.org/Training/NBEMS.html. One of the applications is *flmsg* — a program that wraps a plain text message in formatting and error checking. Also, flmsg includes recognizable, well-established forms such as the ICS–213 and ARRL Radiogram, described earlier, as well as Red Cross forms and many others. You can even create your own forms. This saves a lot of time and mistakes!

The other application, *flamp*, conducts the transmission so that messages are received error-free. Messages can be compressed to save transmission time, as well. Messages can be station-to-station or broadcast for all stations to receive.

NBEMS is particularly useful if you need to communicate over long distances without access to the Internet or the Winlink system of mailboxes. For example, transmissions by a single station using NBEMS on the lower HF frequencies can cover an entire state. NBEMS is a good complement to Winlink and AREDN.

Chapter **11**

Operating Specialties

No matter why you are interested in ham radio, once started you will discover it includes a whole world of interesting specialties. These specialties can be the real attraction of the hobby and keep it endlessly fresh and interesting. More are evolving all the time!

In this chapter, I give you an overview of popular activities and operating styles, cover some of their basic techniques and resources, and demystify some of their specialized jargon. As you follow your interests, don't hesitate to investigate things that catch your interest. Ham radio offers a banquet of choices and you are free to sample as many as you want.

Before we get started, I should reinforce that SDR (software defined radio — see Chapter 8) technology makes traditional operating activities more flexible. For example, instead of having to manually tune from station to station in order to find out "who's on," I can see at a glance how many signals are present, how strong they are (or aren't), and even what type of signal they are. By merging the display with the information from worldwide "spotting networks" on the Internet, I can even tell who the stations are and where they're from. While this changes the ham radio experience, we will always use radio waves to communicate. Analog radio, even using tubes and crystals, will still be around and as enjoyable as it ever was. SDR just helps you "get there" faster and have more fun along the way!

Getting Digital

Operating via digital modes is the fastest-growing segment of amateur radio. Applying the power of *digital signal processing (DSP)* allows hams to communicate keyboard to keyboard, using just a few watts of output power and modest antennas. Hams are innovating like crazy and new modes are popping up all the time.

On HF bands, digital modes must overcome the hostile effect of the ionosphere and atmospheric noise on delicate bits and bytes. Digital modes with error detection and correction mechanisms can overcome those problems. Are you limited in what antennas you can put up? Is the noise level in your neighborhood making it hard to hear signals on CW and SSB? If so, digital mode operating could be just the thing to overcome those challenges and get you on air successfully.

TECHNICAL STUFF

Many of the current HF digital modes can all be used with the same connection between a computer sound card and the transceiver. (See Figure 2-2B in Chapter 2.) The ARRL book *Get On the Air with HF Digital* by Steve Ford, WB8IMY, gives complete instructions for anyone new to this type of operating (www.arrl.org/shop/Get-on-the-Air-with-HF-Digital-2nd-Edition). Some newer radios have a USB connection that supports digital audio connections to a PC. Many vendors also sell data interfaces that have all the necessary cables to connect your computer sound card and transceiver's audio jacks.

On the VHF and higher frequency bands, digital data modes have fewer restrictions on bandwidth. The bands are quieter, with less fading and interference, so data communication works better. At 440 MHz and up, you can use 56 kbaud technology. D-STAR supports an Ethernet bridge mode that connects two computers via the amateur 1.2-GHz band as though they were connected by an Ethernet network cable. Mesh networks like AREDN (see Chapter 10) create full WiFi-like connections on the 2.4-, 3.4-, and 5.6-GHz bands.

TIP

If you are interested in the technical characteristics of amateur radio's many digital modes, the "Digital Protocols and Modes" chapter in *The ARRL Handbook* is an overview of the different modulations and protocols involved. You can hear what some of the various modes sound like at www.arrl.org/HF-Digital. Table 11-1 lists a several current PC software packages that support multiple digital modes. (I mention some Android/iOS apps that support specific modes later.)

WARNING

Some digital modes are *constant amplitude*, meaning that once you start transmitting, your transmitter stays on for the whole transmission at the same power. (This is also how FM voice behaves.) This can be hard on a transmitter! Many of the digital modes have such good performance that full power transmissions aren't required to get through. Nevertheless, when operating the digital modes turn the power down a bit and make sure you keep the radio cool.

TABLE 11-1

Digital Mode Software Resources

Resource	Description
fldigi (www.w1hkj.com)	Free, Multi-mode, Windows
WSJT-X (physics.princeton.edu/pulsar/K1JT/wsjtx.html)	Free, Multi-mode, Windows/Linux/Macintosh
MultiPSK (f6cte.free.fr/index_anglais.htm)	Free, Multi-mode, Windows
MixW (mixw.net)	Multi-mode, Windows
MultiMode (www.blackcatsystems.com/software/cw-rtty-sstv-fax-psk31-packet-decoding-software.html)	Multi-mode, Macintosh

Digital definitions

TIP

These definitions would be good to know for the exam questions!

AFSK — Audio frequency shift keying, using a radio's speech circuits to modulate a signal with audio tones via the microphone input. *Direct FSK* is shifting the frequency of a carrier signal by using a digital signal. Both types of signals sound the same on the air.

baud — The rate at which symbols are transmitted (no need to say "baud rate" since that would be equivalent to "symbol rate rate").

bit — One unit of data, a 0 or a 1.

bps — Bits per second, a measure of the rate at which data is being transferred.

encode (decode) — Change digital 0 or 1 bits into some characteristic of a transmitted signal (recover the 0 or 1 bits from the transmitted signal).

FEC — Forward error correction, the technique of adding redundant information to allow errors in a received signal to be detected and removed.

FSK — Frequency shift keying, modulating a signal with tones that change in a pattern that represents individual characters.

Keyboard-to-keyboard — A mode which sends and receives characters as they are typed by the operators.

keying — Causing a symbol to be transmitted.

MFSK — Multi-frequency shift keying, using more than two tones in combinations or sequences to send digital bits. The group of individual tones are heard as "notes" or "chords" on your receiver.

PSK — Phase shift keying, modulating a signal by shifting the phase of a single carrier signal in a pattern that represents individual characters.

Soundcard mode — A digital mode implemented by PC software using a soundcard (internal or external) as the digital interface for received and transmitted audio.

symbol — A change in the transmitted signal that represents data.

WPM — Words per minute, the speed at which five-character groups are transmitted.

WSJT modes — fast and slow

The software package *WSJT-X* includes several different modes developed by a group led by Joe Taylor, K1JT, and Steve Franke, K9AN, for extremely weak signal operation. Referred to as "WSJT modes" these use AFSK (see the definitions above) with a regular SSB transceiver so they don't need any special supporting equipment.

The WSJT modes use multi-tone FSK or MFSK. Why use so many tones? The "secret sauce" that makes the WSJT modes perform so well are specially coded tone sequences and message structures. Using multiple tones this way establishes a unique pattern that receiving software can detect, even when the signal is hundreds of times weaker than the noise level! The ability to recognize the codes and sequences is called *processing gain* and is what enables *WSJT-X* to recover signals deep in the noise that human ears can't detect.

These modes enable hams with modest stations to make contacts that previously required large antennas and high power. For example, bouncing a signal off the Moon (called *moonbounce* or *EME* for Earth-Moon-Earth, strangely enough) can now be done on 2 meters with a pair of Yagi antennas (see Chapter 12), a good preamp, and 100 watts of RF power. There is a thorough "how to get started" instruction package (*WSJT-X User Guide*) and an online support forum that has helped hundreds of hams get going (www.physics.princeton.edu/pulsar/K1JT and groups.io/g/WSJTX). You can do this, too!

A "NOBLE" HAM RADIO OPERATOR

Joe Taylor, K1JT, received the 1993 Nobel Prize in Physics for his work with Russell Hulse in detecting the first direct evidence of gravitational waves by observing a binary pulsar with the Arecibo Observatory's 1,000-foot dish (en.wikipedia.org/wiki/Joseph_Hooton_Taylor_Jr.). (You may have read about gravity waves finally being detected in 2016 by the LIGO system.) Not bad for a guy who started out with his brother as teenage ham radio operators interested in VHF scatter propagation!

TIP

WSJT-X includes several "slow" modes: JT4, JT9, JT65, QRA65, and FT8/FT4 (see the next section). Each mode is optimized for extremely weak-signal communication using a particular type of propagation. Each type of path affects the radio signal a little bit differently. These modes are considered "slow" because of the long transmit and receive times required — from 7.5 seconds (FT4) to one-minute long for some of the other modes.

The "fast" modes are: MSK144 and ISCAT. These send information very quickly since they are designed for "scatter" propagation from meteors, airplanes, and weather phenomena in the atmosphere.

TIP

A special mode, WSPR for "weak signal propagation reporter," acts as a very low-power beacon to assess worldwide propagation on the HF bands or use the VHF/UHF bands to check your own echoes "off the Moon." Doesn't that sound interesting? It is! I have my own small WSPR transmitter, which generates only one-fifth of a watt, but which was heard one sunset in western Australia, France, and South Africa, all in the span of a few minutes.

You can learn all about the *WSJT-X* modes in a pair of *QST* articles by Taylor, Franke, and Bill Somerville, G4WJS. "Work the World with WSJT-X, Part 1: Operating Capabilities" and ". . . Part 2: Codes, Modes, and cooperative Software Development." Joe's *QEX* articles "The MSK Protocol for Meteor-Scatter Communication" and "The FT4 and FT8 Communication Protocols" give a close look at those modes. All are posted online along with quite a few other interesting articles in the Reference section of Joe's website: physics.princeton.edu/pulsar/k1jt.

FT8 and FT4

Since FT8's introduction in mid-summer 2017, it quickly became the most popular digital mode in ham radio. Its simple message structure, ability to decode signals that are extremely weak or that are buried in noise, and ease of integrating with a SSB transmitter make it very popular with hams don't (or can't) have large antenna systems and full-power transmitters. With HF radio noise being a challenge in urban and suburban settings, FT8's ability to dig out signals is even more appreciated. Most stations use 100 watts or less and simple antennas like dipoles, whips, and small loops to make contacts all over the world. On VHF, particularly 6 meters, the ability to copy very weak signals shows us that propagation is really present a lot of the time, even when CW or SSB signals would be very difficult to copy by ear. (If FT8 signals are strong, CW or SSB is often quite strong, too, and contacts are much quicker to make so keep your analog mode operating skills sharp!)

TIP

The basic structure of FT8 is a series of 15-second transmissions that send and answer a CQ message, exchange location and signal strength, and acknowledge that call signs and other information was received correctly. The transmission periods are synchronized to within a second by using online time sync services. Messages can be configured for casual or contest operation. This type of short exchange is perfectly good for most short DX and contest contacts. Each message is specially structured to enhance reception through special encoding and error-correction techniques. The mode's name comes from the 8 different tones, called 8-FSK, used to create the FSK signal. These basic characteristics are covered by several exam questions.

Figure 11-1 shows the *WSJT-X* software during FT8 operation. At the upper left are all the messages FT8 decoded at least once (left-most frame) and those with my call sign (right-most frame) either received or transmitted. The messages are in different colors depending on whether they are a CQ, a transmission from your station, or a transmission received from another station to yours. Below the messages frames are the FT8 operating controls and configuration settings. You can also see messages 1 through 6 that are created by *WSJT-X*. In the right-hand message frame you can see me working W1ARY near the top and then calling VE2BAE, WJ2N, and AC2BI afterward. Call signs from the U.S., Europe, and the Middle East were all being received on this busy morning.

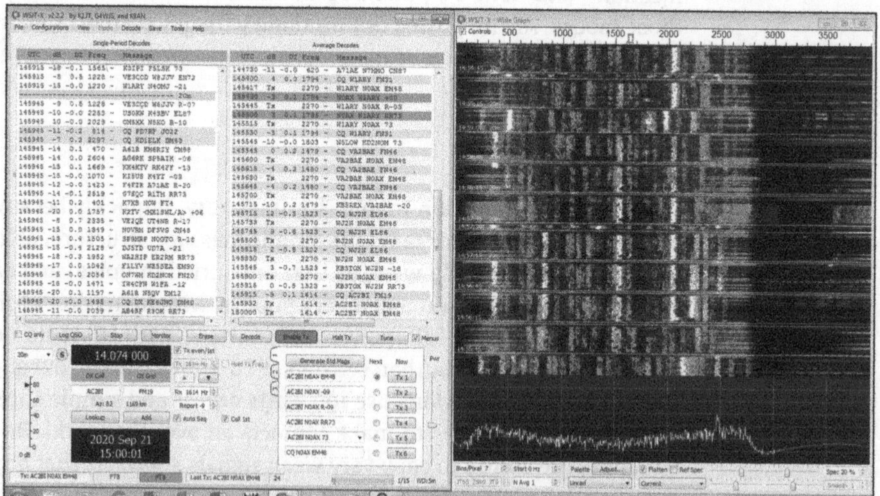

FIGURE 11-1:
The *WSJT-X* windows during FT8 operation on 20 meters.

At the upper right is the FT8 waterfall display showing all of the received signals during each successive listening window. Brighter colors indicate a higher signal strength. The frequency of each received audio tone (not the RF signal) extends from 0 Hz on the left to a configurable frequency on the right. I prefer to set the right-hand frequency to a bit higher than 3000 Hz so I can see every signal in my receiver's pass-band. You can clearly see the 15-second "blocks" of time during which stations are listening or transmitting. Below the waterfall is a spectrum display that shows the amplitude of each signal in the receiver's pass-band.

A set of typical FT8 QSO with me answering a CQ works like this:

1. I see CQ messages in the left-hand decode window and double-click on one. I do this quickly, as soon as the current 15-second receive period is over, so I can answer the CQing station right away.

2. Let's say I see a CQ from K1JT. My message responding to Joe's CQ is "K1JT NØAX EM48" which says "K1JT, NØAX hears your CQ and is calling you from grid square EM48." (See the section on DXing on VHF and UHF for more about grids and grid locators.)

3. If I am copied by Joe's station, he will respond "NØAX K1JT -04" meaning that my signal has a -4 dB signal-to-noise ratio. (My signal is 4 dB weaker than the noise level at Joe's station, but we can still make this short contact.)

4. I respond "K1JT NØAX R+03" to tell Joe that his response was received with a signal level 3 dB greater than the noise level.

5. Joe responds "RRR" and I respond with "73," concluding our contact.

If at any point, either of us don't decode a response, we'll simply retransmit our message until a response is received or a repeat count limit is reached, when we simply give up and quit to try again another time.

The 15-second periods of FT8 are pretty long for contest-style operation when signals are typically quite a bit stronger or for when the bands are wide open. Since those conditions don't require detecting such a weak signal, FT4 was created with fewer tones (4) and a shorter transmit and receive window of 7.5 seconds. You'll find the FT4 calling frequencies to be busy during contests and good band conditions. It is challenging to respond promptly to FT4 calls and keep up with the shorter transmit windows.

JTAlert and *JS8Call* are two extensions of the basic FT8 system created from the WSJT-X open-source software. *JTAlert* (www.dxlabsuite.com/dxlabwiki/Getting StartedwithK1JTModesWithJTAlert) acts as a "wrapper" for the *WSJT-X* program to manages the process of "connecting" FT8 to a regular or contest contact logging program. *JS8Call* (js8call.com) is an independent program that builds on FT8 to provide free-form messaging while still communicating at low signal levels. Both of these apps are typical of ham innovation, building on and extending an original idea.

PSK31 and PSK63

The "31" in PSK31 represents the 31.25-baud rate of the signal — about regular typing speed for this keyboard-to-keyboard mode. (PSK63 operates at twice the bit rate of PSK31.) Invented by Peter Martinez, G3PLX, it also uses a coding system for characters called *Varicode* which has a different number of bits for different characters, not unlike Morse code. Instead of a carrier turning on and off to transmit the code, a continuous tone signifies the bits of the code by shifting its timing relationship (known as *phase*) with a reference signal. A receiver syncs with the transmitter and decodes even very noisy signals because the receiver knows when to look for the phase changes. PSK31 is very tolerant of the noise and other disturbances on HF bands. In fact, you can obtain nearly "solid copy" (receiving 100 percent of the transmission) with signals barely stronger than the noise itself.

TIP

The smartphone apps DroidPSK (Android) and PSK31 (iOS) decode PSK31 signals. DroidPSK can also be used to transmit for full two-way QSOs.

The most common frequencies for PSK31/63 operation in North and South America on the HF bands are 3580, 7070, 10142, 14070, 18100, 21080, 24920, and 28120 kHz. See the calling frequency table at en.wikipedia.org/wiki/PSK31 for other regions and more bands.

If you want to find out more about using PSK31 and PSK63, the PODXS Ø7Ø Club (www.podxs070.com) specializes in this mode, including a good tutorial in its "Frequently Asked Questions" section. For more information on operating PSK31, this online video is pretty good: www.youtube.com/watch?v=wbmXFzmXF00&ab_channel=HamRadioConcepts.

Radioteletype (RTTY)

The first fully automated data transmission protocol was *radioteletype (RTTY)*. RTTY (pronounced "*ritty*" by hams) uses a 5-bit code known as Baudot after its inventor — also the origin of the word *baud*. The Baudot code sends plain-text characters as 5-bit codes that use alternating patterns of two audio frequencies known as *mark* and *space*. Because RTTY uses just two tones, it is sometimes called *binary FSK* to distinguish it from schemes with more than two tones.

The tones 2125 Hz (mark) and 2295 Hz (space) can be transmitted as AFSK using a regular voice SSB transceiver. The separation of the tones, 170 Hz, is the FSK signal's *shift*. RTTY characters are transmitted at a standard speed of 60 words per minute (WPM). There are other variations of tones, shift, and speed but these are the standard configuration.

On the receiving end, the transmission is received as an audio signal. The text characters are typically recovered from the pair of mark and space tones by a computer and sound card running software such as MMTTY (see Table 11-2).

TABLE 11-2 ## RTTY Resources

Resource	Description
AA5AU's RTTY page (www.aa5au.com/rtty)	Tutorial information, links to RTTY programs, troubleshooting, and RTTY contesting
RTTY Contesting (www.rttycontesting.com)	Dedicated to contesting using RTTY
MMTTY (hamsoft.ca/pages/mmtty.php)	Free, RTTY decoder, Windows
2Tone (www.rttycontesting.com/downloads/2tone)	Free, RTTY decoder, Windows
DroidRTTY (www.wolphi.com)	Android app
RTTY email reflector (groups.io/g/RTTY)	International membership email group

Because of the two tones, you will see an RTTY signal as two closely spaced lines on a waterfall display. To tune in the signal, the decoding software will display a *tuning indicator* like one of those in Figure 11-2. The waterfall display at the lower left shows the two vertical lines corresponding to the mark and space tones. Above the waterfall display is a spectrum display showing the strength of each tone. The vertical lines show the frequencies the software is expecting. To the right is an interesting *dual-ellipse* display with each ellipse representing one tone. To decode the RTTY signal, tune your receiver so that the waterfall and spectrum display peaks align with the vertical lines and the ellipses are at right angles.

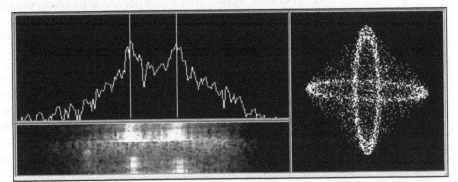

FIGURE 11-2: RTTY tuning indicators of the MMTTY software.

Courtesy American Radio Relay League

TIP

Invented before the transistor, the original teleprinters are mechanical marvels. If you get a chance to watch one of these devices at work, you'll be amazed by its complexity. Here's a YouTube video of a Model 15 teleprinter doing its thing: www.youtube.com/watch?v=bHkNZA28cMA.

RTTY is still a presence on the bands during contests. Tune through the digital signals above the CW stations, and you'll hear lots of two-tone signals "diddling" to each other. A sizable community of RTTY contesters are supported by award programs and contests. Table 11-2 lists several online resources for beginning RTTY operators.

Non-WSJT MFSK modes

When you tune around the digital signal areas of the bands, you're likely to come across MFSK signals that sound like crazy calliopes or steam whistles playing what sound like random melodies.

These modes are all supported by the *fldigi* and *MultiPSK* digital mode software. You can find the technical details on MFSK modulation at www.qsl.net/zl1bpu/MFSK. Here are a few keyboard-to-keyboard MFSK modes you can find on the bands today:

» **MFSK16:** 16 tones, spaced 15.625 Hz apart, sent one at a time at 15.625 baud. Each tone represents four bits of data in a bandwidth of about 316 Hz. The mode includes *forward error correction* (*FEC*) to produce excellent error resistance under even poor conditions. The overall character rate is about 40 WPM, which is a good typing speed for most of us.

» **DominoEX:** With six different modes and FEC that can be turned on or off, this mode can be adapted to different band conditions with speeds ranging from 12 to 140 WPM. Instead of each tone needing to be exactly of a certain frequency, which would make tuning critical, DominoEX uses *incremental frequency keying* (*IFK*) so that only the difference between tones matters. (DominoEX was invented by Murray Greenman, ZL1BPU.)

» **Olivia:** There are quite a number of different types of Olivia signals: 5 different bandwidths and 8 different sets of tones for a total of 8 × 5 = 40 different variations of Olivia! This flexibility allows the mode to recover signals that are as much as 10 dB below the noise or even more. The tradeoff is that the rate of character exchange is quite low, stretching contacts out over a long time. The alternative, though, is no contact at all on less robust modes. (Olivia was invented by Pawel Jalocha, SP9VRC.)

>> **Automatic Link Establishment (ALE):** ALE uses eight different tones, each representing three bits of information. The overall character rate is 375 bps. ALE stations can establish contact with each other automatically using software such as *PCALE* (hflink.com/software).

PACTOR, ARDOP, and VARA

All three modes are used in the Winlink email messaging system which is discussed in Chapter 10. They are *structured modes* that establish connections (called *sessions*) between stations and then begin a *training* process to determine the best data transmission rate for the current conditions. Training can be repeated at any time if the overall data rate gets too low. Maximum data rates approach 6k to 7k bits/sec (not bytes/sec) under excellent conditions. HF radio propagation is not kind to digital modes!

TIP

After training is completed, the protocols begin the process of transferring data as a series of packets. Each packet contains address and error-checking information and is acknowledged as having been received correctly by sending an ACK packet back to the sending station or requesting a retransmission by sending a NAK packet. This type of protocol is called *ARQ* for *Automatic Repeat Request*. If error-free data packets can't be received in an acceptable time, the connection *times out* and the session is closed without accepting any data. Because no data is accepted unless received error-free, this type of data transfer is called *reliable transport*. The exam has a question or two about the process of ARQ protocols.

PACTOR stands for Packet Teleprinting Over Radio and was initially released to the amateur community in 1991. PACTOR 4 is the most recent release of this technology available to hams. PACTOR 1 and 2 can be implemented by software on a PC but PACTOR 3 and later versions are proprietary and are available only in equipment available from SCS (www.scs-ptc.com/pactor).

WARNING

PACTOR 4 is not yet legal for U.S. hams to use because it exceeds the maximum symbol rate allowed by FCC rules. Hams outside the U.S. can use it and U.S. hams can monitor those contacts without transmitting but may not use PACTOR 4 on the air unless the FCC allows it temporarily during disaster recovery by a special authorization. Be sure you have your PACTOR equipment configured properly to operate legally.

ARDOP (Amateur Radio Digital Open Protocol: www.winlink.org/content/ardop_overview) by Rick Muething, KN6KB, replaces an older mode WINMOR. Both are intended to provide a public alternative to PACTOR, which requires using commercial equipment to achieve the best performance. ARDOP can operate in ARQ mode but it can also "broadcast" data with error-correcting information (FEC:

forward error correction) added for the receiver to make the necessary corrections. VARA (www.rosmodem.wordpress.com) is a newer competitor to PACTOR and ARDOP that offers a low-cost license ($70 as of late 2020) for the high-performance software. Developed by Jose Alberto Nieto Ros, EA5HVK, VARA is reported to achieve the best performance of the three protocols but is relatively new. VARA FM is optimized to run over FM voice channels so it can be used with standard VHF/UHF FM radios although at a lower data rate.

Choosing one mode over another is a decision that involves your finances, how much data you plan on sending, how fast that data needs to be sent, what your public service or other communications group uses, what other digital modes you might want to use, and what equipment you already have. This Communications Academy presentation by Scott Currie, NS7C, provides an excellent overview of all three modes so you can begin considering which mode or modes are right for you. Look for this tutorial "Digital Sound Card Modes for WinLink" in the Sunday folder of the 2019 Archives at www.commacademy.org.

Packet radio

Packet radio is a wireless network system based on the commercial X.25 computer-to-computer protocol. Developed by the Tucson Amateur Packet Radio group (TAPR: www.tapr.org), packet can send error-corrected data over VHF links. The protocol that controls packet construction, transmission control, and error correction is called AX.25 (for Amateur X.25).

TIP

In packet radio, ordinary VHF/UHF FM transceivers transfer data as audio tones. An external modem called a *terminal node controller (TNC)* provides the interface between the radio and a computer or terminal. Packet radio can also be implemented as a soundcard mode, using PC software instead of a TNC. Data is sent at 1,200 or 9,600 baud as packets of variable length up to about 1,000 bytes.

Packet stations are organized into networks in which a packet controller is called a *node.* Connections between nodes are *links.* Connecting to a remote node by relaying packets through an intermediate node is called *digipeating.* A node that makes a connection between two packet networks or between a packet network and the Internet is called a *gateway.*

APRS and tracking

The Automatic Packet Reporting System (APRS: www.aprs.org) is an amateur invention that combines GPS positioning and packet radio. APRS was developed by Bob Bruninga, WB4APR.

TIP

The most common use of APRS is to relay location data from GPS receivers via 2 meter radio as an "I am here" service. APRS packets are received directly by other hams or by a packet radio digipeater. If they're received by a digipeater, the packets may also be relayed to a gateway station that forwards the call sign and position information to an APRS server accessed through the Internet. After the information is received by a server, it can be viewed through a web browser or an APRS viewing program.

The most common frequency for APRS is 144.39 MHz, although you can use 145.01 and 145.79 MHz. You can find a group of HF APRS users using LSB transmission on 10.151 MHz. The actual tones are below the carrier frequency of 10.151 MHz and fall inside the 30 meter band.

The map in Figure 11-3 shows the location of WA1LOU-8. *WA1LOU* is a call sign, and *-8* is a secondary station ID (SSID), allowing the call sign to be used for several purposes with different SSIDs. The figure shows that WA1LOU is in Connecticut, 7.9 miles northeast of Waterbury. You can zoom in on WA1LOU's location. If the radio changes location, this change is updated on the map at the rate at which the operator decides to have his APRS system broadcast the information.

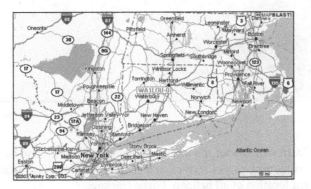

FIGURE 11-3:
WA1LOU's
position reported
by APRS.

TIP

Lots of people have developed maps based on APRS data. Some of the best known are www.findu.com, www.aprs.fi, and www.aprsdirect.com. Enter **aprs map display** to obtain a long list of available mapping services.

Kenwood and Yaesu make APRS-ready 2 meter radios that include GPS receivers or have direct GPS data interfaces. Icom radios with GPS and D-STAR can send data to the APRS network via the D-PRS app.

You don't need a handheld or mobile transceiver to send information to the APRS servers. An APRS *tracker* is a combination low-power 2 meter transceiver, GPS receiver, and a microcontroller to handle the packet radio protocol. These are available in various power levels, sizes, and configurations. The Byonics Micro-Trak product line (www.byonics.com/microtrak) includes the most common types of trackers. You can put trackers in your car (this is where the commercial LoJack products came from), on your bicycle, or carry them with you on a hike.

TIP

Check out the apps APRSdroid (Android) or aprs.fi (iOS) for smartphone access to the APRS network.

You can use APRS trackers in high-altitude balloons, model rockets, and drones. Hams have sent balloons to altitudes of more than 100,000 feet — the edge of space! Figure 11-4 shows the track of WB8ELK's Skytracker balloon that went around the world six times in 75 days. The solar powered tracker provided updates on the balloon's position the whole way.

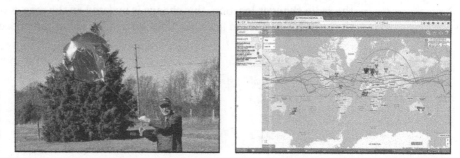

FIGURE 11-4:
The WB8ELK Skytracker balloon tracked via APRS.

Courtesy American Radio Relay League

You can learn more about the interesting blend of science and ham radio in high-altitude ballooning at www.arhab.org. The website tracker.habhub.org shows the current location of all balloon-borne trackers.

You can do a lot more with APRS than just report location. APRS supports a short-message format similar to the Short Message Service (SMS) format that's used for texting on mobile phones. Popular mapping software offers interfaces so that you can have street-level maps linked to your position in real time by ham radio. Race organizers use APRS to keep track of far-flung competitors. You can also add weather conditions to APRS data to contribute to a real-time automated weather tracking network. To find out more, including detailed instructions on configuring equipment, start with the resources listed in Table 11-3.

TABLE 11-3   **APRS Resources**

Resource	Description
APRS (www.aprs.org)	Website describing the current state of the technology, with useful articles and links
APRS introductions (www.arrl.org/aprs-mode)	Primer on APRS technology and use
APRS technical details (www.aprs-is.net/presentations/how%20aprs%20works.pdf)	Detailed information on APRS networks and functions

DXing — Chasing Distant Stations

Pushing your station to make contacts over greater and greater distances, or *DXing* (*DX* means *distance*), is a driving force that fuels the ham radio spirit. Somewhere out in the ether, a station is always just tantalizingly out of reach; the challenge of contacting that station is the purpose of DXing. The history of ham radio innovation is tightly coupled with DXing which drove improvements in many types of equipment, especially antennas and receivers. If this sounds like fun, you might be a *DXer*!

TIP

Listen for and contact *DXpeditions* — special trips made to remote or unusual locations by one or more hams just for the purpose of putting them on the air.

VHF/UHF DXing with a Technician license

Once you get your first license, you can begin DXing! The VHF and UHF bands have undeserved reputations for being limited to line-of-sight contacts. As I discuss in Chapters 8 and 9, CW and SSB and the new digital modes have much longer ranges than FM. By taking advantage of well-known modes of radio propagation, you can extend your VHF and UHF range dramatically beyond the horizon. You also have operating privileges on 10 meters which I discuss in the HF DXing section below.

You will find distant stations at the lowest frequencies on the band in the so-called weak signal segments. On the 6 meter band, for example, 50.0–50.4 MHz — a 300-kHz segment as large as most HF bands — is where the CW, SSB, and digital mode calling frequencies are located. Similar segments exist on all VHF and UHF bands through the lower microwave frequencies.

TIP

Calling frequencies on *WSJT-X* modes like FT8 (50.313 MHz) and MSK144 (50.260 MHz) are built in to the software. SSB on 6 meters is often near 50.125 MHz. The calling frequency on 2 meters is 144.200 and on 70 cm it's 432.100 MHz (www.arrl.org/band-plan).

When you're DXing on VHF or UHF, stay close to the calling frequencies. Propagation between widely separated points may be short-lived so don't wait for email or an Internet "spot" about an opening! Set your squelch control so that the radio is barely muted. If anything shows up on the frequency, the radio springs to life. This way, you (and whoever else is in earshot) don't have to listen to continuous receiver noise.

For DXing on VHF and UHF, use a small beam antenna (see Chapter 12). A beam antenna is easy to build, is relatively small compared with HF antennas, and is a terrific homebrew project. Mount the antenna for horizontal polarization, with the antenna elements parallel to the ground. You should be able to rotate the antenna to point in any horizontal direction, because signals may appear from nearly anywhere at any time.

> **TIP** To find out more about VHF/UHF propagation, join one of the many VHF/UHF contest clubs. These helpful, energetic groups make a lot of expertise available through their websites and at meetings. Watching a QSO mapping site like DXmaps.com will also help you learn when and where propagation is present on the VHF and UHF bands.

VHF and UHF DX propagation

Although HF propagation depends most strongly on the ionosphere's upper F layers, the lower E layer comes into its own on the VHF and UHF bands. Even the lower atmosphere region, called the *troposphere*, gets into the act above 50 MHz with weather-related ways of bending and reflecting signals for long-distance contacts.

>> **Sporadic E:** The term *sporadic E* refers to an interesting property of one of the lower ionospheric layers: the E layer. Somewhere around 65 to 70 miles above the Earth, small, highly ionized regions form that reflect signals from the 10 meter, 6 meter, 2 meter, 1-1/4 meter, and (rarely) 70 cm bands back to Earth. These regions usually don't last more than an hour or two as they drift, usually to the west or southwest. While they're available, hams can use them as big radio reflectors. Their unpredictable nature has led to the name *sporadic E* which is abbreviated Es or "E skip."
>
> Sporadic E propagation occurs throughout the year but is most common in the early summer months and the winter. When sporadic E is present, signals appear to rise out of the noise over a few seconds as the ionized patch moves into position between stations. The path may last for seconds or for hours, with signals typically being very strong in both directions. Working Es with only a few watts and simple antennas is possible. Most VHF and UHF DXers get their start working Es openings on 6 meters; certainly, more people are actively DXing in that way than in any other.

The dramatic appearance and disappearance of signals propagating by sporadic E has led to 6 meters getting nicknamed the "magic band."

>> **Meteor scatter:** The most fleeting reflectors of all result from the tens of thousands of meteors that enter the Earth's atmosphere each day, traveling at thousands of miles per hour. The friction that occurs as the meteors burn up ionizes the gas molecules for several seconds or longer. These ionized trails can reflect radio signals, so two lucky stations with the meteor trail between them can communicate for a short period — from less than a second to several seconds. Six meters is the easiest band for beginners to use for making contacts via meteor scatter.

Meteor scatter operators are called *ping jockeys* because the many short reflections off small meteors make a characteristic pinging sound. (Longer pings are called "burns.") As you may imagine, ping jockeys go into high gear around the times of meteor showers, large and small. Because of meteor scatter, hams can enjoy meteor showers even during daylight hours with the best opportunity before and after sunrise as the atmosphere enters the meteor stream straight on. Check out www.pingjockey.net/cgi-bin/pingtalk for the latest activity and contact reports.

Check out Stan Gibilisco's video explanation of meteor scatter at www.youtube.com/watch?v=7RDasS20FP0&ab_channel=StanGibilisco. See the *WSJT-X User Guide* for tutorial information about using MSK144 to work meteor scatter. You can hear an MSK144 QSO being made in this YouTube video from VE2XK: www.youtube.com/watch?v=i5JfpgPyD14&t=93s&ab_channel=MichelXK.

Airplanes reflect the microwave signals of radar and they'll reflect your VHF and UHF signals, too! A growing number of hams use airplane tracking software to aim their antennas at them, making MSK144 contacts over hundreds of miles. Don't worry — even a high-power ham signal can't cause any problems on board an aircraft.

>> **Aurora:** Another large ionized structure in the ionosphere is the *aurora borealis* or *aurora australis.* It is created by charged particles from the solar wind entering the Earth's atmosphere. The aurora is oriented vertically instead of horizontally like sporadic E but still reflects signals. Both *auroral zones* are centered near the magnetic North and South Poles.

One of the neatest things about auroral propagation is that its characteristic movement adds its own audible signature to the reflected signals. Signals reflected by an aurora have a characteristic rasp or buzz impressed on the Morse tone or the spoken voice. A very strong aurora can turn Morse transmissions into bursts of white noise and render voices unintelligible. After you hear the auroral signature, you'll never forget it.

>> **Tropospheric:** Tropospheric propagation (also known as *tropo*) occurs in the atmospheric layers closest to the Earth's surface, known as the *troposphere*. Any kind of abrupt change in the troposphere, such as temperature inversions or weather fronts, can serve as a long-distance conduit for VHF, UHF, and even microwave signals. If your region has regular cold or warm fronts, you can take advantage of them to reflect or guide your signals.

Tropo supports surprisingly regular communications on 2 meters and 1-1/4 meters between stations in California and on the upper slopes of Hawaiian volcanoes. A stable temperature-inversion layer forms over the eastern Pacific Ocean most afternoons, so a properly located station on the slope of a volcano at the right altitude can launch signals along the inversion. As the inversion breaks up near land, the signals disperse and are received by mainland amateurs. When conditions are right, mainlanders can send signals back along the same path — more than 2,500 miles!

>> **Mountaintopping:** What do you do when all the popular DXing methods fail to provide you an over-the-horizon path? Move your horizon! Because VHF/UHF radios are lightweight and the antennas are small, you can drive, pack, or carry your gear to the tops of buildings, hills, ridges, fire lookouts, and even mountaintops.

The higher the elevation of your station, the farther your signal travels without any assistance from the ionosphere, weather, or interplanetary travelers. Camping, hiking, and driving expeditions can take on a ham radio aspect, even if you're just taking a handheld radio. From the tops of many hills, you can see for miles, and a radio can see even better than you can. These expeditions are particularly popular in VHF contests, discussed in "Taking Part in Radio Contests," later in this chapter. All you have to do is pick up a book of topographic maps of your state, load the car with your radio gear, and head out.

TIP

A special award is available for working mountaintop stations. The Summits On the Air (SOTA) program encourages activity by these sociable climbers, who often use ultra-low-power gear, which makes contacts with them challenging and fun. You can read up on SOTA at www.sota.org.uk.

Earning VHF and UHF DX awards

TIP

Because VHF DX contacts generally aren't as distant or dispersed as their shortwave cousins are, VHF DX awards deal with geographic divisions on a smaller scale, called *grid squares*. Grid squares are the basis for the Maidenhead Locator System, in which one grid square measures 1° latitude by 2° longitude. Each grid square is labeled with two letters (called the *field*) and two numbers (called the *square*). A location near St. Louis, Missouri, for example, is in the EM48 grid square. Grid squares are divided even further into subsquares, which are denoted by two additional lowercase letters, as in EM48jw. The exam will include a question about grid squares.

TIP

Find your grid square by using one of the grid-square lookup utilities listed on the ARRL's Grid Squares page (www.arrl.org/grid-squares).

In North America, where countries tend to be large (except in the Caribbean), the primary VHF/UHF award program is the ARRL's VHF/UHF Century Club (VUCC; www.arrl.org/vucc). The number of grid squares you need to contact to qualify varies by the band, due to the degree of difficulty. As an example, on the lowest two bands (6 meters and 2 meters) and for contacts made via satellites, contacts with stations in 100 different grid squares are required. The ARRL's Worked All States (WAS) program (refer to Table 11-4, later in this chapter) has a vigorous VHF/UHF audience as well.

In Europe, where more countries are within range of conventional VHF/UHF propagation, many of the shortwave DX awards have VHF/UHF counterparts. Many of those awards are based on contacting different countries, too.

Finally, what would DXing be without a distance record? On shortwave bands, with signals bouncing all the way around the world, the maximum terrestrial distance records were set long ago. In VHF/UHF, though, many frontiers are still left. Al Ward (W5LUA) has put together a VHF/UHF/Microwave record list at www.arrl. org/distance-records, and new records are added all the time. Maybe your call sign will be there one day.

HF DXing with a General license

After you upgrade your Technician license to a General, the HF bands are all available to you! Signals routinely travel long distances at frequencies below 30 MHz, bouncing between the ionosphere and the planet's surface as they go. Even low-power signals can be heard over long distances on the shortwave bands, so HF DXing is often a worldwide event, with stations calling from several continents. Even if you have a very modest home or mobile HF station, you can work DX.

REMEMBER

Don't forget that Technicians have access to CW, digital, and voice modes on 10 meters from 28.0 to 28.5 MHz. They also have limited CW privileges on 80, 40, and 15 meters.

TECHNICAL STUFF

For in-depth information about HF DXing from the perspective of an experienced and dedicated DXer, I recommend *The Complete DXer*, by Bob Locher (W9KNI), published by Idiom Press. Now in its third edition, Locher's book has mentored legions of beginning DXers.

Receiving HF DX signals

HF signals coming from far away have to make one or more hops off the ionosphere and take several paths to your station. (This is called *skip* propagation.) The signals mix with one another as they arrive at your antenna, spreading out in time and changing strength rapidly. You will soon recognize accents and signals with a curious, hollow, or fluttery "DX" sound.

Start tuning from the bottom of the band, and keep listening, noting what you hear and at what times. In DXing, experience with the characteristics of a band's propagation is the best teacher. Try to detect a pattern when signals from different population centers appear and see how the seasons affect propagation on the different bands.

If you plan on doing a lot of DXing, bookmark the ARRL's DXCC Award program website at www.arrl.org/dxcc, and download the ARRL DXCC List (www.arrl.org/files/file/DXCC/2016%20DXCC%20Current.pdf). A ham radio prefix map of the world is great for reference. Maps are often available for free from manufacturers like Yaesu and Icom, or you can use the software maps described in the nearby Tip. These tools help you figure out what countries correspond to the call signs you hear. Another handy tool is a prefix list, such as www.arrl.org/international-call-sign-series which helps you figure out call signs' countries of origin. Your logging software will probably use a list like the CTY files available at www.country-files.com to keep track of currently active and new station calls.

While you're collecting resources, here's another suggestion: Centered on your location, an *azimuthal-equidistant* (or *az-eq*) map, such as the one in Figure 11-5, tells you the direction from which a signal is usually coming. Because signals typically travel along the "Great Circle" paths between stations (imagine a string stretched tightly around a globe between the stations), the path for any signal you hear follows the radial line from the middle of the map (your location) directly to the other station. If the path goes the long way around, it goes off the edge of the map, which is halfway around the world from your station, and reappears on the other side. Most signal paths stay entirely on the map because they take the short path. Some az-eq maps are available in the *ARRL Operating Manual*; you can also generate a custom map at sites such as www.wm7d.net/azproj.shtml. You may be surprised to see what the world looks like when an az-eq map is centered on your exact location!

Two great collections of online ham radio maps and mapping software are DX Atlas by Alex Shovkoplyas (VE3NEA), at www.dxatlas.com/DxAtlas, and Mapability, by Tim Makins (EI8HC), at www.mapability.com/ei8ic. Either of these packages is a tremendous asset to ham radio operating and enjoyment.

>> **Daytime signals:** Depending on the hour, the ionosphere absorbs a signal or reflects it over the horizon as described in Chapter 8. In the daytime, the 14, 18, 21, 24, and 28 MHz bands, called the *high bands,* tend to be open to DX stations — that is, they support long-distance, multi-hop propagation.

FIGURE 11-5:
An azimuthal-equidistant map centered on the Midwestern United States.

Before daylight, signals begin to appear from the east, beginning with 14 MHz and progressing to the higher-frequency bands over a few hours. After sunset, depending on solar activity the signals may linger from the south and west, with the highest-frequency bands closing first, in reverse order.

Daytime DXers tend to follow the *maximum usable frequency* (MUF), the highest signal the ionosphere reflects. Reflections or *hops* at a low angle travel the longest distance in one hop. A signal that gets to where it's going in the fewest hops will have a higher signal strength.

>> **Nighttime DXing:** The nighttime bands of 10, 7, 5, 3.5, and 1.8 MHz are known as the *low bands*. These bands are throttled for long-distance communication during the daytime hours by absorption in the lower layers of the ionosphere. At sunset, these bands start to come alive.

First, 10, 7, and 5 MHz may open in late afternoon and stay open somewhat after sunrise. The 3.5 and 1.8 MHz bands, however, change fairly rapidly around dawn and dusk. Signals between stations operating on 3.5 and 1.8 MHz often exhibit a short (15- to 30-minute) peak in signal strength when the easternmost stations are close to sunrise, a peak known as *dawn enhancement*. This time is good for stations with modest equipment to be on the air and to take advantage of the stronger signals on these more difficult DX bands.

REMEMBER

The 160 meter (1.8 MHz) band is known as the *top band* because for a long time it had the longest wavelength of any authorized amateur band. This long wavelength requires larger antennas. Add more atmospheric noise than at higher frequencies, and you have a challenging situation, which is why some of the most experienced DXers love top-band DXing. Imagine trying to receive a 1-kilowatt broadcast station halfway around the world. That's what top-band DXers are after, and many of them have managed it.

TIP

Hams have two new bands in the U.S.! The 630 meter band at 472–479 kHz and the 2200 meter band at 135.7–137.8 kHz. Both of these bands have special rules as explained at www.arrl.org/frequency-allocations. You may also see references to the "4 meter" or "70 MHz" band. Those frequencies are only available to hams in Europe and Africa at the moment.

WATCHING THE SUN AND ITS SPOTS

Because the Sun is so important in determining what bands are open and in what direction, you need to know what portions of the Earth are in daylight and darkness. You can use a variety of tools to keep track of the Sun. The figure below shows the handy map available at dx.qsl.net/propagation/greyline.html.

Solar activity is also very important, strongly affecting when and to where the HF bands are open. The Sun's sunspot cycle lasts for about 11 years with years of peak activity corresponding to the best conditions on the higher HF bands. As of late-2020 a new sunspot cycle, Cycle 25, is beginning. Watch for more spots on the Sun in the coming months and years! You can get the latest news on solar activity at www.spaceweather.com and in the weekly videos from Dr. Tamitha Skov, WX6SWW, who is also known as "Spaceweather Woman," at www.spaceweatherwoman.com.

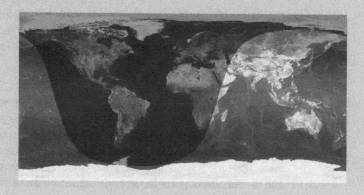

Contacting a DX station

Making a call to a DX station requires a little more attention to the clarity of your speech and the quality of your sending than making a call to a nearby ham does. Your signal likely has the same qualities to the DX station as the DX station's signal does to you — hollow or fluttery and weak — so speak and send extra carefully. When responding to a DX station CQ, give the DX station's call sign, using the same phonetics that they are using; then repeat yours at least twice, using standard phonetics. When using CW, match your sending speed to that of the DX station.

TIP

To find out when DX stations are active, particularly expeditions to rare places and hams taking a holiday in some exotic location, subscribe to one of the online DX newsletters, such as The Daily DX (www.dailydx.com) or the weekly OPDX Bulletin (www.papays.com/opdx.html). Watch these publications for upcoming or current DX station activity.

DX contacts tend to be shorter than contacts with nearby stations. When signals are weak or the other station is rare, a contact may consist of nothing more than a confirmation that both stations have the call signs correct, as well as a signal report (see Chapter 8). To confirm the contact, both you and the DX station must get each other's call signs correct. To do that, use standard phonetics (on voice transmissions), speak clearly, and enunciate each word.

When it's time to conclude the contact, the DX station will probably let you know whether to confirm the contact by using an online system like Logbook of the World (www.arrl.org/logbook-of-the-world) or by sending a QSL card. Collecting QSL cards, like those shown in Figure 11-6, is a wonderful part of the hobby — and beautiful, too! See Chapter 14 for more info on these cards and on QSLing systems.

TIP

If you call and call and can't get through, or if the stations you contact ask for a lot of *repeats* and *fills* (in other words, if they often ask you to repeat yourself), you may have poor transmitted audio quality. Have a nearby friend, such as a club member, meet you on the air when the bands are quiet, and do some audio testing. The radio's power-meter output alone may not tell you when you have a problem, so an on-the-air check is necessary to find it.

Navigating pileups

A *pileup* is just that: a pile of many signals trying to get through to a single, often quite rare station. Pileups can sound like real messes, but if you listen carefully, you'll notice that some stations get right through. How do they do it? They listen to the rare station's operating technique, follow instructions, and carefully time their calls. If they don't get through the first time, they stop calling and listen until they have the pattern figured out. These smooth operators use their ears instead of their lungs or amplifiers to get through. You can, too, by listening first and transmitting second. Here are some common tricks to listen for and try yourself:

>> Time your call a little bit differently from everybody else. Wait a second or two before beginning, or wait for the short lull when most hams have given their call signs once and are listening.

>> Make your signal sound a little bit different — higher or lower — by offsetting the transmit frequency by 200 Hz or 300 Hz.

>> Give your call once or twice before listening for the DX station. Some folks never seem to stop calling; how would they know if the DX station did answer them?

>> Use phonetics that the DX station has used.

>> Try to figure out what the DX station hears well and do that.

FIGURE 11-6: DX QSL cards from (clockwise from upper left) Reunion Island, Somalia, Pitcairn Island, and Egypt.

TIP

YOU DON'T NEED TO SHOUT INTO THE MICROPHONE! Shouting doesn't make you any louder at the other end. By adjusting your microphone gain and speech processor, you can create a very understandable signal at normal voice levels. Save the shouting for celebrating your latest DX contact; your contacts and family members will thank you for doing so.

Working split: Split-frequency operation

A *split* refers to a station that's transmitting on one frequency but listening on another. This procedure, called *working split*, is common when many stations are trying to get through to a single station, such as a rare DX station. You can tell that a station is working split when you hear the station contacting other stations but can't hear those stations' responses.

The process can also work the other way around: Sometimes, you tune in a pileup of stations trying to contact a DX station, but you aren't able to hear the DX station's responses. Typically, the DX station's split listening frequency is a few kHz above the transmitting frequency. The station being called gives instructions such as "Listening up 2" or "QRZed 14205 to 14210." The former means that the DX station is listening for stations 2 kHz above the transmit frequency; the latter means that the station is listening in the range between the two frequencies (in this case, probably 14205–14210 kHz). Your radio's instruction manual can show you how to configure your radio to receive and transmit on different frequencies.

WARNING

Don't bother trying to spin the dial back and forth between the receive and transmit frequencies; you won't be quick enough. If you aren't experienced with working split, practice with a nearby friend, using low power, until you're comfortable with using your radio's controls that way.

Using spotting networks

DXers share the frequencies and call signs of DX stations that they discover on the air through an extensive worldwide system of websites. A message that describes where you can find the DX station is called a *spot*, and websites that link up to provide and relay the spots form what is called a *spotting network* or the *DX cluster*. Information from the spotting network is usually obtained with a web browser or by logging software opening a TELNET connection (a type of online protocol). A directory of spotting network sites is available at www.dxcluster.info/telnet/ index.php. Your logging software may have a built-in list, as well.

Here is an example spot received from the popular DX Summit website (www. dxsummit.fi):

W5VX 7003.7 **A61AJ** 0142 05 Nov Up 1 to 3U.A.E.

This spot means that W5VX is hearing A61AJ from the United Arab Emirates (A6 is the prefix of call signs for amateurs in the U.A.E.) on a frequency of 7003.7 kHz at 0142UTC (01:42 a.m. in London) on November 5. A61AJ is operating split (see the previous section) and listening up 1 to 3 kHz.

TIP

UTC (Universal Time Coordinated) is also referred to as *world time* and is local time on the Prime Meridian in London. UTC used to be referred to as Greenwich Mean Time (GMT) and was abbreviated by Z following the 24-hour time.

Although jumping from spot to spot can be a lot of fun, maintaining your tuning and listening skills is still important. Be sure that you have the station's call sign correct before you put it in your log. It's disappointing to contact what you think is a rare station, only to find out that due to a busted spot and your fabulous DX contact isn't so fabulous after all. Because spotted stations attract quite a crowd, working DX by not chasing the spotted stations and tuning for them yourself may be easier. Don't become dependent on the spotting networks.

Earning awards

Many DXing award programs are available, and the most popular are listed in Table 11-4.

TECHNICAL STUFF

CAN THEY HEAR ME NOW?

In Chapter 8, you learn about propagation beacon stations that transmit from a known location. When you can hear the beacon, propagation is available between your location and that of the beacon. But what about the other way 'round? Where can *your* signal be heard? The Reverse Beacon Network or RBN (www.reversebeacon.net) and PSKReporter (www.pskreporter.info/pskmap.html) networks answer that question.

These networks accept spots from individual stations with automated receivers decoding CW and RTTY signals (Reverse Beacon) and digital mode signals (PSKReporter). Stations set up to receive WSPR signals also report what they hear at www.wsprnet.org/drupal. Spots can be displayed on a map or as a list, which can be downloaded, as well.

If you want to know where you signal can be heard, get on the air using CW or RTTY and call CQ two or three times at a nice even speed, repeating your call sign at least twice with each CQ. Within seconds, any RBN network receiver that hears and decodes your call will relay its spot to the RBN server, including a *signal-to-noise ratio* (*SNR*). Not only can you assess propagation but you can easily do antenna comparisons, too: Call CQ on antenna A and be spotted. Immediately switch to antenna B and call CQ again after moving a few kHz so the RBN receivers will spot you again. Then you can compare signal strengths at the same receiver under the same conditions. The ionosphere changes quickly, so make several comparisons to get a rough average of the differences.

TABLE 11-4   **Popular DX Awards Programs**

Sponsor	Awards Program	Achievement
ARRL (www.arrl.org/awards)	IARU Worked All Continents (WAC)	Confirm a contact in each of six continents.
	Worked All States (WAS)	Confirm a contact in each of the 50 U.S. states.
	DX Century Club (DXCC)	Confirm a contact with 100 of the DXCC entities (currently, 340 countries, islands, and territories).
CQ Magazine (www.cq-amateur-radio.com)	Worked All Zones (WAZ)	Confirm contacts with all 40 of the world's CQ-defined zones.
	Worked Prefixes (WPX)	Confirm contacts with stations with different prefixes in the call signs to receive awards.
Radio Society of Great Britain (www.iota-world.org)	Islands on the Air (IOTA)	Confirm contacts with saltwater islands around the world to achieve various levels of awards.

Most DX awards programs reward achievement in the same manner. First, you must qualify for the basic award (100 entities, 100 islands, 300 prefixes, and so on). You receive a certificate and your first *endorsement* (a sticker or other adornment signifying a level of achievement). From that point, you can receive additional endorsements for higher levels of achievement: more contacts on one band, more contacts in one geographic region, and so on. For more information on awards, see "Chasing Awards," later in this chapter.

Taking Part in Radio Contests

TIP

Radio contests, or *radiosport*, are competitions between stations to make as many contacts as possible with as many stations as possible during the contest. Contest lengths range from a couple of hours to a weekend with rules specifying who can contact whom and on what bands, and what information must be exchanged. Each contest has a theme for which stations to contact, such as stations in different countries, grids, or states. The exam has a question or two about contest operation.

When the contest is over, participants submit their logs to the contest sponsor via email or a web page. The sponsor performs the cross-checking between logs to confirm that the claimed contacts actually took place. Then the final scores are computed, and the results are published online and in magazines. Winners receive certificates like those in Figure 11-7, plaques, and other nonmonetary prizes.

2018 Missouri QSO Party
April 7-8, 2018
The Boeing Employees'
Amateur Radio Society – St. Louis
Presents this
Certificate of Accomplishment to:
Ward Silver
KØS (NØAX, KDØPES)
First Place,
Missouri Mobile
Multi-Op Low Power
Mike Heitmann, NØSO
Mike Heitmann – NØSO, President
George Mackus – ABØRX, Contest Coordinator
Saint Louis

FIGURE 11-7: Participating in a contest can result in an attractive certificate.

What's the point of such contests? Well, for one thing, with many stations all on the air at the same time, trying for rapid-fire short contacts, they are pretty exciting. In the big international contests, such as the CQ World Wide DX Contest (`www.cqww.com`), thousands of stations around the world are on the air on the bands from 160 meters through 10 meters. In a few hours, you can find yourself logging a Worked All Continents (WAC) award and being well on your way to earning some of the DX awards I mention later in this chapter.

Contests are also great ways to exercise your station and your operating ability to their limits. Test yourself to see whether you can crack contest pileups and copy weak signals through the noise; find out whether your receiver is up to the task of handling strong signals. If you want to increase your Morse code speed, spend some time in a contest on the CW sub-bands. Just as in physical fitness, competitive activities make staying in shape a lot more fun.

TIP

If contesting sounds like fun (hint — it is, that's how I got started in ham radio!) you should get a copy of Doug Grant, K1DG's *Amateur Radio Contesting for Beginners* from the ARRL (`www.arrl.org/shop`). Doug is a world-champion operator many times over and a member of the CQ Contest Hall of Fame. He knows his stuff!

Choosing a contest

Contest styles run the gamut from low-key, take-your-time events occurring on a few frequencies to band-filling events involving hectic activity in all directions. Table 11-5 lists some popular contests that you can try.

TABLE 11-5   **Popular Contests for Beginners**

Contest Name	Sponsor
ARRL VHF Contests (Jan, June, Sep)	ARRL (www.arrl.org/contests)
ARRL International DX Contest	
ARRL RTTY Roundup	
ARRL Field Day	
ARRL November Sweepstakes	
ARRL 10 and 160 Meter Contests	
CQ World Wide DX and WPX Contests	CQ Magazine (www.cq-amateur-radio.com)
CQ World Wide VHF Contest	
IARU HF Championship	IARU (www.arrl.org/contests)
North American QSO Parties	National Contest Journal (www.ncjweb.com)
World Wide Digi DX Contest	WWROF and SCC (www.ww-digi.com)

Most contests run annually, occurring on the same weekend every year. The full-weekend contests generally start at 0000 UTC (Friday night in the United States) and end 48 hours later (Sunday night in the United States) at 2359 UTC. You don't have to stay up for two days, but some amazing operators do. Most contests have time limits or much shorter hours.

TIP

The Rookie Roundup (www.arrl.org/rookie-roundup) is a six-hour contest just for new hams. There are three each year: April (SSB), August (RTTY), and December (CW). Report your score online and the results are published a few days later. The old-timers will be lining up to call *you* for a change! If you're a student, the School Club Roundup (www.arrl.org/school-club-roundup) is another great contest to get your feet wet. There's one in October and one in February.

Start by finding out what contests are coming up. Use Table 11-6 to locate several sources of information, or enter *contest calendar* in a web search engine. The list of contests will also help you identify which contest a station is participating in if you hear it on the air.

TABLE 11-6 Contest Calendars

Calendar	URL
ARRL Contest Calendar	`www.arrl.org/contests/calendar.html`
ARRL Contest Update (biweekly email newsletter, free to ARRL members)	`www.arrl.org/contest-update-issues`
WA7BNM Contest Calendar	`www.contestcalendar.com`

TIP

Many countries sponsor a contest that encourages foreign stations to work in-country stations. Examples include the Worked All Germany (WAG), Japan International DX (JIDX), and ARI (Amateur Radio Italia) contests. This is a great way to work stations from those countries for award programs and maybe get a certificate of your own!

TIP

Each state usually sponsors its own contest, called a *state QSO party,* once a year. These are great ways to work stations in states you need for your Worked All States award! They are typically more relaxed than the bigger contests and great ways to practice contest operating (and maybe pick up a certificate). You can find when they are held at `www.contestcalendar.com/stateparties.html`.

Operating in a contest

You needn't have a huge and powerful station to enjoy contesting; most contesters have a simple setup. Besides, the most important part is the operator. If you listen, know the rules, and have your station ready to go, you're all set. Remember that all participants, including the big guns, need and want to talk to you! By reading the rules or simply listening, you'll know what's required and the order in which you need to send your information. When you're ready, give the contest a try.

REMEMBER

Contests emphasize efficiency — this is a good opportunity to work on eliminating time-wasting habits. Unnecessary phrases like "Please copy" and "You are" slow things down when time is at a premium. Don't repeat the other station's information back to them — they already know it! Once they have your call sign, just give the minimum exchange information once. If a repeat is needed, it will be requested. Give your exchange in the same order the rules suggest. These are good practices for any kind of operating!

Making contest contacts

After you know the rules for a particular contest, listen to a participating station. The most important part of each contact is the information passed between stations, known as the *exchange.* For most contests, the exchange is short — a signal report and some identification such as a *serial number* (the count of contacts you made), name, location, or club membership number.

Here's an example of a contact in a typical contest: the Washington State Salmon Run. In this scenario, I'm W7VMI in King County, calling CQ to solicit contacts, and you're W1AW in Connecticut, tuning around the band to find Washington stations. The information we exchange is a signal report (see Chapter 8), my county and your state, because (at least for this example) you're not in Washington.

Me (W7VMI): CQ Salmon Run CQ Salmon Run from Whiskey Seven Victor Mike India.

You (W1AW): Whiskey One Alfa Whiskey.

(**Note:** You send or say your call sign once, using standard phonetics on voice transmissions.)

W7VMI: W1AW, you're five-nine in King County.

W1AW: QSL, W7VMI, you're five-nine in Connecticut.

W7VMI: Thanks, QRZed Salmon Run Whiskey Seven Victor Mike India.

The whole thing takes about ten seconds. Each station identifies and exchanges the required information. "Five-nine" is the required signal report signifying "loud and clear." That's an efficient contest contact, and most contacts are similar.

What if you miss something? Maybe you've just tuned in the station, and the band is noisy or the signal is weak. To continue the preceding example, my response to your call might sound like this:

Me (W7VMI): W1AW, you're five-nine in BZZZZTCRASH@#$%^&*.

You (W1AW): Sorry, please repeat your county.

W7VMI: Kilo India November Golf, King County.

W1AW: QSL, W7VMI, you're five-nine in Connecticut.

You're probably thinking, "But I missed the county. How can the signal report be five-nine?" By convention, most contesters say "Five-nine," type 599, or send 5NN in Morse code (the *N* represents an abbreviated 9). Since the signal report doesn't affect the score unless it's miscopied, contesters have naturally decided not to make more work for themselves and send the same signal report.

WARNING

To use FT8, FT4, or MSK144 in a contest, be sure to configure *WSJT-X* so that it uses the right information in its messages. There are several options available on the **Settings > Advanced** tab. Check the box next to **Special operating activity: Generation of FT8 and MSK144 messages.** Then select the appropriate option for your contest. If you don't do this, the right information won't be included in your messages and you may not be able to complete contest QSOs.

TIP

If you're unsure of yourself, try "singing along" without actually transmitting. Make a cue card that contains all the information you need to say or send in a script or list. If you think you may get flustered when the other station answers your call, listen to a few contacts, and copy the information ahead of time. Serial numbers advance one at a time, so you can have all the information before your contact.

Contesting is no more complicated than getting your sandwich order taken at a busy deli counter during lunch hour. Contesting has a million variations, but you'll quickly recognize the basic format.

TIP

If you don't know the rules of a contest but want to help a station calling "CQ contest" with a contact, wait until the station doesn't have anyone calling and then ask what information is needed. Stations in the contest want your contact and will help guide you through whatever they need.

Your score for almost all contests is made up of *QSO points* and *multipliers*. Each contact counts for one or more QSO points, sometimes depending on the mode, band, or other special consideration. Multipliers — so named because they multiply QSO points for the final score — are what make each contest an exciting treasure hunt. Depending on the theme, you may be hunting for states, grids, counties, lighthouses, islands, or anything else. Read the rules carefully to find out how the multipliers are counted: only once, once per band, once per mode, and so on. Special bonus points may be awarded for working certain stations or multipliers.

REMEMBER

You don't have to be a speed demon; just be steady. Good contest operators are smooth and efficient, so send your full call sign once. If the station answers with your call sign, log the exchange and send your information only once, even if you're using a small station. The other operator will ask you to repeat yourself if he or she misses some of the information.

WARNING

Because hams must keep their signals entirely within our allocated bands or segments, you need to remember where your signal is actually transmitted. Most voice signals occupy about 3 kHz of bandwidth. If the radio is set to USB, your signal appears on the air starting at the radio's displayed frequency to about 3 kHz higher. Similarly, on LSB your signal extends below the displayed frequency. When you're tuning close to the band edges, make sure that your signal stays entirely in the band. For example, the 20 meter band upper limit is 14.350 MHz so you shouldn't transmit if your radio is displaying a frequency above 14.347 MHz.

Logging your contacts

Manual logging (with pencil and paper) is the easiest method when you're a beginner. Often, the contest sponsor has a log sheet that you can download or print from a website. After the contest, you can convert your written entry to electronic form by using logging software or an online converter such the one at www.b4h.net/cabforms.

If you're a more experienced contester, using a general-purpose logging program or special contesting software makes contesting much easier. The software keeps score, maintains a *dupe list* (a list of stations you've already worked), shows needed multipliers, connects to spotting networks, and creates properly formatted logs to submit to the sponsors.

Table 11-7 lists some popular software programs. Entering *contest logger* in a search engine also turns up many useful programs.

TABLE 11-7

Popular Contest Logging Software

Software	URL
CQ/X (for mobile operation in state QSO parties)	www.no5w.com
N1MM Logger+	n1mmwp.hamdocs.com
WriteLog	www.writelog.com
N3FJP contest loggers	www.n3fjp.com
Rover Log	code.google.com/p/roverlog
Win-Test	www.win-test.com
SD by EI5DI	www.ei5di.com

TIP

If you want to operate away from a home station PC there are logging programs for phones and tablets, too! Go to the Android or Apple app stores and search for *ham radio logging*. You'll turn up several, like *Ham Radio Logger* (Android) or *HamLog* (iOS). Once you're done with the contest, you can export a file to send to the sponsors or import into another logging program.

While you are operating, most contest logging software can send your score to one of the "real-time scoreboards" like www.contestonlinescore.com or www.cqcontest.net/view/readscore.php. (To send your score to all of the scoreboards, use the Score Distributor service at www.scoredistributor.net.) You can see how you're doing compared to friends — how about a friendly challenge next time?

WRTC – THE HAM RADIO WORLD CHAMPIONSHIP

The World Radiosport Team Championship (WRTC) is a true "world series of radio contesting." Held every four years, the event was first held in Seattle, Washington in 1990. Since then, WRTC events have been organized in San Francisco (1996), Slovenia (2000), Finland (2002), Brazil, (2006), Moscow (2010), Boston (2014), and Germany (2018). As this book is being written, preparations are underway for WRTC 2022 in Bologna, Italy (www.wrtc2022.it). This event will be headquartered in the family home of the inventor of radio, Guglielmo Marconi! To find out more about WRTC and really get the flavor of radio contesting, get a copy of frequent competitor Jim George, N3BB's excellent book, *Contact Sport*.

Most contests expect logs in *Cabrillo format*, which is nothing more than a method of arranging the information in your log so that the sponsor's log-checking software can read it. Contest-logging software inserts Cabrillo-formatted logs for you. To find out more about Cabrillo, visit www.arrl.org/cabrillo-format-tutorial.

Many sponsors post a Logs Received web page that you can check to make sure that your log was received. Don't miss the deadline — usually a few days after the contest — for submitting logs. Even if you're not interested in having your score posted in the results, submitting your log just for the sponsor to use in checking other logs helps improve the quality of the final scoring. You can also post a summary of your score on the www.3830scores.com website. Include a few comments in the "Soapbox" window and see how your score stacks with others.

Calling CQ in a contest

To make a lot of contacts, you have to call CQ. In any contest, more stations are tuning than calling. You can turn those numbers to your advantage. Find a clear frequency (see "Being polite," later in this chapter), and when you're sure that it's not in use, fire away.

Following are a few examples of appropriate ways to call CQ in a contest. (Replace *Contest* with the name of the contest or an abbreviation of the name.)

> **Voice transmissions:** CQ Contest CQ Contest from Whiskey One Alpha Whiskey, Whiskey One Alfa Whiskey, Contest

> **Morse code or digital modes:** CQ CQ TEST DE W1AW W1AW TEST

>> **FT8, FT4, or MSK144: CQ W1AW FN31** (include your grid locator)

>> **VHF/UHF transmissions:** CQ Contest from W1AW FN31

Keep transmissions short, and call at a speed at which you feel comfortable receiving a reply. Pause for two or three seconds between CQs before calling again. Other stations are tuning the band and can miss your call if you leave too much time between CQs.

If you get a stream of callers going, keep things moving steadily. Try to send the exchange the same way every time. On voice, don't say "uh" or "um." Take a breath before the exchange, and say everything in one smooth sentence. As you make more contacts, your confidence builds. An efficient rhythm increases your *rate* — the number of contacts per minute.

Contesting being what it is, you'll eventually encounter interference or a station that begins calling CQ on your frequency. You have two options: Stick it out or move. Sometimes, sending a simple "The frequency is in use, CQ contest . . ." or "PSE QSY" (which means "Please change your frequency") on CW and digital modes does the trick. Otherwise, unless you're confident that you have a strong signal and good technique, finding a new frequency may be more effective. The high end of the bands is often less crowded, and you may be able to hold a frequency longer.

Searching and pouncing

Searching and pouncing (S&P) is usually accomplished by tuning across the band and finding stations manually. Another popular method is to connect to the spotting network and use logging software to create a list of stations and their frequencies, called a *band map*. If your logging software can control your radio, all you have to do is click the call signs to jump right to their frequencies. If you're using SDR software, you can just click on the individual peaks for each signal.

If you call and get through right away, terrific. Sometimes, though, you won't get through right away. Use your radio's memories or alternate between the radio's VFOs. By saving the frequencies of two or three stations, you can bounce back and forth among several pileups and dramatically improve your rate.

WARNING

Using information from the spotting networks during the contest usually requires you to enter in an assisted or multiple-operator category. Know the rules of the contest regarding spotting information, and be sure to submit your score and log in the proper category. If you do use information from the spotting networks, listen to make sure that it's correct because many spots are incorrect (*busted*). If you log the wrong call sign, you'll not only lose credit for the contact, but also incur a small scoring penalty for the mistake, like jumping offside in a football game.

Being polite

Large contests can fill up most or all of an HF band, particularly during voice-mode contests, and often cause friction with noncontest operators. As in most conflicts, each side needs to engage in some give-and-take to keep the peace. If you're participating in a contest, be courteous, and make reasonable accommodations for noncontesters. If you're not contesting, recognize that large competitive events are legitimate activities and that you need to be flexible in your operating expectations. Check the contest calendars to avoid particularly busy weekends or when making plans.

That said, how can you get along with everyone? Here are a few tips:

>> **Make sure that your signal is clean.** *Clean,* in this context, means not generating key clicks or splatter from overmodulation. A clean signal gets more callers every time and occupies less bandwidth.

>> **Make sure that your receiver isn't overloaded.** Keep your noise blanker and preamp off, and turn on the attenuator if signals are particularly strong.

FINDING OUT MORE ABOUT CONTESTING

The best way to find out more about contesting is to work with an experienced contester. You'll probably find that one or two multiple-operator stations in your region are active in the big contests. Look through the results of previous contests for their call signs. Contact the station owner, and volunteer to help out; most operators are eager to have you on board or can help you find another team. As a rookie, expect to listen, log, or spot new multipliers, all of which are valuable learning opportunities. When you know the ropes, you can fill in on the air more and more.

You can also find many contest clubs around the country. Look at the club scores in the results, and contact them.

QST and *CQ* magazines feature contest results and articles on technique. The ARRL also publishes the *National Contest Journal* (www.ncjweb.com), which sponsors several HF contests every year and features interviews with contesters, as well as articles and columns on contesting. ARRL members can receive the biweekly email newsletter Contest Update (www.arrl.org/contest-update-issues) without charge. The CQ-Contest email list is a source of many good ideas; subscribe to it at lists.contesting.com/mailman/listinfo/CQ-Contest.

>> **Listen before you leap.** Noncontest contacts are relaxed, with long pauses, so a couple of seconds of dead air don't mean that the frequency is clear. Asking "Is the frequency in use?" (QRL? in Morse code) before calling CQ is the right thing to do, whether you're in a contest or not. Make your first CQ a short one in case another station was busy on the frequency.

When you're participating in a contest, keep a minimum of 1.5 kHz between you and adjacent contest contacts on phone and 400 Hz on CW or radiotele-type (RTTY). Don't expect a perfectly clear channel.

>> **Avoid major net frequencies.** Examples include the Maritime Mobile Service Net on 14.300 MHz. Also, be aware of any emergency communications declarations or frequencies where regional emergency nets meet.

Chasing Awards

If the awards I mention in the section on DXing pique your interest, the following sections may satisfy that interest by discussing awards in greater detail. Seeking awards is one of the most fulfilling activities in the hobby of ham radio. Certificates (often called *wallpaper*) are the usual rewards. Some radio shacks that I've visited are literally papered (ceilings, too) with certificates and awards in all shapes, sizes, and colors. Some wallpaper is plain, but it can be as colorful and as detailed as paintings or photographs.

If awards sound interesting, you may be a member of the species of ham known as the paper chaser or wallpaper hanger. Believe me, a lot of them are out there!

Finding awards and special events

You can find awards almost everywhere you look. *CQ* magazine, for example, runs a column featuring novel awards every month, and The K1BV DX Awards Directory (www.dxawards.com) lists more than 3,300 awards from nearly every country. Want to try for the "Tasmanian Devil" award? Contact VK7 (Tasmania's prefix) amateurs. Or pursue the South African Relay League's "All Africa" award for contacting the six South African call areas and 25 other African countries. Although the website is no longer actively maintained, it is a rich archive of currently available DX awards.

Most awards have no time limit, but some span a given period, often a year. Whatever your tastes and capabilities, you can find awards that suit you. For example, during the ARRL's centennial year, its famous call sign W1AW was activated in each of the 50 U.S. states and many of the territories. If you "worked 'em all" you received the nice certificate shown in Figure 11-8.

FIGURE 11-8:
Special events and achievement award certificate.

Along with ongoing awards programs, you can find many special-event stations and operations, which often feature special call signs with unusual prefixes (of great interest to hams who chase the WPX award; refer to Table 11-5) and colorful, unusual QSL cards. Ham stations often operate at large sporting events and public festivals, such as international expositions and the Olympic Games. The larger special-event stations are well publicized and are listed on web pages such as www.arrl.org/special-event-stations, in the DX bulletins mentioned earlier in this chapter, and on the ham Internet portals I discuss in Chapter 3. Other special-event stations just show up unannounced on the air, which makes finding them exciting.

TIP

There is always a station at the Scouting national jamboree, held each year on the third weekend of October. Referred to as Jamboree On The Air (JOTA: www. scouting.org/jota.aspx), thousands of scouts around the world get a taste of ham radio. Hosting scout-age teens and preteens at your station is a great way to introduce them to ham radio by making contacts with the scouts at the event.

Logging contacts for awards

First, verify that the program is active by checking the sponsor's website. Determine if the award requires you to submit QSL cards. Overseas sponsors may allow you to submit a simple list of contacts that comply with their General Certification Rules (GCR) instead of requiring actual cards.

When you make an eligible contact, be sure to log any information that the award may require. If you're working Japanese cities, for example, some awards may require you to get a city number or other ID. This information may not be on the QSL card that the other station sends you, so ask for the information during the contact itself. Grid square information isn't always on QSLs, either. Be sure to ask during the contact or in a written note on the QSL you send to the other station.

TIP

If you make a contact for an award that favors a certain geographic area, ask the station's operator whether he or she will let others know you're chasing the award, which may even generate a couple more contacts for you right on the spot. This can help with difficult awards or contacts in remote areas.

Applying for awards

When applying for an award, use the proper forms, addresses, and forms of payments. Follow the instructions for submitting your application to the letter. If you aren't certain, ask the sponsor. Don't send your hard-earned QSLs or money before you know what to do.

WARNING

When you do apply for the awards, you may want to send the application by registered or certified mail, particularly if precious QSLs or an application fee are inside. Outside the developed countries, postal workers are notorious for opening any mail that may contain valuables. Make your mail look as boring and ordinary as possible; keep envelopes thin, flat, and opaque.

Mastering Morse Code (CW)

Mastering Morse code, also known as *CW* for "continuous wave," is a personal thing, such as playing an instrument or achieving a new athletic maneuver. Many people liken it to studying another language because it involves the same sudden breakthroughs after periods of repetition. Becoming a skilled Morse code operator results in a great sense of accomplishment, and you'll never regret taking the time to learn it.

WARNING

Avoid learning Morse code as a table of character patterns. It's hard to go much faster than five words per minute while looking up each character in your mind. You'll find it hard to move to the higher speeds that make Morse code fun.

Learning Morse correctly

The methods that most hams are successful with are the *Farnsworth method* or the *Koch method*. Both start you out with code at the speed you want to achieve. In Farnsworth, the dits and dahs of each character are sent at the code speed you want to achieve (measured in words per minute [wpm]), but the individual characters are spaced far enough apart in time that the overall word speed is low, a technique called *character spacing*. The Koch method is similar, starting at the desired speed but only working on a couple of characters at first. Both keep you from thinking of the code as a table of characters to memorize.

Several groups are devoted to teaching other ham how to send and receive Morse code. The FISTS Club (www.fists.org) offers low-cost training software from K7QO and on-the-air assistance and training. (A person's style of sending is known as their *fist*.) The CW Ops group (www.cwops.org/) features free training classes as part of its CW Academy program, too. The Straight Key Century Club (www.skccgroup.com) encourages on-the-air operation and sponsors several low-speed operating events.

CW is a lot easier to learn and copy if you're equipped to listen to it properly. For starters, headphones (cans) really help because they block out distracting noise. When you're copying code, your brain evaluates every little bit of sound your ears receive, so make its job easier by limiting non-code sounds.

TIP

CW practice and decoding apps are among the most popular for hams on either Android or iOS smartphones. Enter *ham radio morse code* into the app store's search window and browse to find one with the features you want.

While you're listening to Morse, you may hear some odd characters that don't make up a word or abbreviation, especially as the operators start and stop sending. These characters are *prosigns* (short for *procedural signs*) used to control who sends and who doesn't. On voice, "over" is the same as the prosign K. Some prosigns, such as BK (break), are two letters sent together. Others include SK, which means "end of contact," and KN, which means "only the station I am in contact with start transmitting." You can find out more about prosigns and other Morse conventions and abbreviations at www.hamuniverse.com/qsignals.html.

TIP

A great way to gauge your Morse code proficiency is with live code practice, which enhances your taped or computer-generated studies. The most widely received code-practice sessions are transmitted by the ARRL's station W1AW in Newington, Connecticut (www.arrl.org/w1aw). Code practice may be available on a VHF or UHF repeater in your area, too. Check with your local radio clubs to find out.

The secret of mastering Morse code is keeping at it. You'll have days when conquering new letters and higher speeds seem to come effortlessly; then you'll have days when progress is elusive. Those plateau days are the most important times to keep going, because they're the times when your brain is completing its new wiring.

TIP

You'll be a better Morse operator if you learn how to send and receive it yourself, but programs are available that allow you to send Morse directly from your keyboard. A simple interface from a COM or USB port is all that's required to key your radio. Some of these programs include plug-ins that can copy code, too, as long as the frequency isn't too crowded or noisy.

Copying the code

To get really comfortable with CW, you need to copy in your head. Watching good operators having a conversation without writing down a word is an eye-opener. How do they do that? The answer is practice.

As your code speed increases during the learning process, you gradually achieve the ability to process whole groups of characters as one group of sound. Copying in your head just takes that ability to another level. To get there, spend some time just listening to code on the air without writing anything down. Without the need to respond to the sender, you can relax and not get all tensed up, trying not to miss a character. Soon, you'll be able to hold more and more of the contact in your head without diminishing your copying ability.

"Read the mail" (listen to contacts) on the bands, trying to relax as much as possible without staying right up with each character. Don't force the meaning; let your brain give it to you when it's ready. Gradually, the meaning pops into your head farther and farther behind the characters as they're actually received. What's happening is that your brain is doing its own form of error correcting, making sure that what you copy makes sense and taking cues from previous words and characters to fill in any blanks.

Be sure that your radio is set up properly. Most radios come with a filter modules or DSP settings intended for use with voice signals. Typically, a voice filter is 2.4 kHz wide (meaning that it passes a portion of radio spectrum or audio that spans 2.4 kHz) to pass the human voice clearly. CW doesn't need all that bandwidth. A filter 500 Hz wide is a better choice, and you can purchase one as an accessory for an older radio without adjustable software filters.

Narrower isn't necessarily better below 400–500 Hz. A very narrow filter, such as 250 Hz models, may allow you to slice your radio's view of the spectrum very thin, but the tradeoff is an unnatural sound, and you'll be less able to hear what's going on around your frequency. Extra-narrow filters are useful when interference or noise is severe, but use a wider filter for regular operation.

Settling on one preferred pitch is natural, but over long periods, you can wear out your ear at that pitch. Keep the volume down and try different pitches so you don't fatigue your hearing.

Be sure to read the sections on CW operation in your radio's operating manual. Find out how to use all the filter adjustment controls, such as the IF Shift and Passband Tuning controls. Most CW operators like to set the AGC control to the FAST setting so that the radio receiver recovers rapidly. Being able to get the most out of your receiver is just as important on CW as on voice.

Pounding brass — sending Morse

The basic telegraph key, or *straight key*, shown in Figure 11-9, is used on the bands every day, but sending good code at high speed with one is challenging. The straight key tops out at somewhere between 20 and 30 words per minute (wpm). At these speeds, sending becomes a full-body experience and you have to be really skilled to make it sound good.

Combination paddle and key

Straight key

FIGURE 11-9:
My paddle-key combo and the venerable straight key.

Electronic *keyers* are the most common way to send CW. These devices generate dots and dashes electronically, controlled by the keying *paddles* (also shown in Figure 11-9), referring to the flat ovals touched by the operator. Electronic keyers keyers make sure that the length of and spacing between dits and dahs is correct. *Iambic* keyers will send alternating di-dah-di-dah-di-dah patterns if both the dot and dash paddle are closed. Many radios have a keyer built in that you can control from the front panel. I recommend that you start with or soon change to a paddle and keyer so that you don't have to change techniques as your speed increases.

No matter whether you decide to start with a straight key or a paddle, use a good-quality instrument, for that's what it is: an instrument. See whether you can borrow or try one out. Experiment with different styles, and eventually you'll find one that feels just right. Key and paddle manufacturers are listed in *QST* magazine's advertising pages.

Making code contacts

As you start, find an operator sending code at a speed you feel comfortable receiving. Slow-speed code operators are often found well above the digital signals — around 3.600, 7.100 to 7.125, 21.100 to 21.175, and just above 28.100 MHz. These are the old Novice license bands where Technicians have operating privileges. When you're sending a Morse code CQ, don't send faster than you can receive. An answering station will assume you can copy code at the speed you're sending.

REMEMBER

After you begin a contact and exchange call signs, giving your call sign every time you turn the transmission over to the other station isn't necessary, but you must include it once every ten minutes, as required by FCC rules. Send your information and end with the "BK" prosign to signal the other station that he or she can go ahead. This method is much more efficient than sending call signs every time.

At the conclusion of a Morse code contact, after all the 73s (best regards) and CULs (see you later), be sure to close with the appropriate prosign: SK for "end of contact" or CL if you're going off the air. You may also hear the other station send "shave and a haircut" (dit-dididit-dit), and you're expected to respond with "two bits" (dit dit). These rhythms are deeply ingrained in ham radio. Yeah, it's a little goofy, but have fun!

QRP (Low Power) and Portable Operating

Just as some folks like fishing with the lightest possible tackle or hike with as little gear as possible, some hams have a similar outlook on operating. It's fun to make contacts with a minimum of power, called *QRP*, relying on antennas and skill to get through. There are plenty of hams who enjoy QRP and manufacturers have created special radios, such as those in Figure 11-10, optimized for low-power operating.

FIGURE 11-10:
The FT-817 and
KX3 are two
popular
commercial QRP
HF radios.

This enthusiasm also applies to operating from portable stations in parks, camp-grounds, and on the trail. Choosing a high spot with a world-class view is a great location to launch low-power radio signals, too. You'll find lots of effective anten-nas, and the latest batteries complete a capable station that fits in a trunk or backpack with room to spare. If operating in the great outdoors sounds interest-ing, this section is for you!

Getting started with QRP

QRP is primarily an HF activity, and most QRP contacts are in Morse code due to the efficiency and simplicity of that mode. Digital operation on weak-signal modes like FT8 is growing rapidly. QRP is up to 5 watts of transmitter output power on Morse code or digital transmissions and 10 watts of peak power on voice, usually SSB. The quality of your antenna or location isn't considered, just transmitter power. If you choose to turn the power down below 1 watt, you're milliwatting.

TIP

While it might seem natural to start with low power and work your way up, that actually makes being successful more difficult for beginners. Making QSOs at QRP levels is more difficult than at higher power. I recommend that you get comfort-able with operating at 100 watts or so, then try turning the power down to see how that works out.

QRPers often hang out around their calling frequencies, shown in Table 11-8. For digital modes with QRP, use the digital calling frequencies listed in the ARRL band plan referenced earlier. The usual frequencies for modes in the *WSJT-X* package are built into the software.

TABLE 11-8

North American QRP HF Calling Frequencies

Band (Meters)	Morse Code (MHz)	Voice (MHz)
160	1.810	1.910
80	3.560, 3.710	3.985
40	7.030, 7.110	7.285
30	10.106	no voice modes
20	14.060	14.285
17	18.096	None
15	21.060, 21.110	21.385
12	24.906	none
10	28.060, 28.110	28.385, 28.885

If you're just getting started, tune in a strong signal, and give that station a call with your output power turned down to QRP levels. (Check your radio operating manual for instructions.) Make sure that your transmissions are clear, which allows the other station to copy your call sign easily. If you decide to call CQ, you don't need to call CQ QRP unless you specifically want to contact other QRPers.

TIP

Some low-power stations send their call with /QRP tacked onto the end to indicate that they're running low power. This procedure isn't necessary and can be confusing if your signal is weak. After all, that's more characters for the other station to copy, isn't it?

When you have some experience with QRPing, you may want to try some of the following:

>> **Building your own QRP gear:** Many QRPers delight in building their own equipment — the smaller and lighter, the better. You can find lots of kits and homebrew designs for hams who have good construction skills.

QRPers probably build more equipment than those in any other segment of the hobby, so if you want to find out about radio electronics, you might consider joining a QRP club (discussed later in this list) and one of the QRP email mailing lists.

>> **Join a QRP organization:** QRPers are enthusiastic and helpful types. Their clubs and magazines are full of "can-do" ham spirit and ideas.

>> **Entering QRP contests:** You'll find some QRP-only contests, and nearly all the major contests have QRP categories. Many awards have a special endorsement for one-way and two-way QRP. The QRP clubs themselves have their

own awards, including my all-time favorite, the 1,000 Miles Per Watt award (www.qrparci.org). Some stations make contact with so little power that their contacts equate to millions of miles per watt!

TIP

Look for special gatherings of QRP enthusiasts at conventions, such as the internationally attended Four Days in May (www.qrparci.org/fdim), which coincides with the Dayton Hamvention (see Chapter 3). Other QRP gatherings occur around the United States throughout the year.

Portable operating

You should consider trying operating from "out there"! The capabilities of low-power and portable transceivers today range from minimalist designs to full-scale state of the art transceivers. Coupled with effective, lightweight antenna designs and some interesting award programs, portable operation has really taken off.

If you like hiking, try the Summits On the Air program (SOTA: www.sota.org.uk) which encourages operating from mountainous areas. Figure 11-11 shows Brian Jester, KB8UIP, operating from an Alpine, California mountaintop using a hand-held FM radio and a small beam for 2 meters.

FIGURE 11-11: KB8UIP enjoys activating peaks on the VHF bands for the Summits On the Air program.

Photo courtesy Brian Jester, KB8UIP/VE3SPG

Another popular program that doesn't require hiking is the Parks On the Air (POTA: parksontheair.com) program. Park "activators" travel to all kinds of parks and recreational areas, each assigned their own tracking ID number. Points are awarding both for putting parks on the air and for contacting the activators. The portable stations range from lightweight QRP gear to full-power mobile and temporary station setups. It's a great way to combine travel to new parks with ham radio!

Table 11-9 lists some QRP and portable operating resources, including large QRP organizations and online groups.

TABLE 11-9 **QRP and Portable Operating Resources**

Organization	URL	Resources/Information
Adventure Radio Society	www.arsqrp.blogspot.com	Portable operation
eHam.net QRP forums	www.eham.net/ehamforum/smf/index.php www.eham.net/ehamforum/smf/index.php/board,23.0.html	Discussions on a wide variety of topics
GQRP Club	www.gqrp.com	*Sprat* magazine, and building and operating information
Parks on the Air (POTA)	parksontheair.com	An award program for activating parks and other recreational areas
QRP Amateur Radio Club International (ARCI)	www.qrparci.org	*QRP Quarterly* magazine and numerous awards
QRP-L email reflector	www.mailman.qth.net/mailman/listinfo/qrp-l	Best-known QRP email reflector; archives for email, files, and articles
Summits On the Air (SOTA)	www.sota.org.uk	Active group of QRP operators who hike to mountaintops
U.S. Islands	usislands.org	Award program for activating freshwater islands in U.S. rivers and lakes

Direction-finding (ARDF)

TIP

Although it's not a transmitting event, the popular outdoor activity known as *direction-finding* has a similar philosophy to portable, low-power operating. (en. wikipedia.org/wiki/Amateur_radio_direction_finding). In these events, sometimes called *foxoring*, low-power transmitters (the foxes) operating on 3.5 MHz and the 2 meter band are hidden around a park or other site. Competitors like those in Figure 11-12 then use a directional antenna or *beam* to find all the transmitters as quickly as they can, combining orienteering and map-reading skills with radio expertise to use directional antennas and lightweight receivers. There are entry classes for kids through seniors and you can participate at a walk or on the run.

FIGURE 11-12: Competitors in a direction-finding meet use lightweight portable antennas.

Courtesy American Radio Relay League

The Amateur Radio Direction Finding organization hosts regular meets around the U.S. and worldwide. The culmination is a world championship and the most recent was in Bulgaria in 2016. The 2019 U.S. ARDF championships were held in North Carolina (backwoodsok.org/2019-ardf-us-and-region-ii-championships).

TIP

An interesting variation on ARDF combines geocaching (www.geocaching.com — finding small caches of trinkets or a logbook from GPS coordinates) with direction-finding. This is often referred to as *geo-foxing* where "fox" is another name for the hidden transmitter. Begin with finding a temporary geocache where you find the frequency and call sign of a hidden transmitter. At the transmitter, you find the coordinates of the next geocache. The process can be repeated for as many geocaches and transmitters as you like — or can find! Be sure your transmitter is located in the beacon segment of the band so that unattended operation is legal.

Operating via Satellites

Nonhams usually are pretty surprised to find out about ham radio satellites. Imagine — do-it-yourself satellites! The first amateur satellite, OSCAR-1 (Orbiting Satellite Carrying Amateur Radio), was built by American hams and went into orbit in 1961, just a couple of years after the Soviet Union launched Sputnik and ignited the space race. Typically, about a dozen satellites capable of providing ham-to-ham communications are active. More than a dozen more support the scientific experiments of student teams by sending telemetry back to Earth.

TIP

The main organization for amateur satellite activities is AMSAT (www.amsat.org). AMSAT coordinates the activities of satellite-building teams around the world and publishes the *AMSAT Journal*, which contains some interesting high-tech articles. There are AMSAT organizations in Japan, the United Kingdom, Germany, and other countries, as well. AMSAT sponsors the "Fox" and "Golf" series of small satellites that conduct most of the ham-to-ham communications.

Getting grounded in satellite basics

Most amateur satellites are located in near-circular low Earth orbit, circling the planet several times each day. For practical and regulatory reasons, satellite transmissions are found on the 21 and 28 MHz HF bands, the VHF/UHF bands at 144 and 432 MHz, and microwaves at 1296 MHz and higher. The ionosphere doesn't pass signals reliably at lower frequencies, and satellite antennas need to be small, requiring shorter wavelengths.

The satellite's input frequencies are called the *uplink*, and the output frequencies are called the *downlink*. The pieces of information that describe a satellite's orbit (and allow software to determine where it is) are called the *orbital* or *Keplerian elements*. Knowing where a particular satellite is in space is required for you to operate through it.

TIP

If satellite operating sounds interesting to you, Sean Kutzko, KX9X, has produced "Operating Amateur Radio Satellites," a six-part series of videos that show you how to do it! You can watch Sean's videos at www.onallbands.com. Just search for *satellite* and you'll find them easily.

There are four common types of satellites:

>> **Transponder:** A transponder listens on a range of frequencies on one band, translates those signals to a different band, and retransmits them in real time.

>> **Repeater:** Just like terrestrial repeaters, repeater satellites listen and receive on a specific pair of channels. Satellite repeaters are *crossband,* meaning that their input and output frequencies are on different bands.

>> **Digital:** Digital satellites usually act as store-and-forward systems using regular packet radio protocols and equipment. The International Space Station (ISS) has a digital packet node that's available to hams on the ground, as well as an onboard APRS digipeater.

>> **Telemetry:** Many student teams and other noncommercial groups (whose members have licenses, like all other hams) use amateur radio frequencies to build small satellites called *CubeSats,* which are launched into low Earth orbit as a group when a commercial satellite launch has spare payload capacity. Each CubeSat measures something or performs some interesting function and then sends a stream of digital data *(telemetry)* back to Earth. A CubeSat may or may not be controllable by telecommand from a ham station on Earth. CubeSats typically operate for less than a month; then they gradually reenter the atmosphere and burn up.

TIP

CubeSats and other non-commercial or academic satellite teams need the support of ham radio operators. If you're interested in supporting or working with a Cube-Sat team, check out the NASA CubeSat initiative (www.nasa.gov/directorates/heo/home/CubeSats_initiative.html). These satellites broadcast telemetry that can be received by anyone and transferred to the operating group, such by the automated stations of SatNOGS (Satellite Network Open-Source Ground Station: www.satnogs.org).

Accessing satellites

The best place to find out which satellites are active and in what mode is the AMSAT home page (www.amsat.org). Click the Satellite Info⇨ Current Status menu items to get complete information about what each satellite does and its current operational status.

TIP

After several years of hard work by AMSAT and other groups, a new package of radio equipment was installed on the International Space Station. You can now use the FM repeater aboard the ISS as well as packet radio operation as of late-2020. The most common call signs in use are RSØISS and NA1SS. More information and current status is available at www.ariss.org/contact-the-iss.html.

To access satellites, you also need a satellite tracking program. Several of these programs, including free and shareware trackers, are listed at www.dxzone.com/catalog/Software/Satellite_tracking. AMSAT also provides several professional-quality tracking and satellite operation programs. There are plenty of satellite tracking apps for both Android and iOS — search for *ham radio satellite tracker.*

When you have the tracking software, obtain the *Keplerian* elements for the satellite you're seeking from the AMSAT home page (click the Satellite Info⇨TLE/ Keplerian Elements menu items). Enter this information into your software program, and make sure that your computer's time and date settings are correct.

A complete set of instructions on using satellites is beyond the scope of *Ham Radio For Dummies*, but a set of instructions is available from AMSAT at www.amsat.org/ introduction-to-working-amateur-satellites.

TIP

One satellite *everybody* can see is the Moon! Your radio signals can see it, too, so hams have naturally tried to bounce signals off the ol' orb — successfully! As mentioned in the "Getting Digital" section, new digital modes and computer processing power make it possible for hams using modest equipment to make *moonbounce* or *Earth-Moon-Earth* (*EME*) contacts. The Moon is definitely within your reach!

Seeing Things: Image Communication

All the ham transmissions I've covered so far in this chapter have been voice, data, or codes. Don't hams care about pictures and graphics? They do! With the increasing availability of excellent cameras and computer software, getting on one of the amateur image modes has never been easier.

The ease of image communication has resulted in several really interesting uses, such as sending images from balloons and radio-controlled vehicles. In addition, emergency communications teams are starting to use images as tools for assessing damage after a disaster or managing public events. *The ARRL Handbook* and *The ARRL Operating Manual* both have a chapter on image communication. The next few sections discuss these image modes. Figure 11-13 shows examples of images sent on each mode.

Slow-scan television

You can find slow-scan television (SSTV) primarily on the HF bands, where SSB voice transmission is the norm. The name comes from the fact that transmitting the picture over a narrow channel made for voice transmissions takes several seconds. Usually, you can hear slow-scan signals in the vicinity of 14.230 and 21.340 MHz by using USB transmissions.

FIGURE 11-13: Pictures typical of those sent via amateur radio image modes.

NARROW-BAND TELEVISION

AMATEUR TELEVISION

SLOW-SCAN TELEVISION

FACSIMILE

Courtesy American Radio Relay League

SSTV enthusiasts start with a webcam or video camera and a sound card. They use frame-grabber software to convert the camera video to data files. Graphics files from any source can be used. SSTV software encodes and decodes the files, which are exchanged as audio transmitted and received with a voice SSB transceiver. You can use analog SSTV, in which the picture is encoded as different audio frequencies, or digital SSTV, in which the picture is broken into individual pixels and transmitted via a digital protocol.

TIP

Facsimile over radio is still a widely used method of obtaining weather information from land-based and satellite stations. Hams rarely transmit fax signals anymore, but it's handy to be able to receive fax transmissions from weather satellites.

You can find links to detailed information about SSTV operation at www.qsl.net/kb4yz.

Fast-scan television

You can also send full-motion video, just as regular broadcasters do, with fast-scan video transmissions. Fast-scan uses the same video standards as analog broadcast and consumer video, so you can use regular analog video equipment. This mode is usually called *amateur television* (ATV) and is most popular in metropolitan and suburban areas, where transmission distances are relatively short. ATV even has its own repeaters.

The Ham TV website (www.hamtv.com) has lots of resources for ATV. It includes sections on using ATV to beam back photos from RC-controlled planes and drones, balloons, and rockets. Another excellent source of information, particularly about digital amateur TV or DATV, comes from Boulder, Colorado at www.kh6htv.com.

TIP

ATV transmissions are restricted to the 440 MHz band and higher frequencies because of their wide bandwidth — up to 6 MHz. You won't be able to use your regular 70 cm transmitter to handle that bandwidth, so you must construct or purchase a transmitter designed specifically for ATV. The transmitters are designed to accept a regular video camera signal, so little extra equipment except a good antenna is required to use ATV.

TIP

Many ATV transmissions use the same video transmission format (called NTSC) that analog TV broadcasts did. That means old analog television receivers can be used as receivers. A frequency converter is used to transfer the ham band ATV signals to one of the higher UHF TV channels, where they're received just like any other TV signals. Cable-ready monitors can receive on the 440 MHz band at Channel 59, as well.

Hams are also using the same types of signals as digital TV broadcasts or DTV as hams migrate to DTV formats. There are repeaters for DTV, too. The DATV-Express project (www.datv-express.com) is working to develop and produce inexpensive DTV equipment that doesn't rely on the availability of broadcast industry surplus. This is an active area of experimentation so expect the situation to change.

4

Building and Operating a Station That Works

Chapter **12**

Getting on the Air

W eb pages and social media for ham radio organizations and vendors present many colorful photos of radios with digits winking, lights blinking, signals spread across every band, and every gadget imaginable. Antennas are even more numerous, their elements sticking out every which way, doodads dripping off them, and all manner of claims made about performance. Then you have to sort through nearly an infinite number of accessories and apps. Deciding on gear to build your first station can be overwhelming. How do you choose?

This is an exciting time for you and for ham radio. Technology continues to change what a radio is and how you interact with it. New digital modes are opening the bands in ways not possible until now. The ever-present Internet gives you the ability to use your radio from anywhere on the planet (remote operating). New operating awards and styles arise all the time. These developments are just a few of the new tools to use and enjoy, with more becoming available every day.

What Is a Station?

Everybody has a different idea of what a station should be. Some think of racks of equipment like in a broadcast station. Some imagine a desktop full of equipment. Others might think of putting a whiz-bang setup in their vehicle or a backpack. They're all in agreement and all correct — these are ham radio stations.

All stations have a few basic parts in common:

>> **Transceiver:** Nearly all hams use the combined receiver-transmitter, but there are some with separate receivers and transmitters.

>> **Power source:** AC line, batteries, generators, solar panels, wind turbines.

>> **Antenna system:** Including all the feed lines, switches, tuners, masts, towers, ropes, and so forth.

>> **Control:** This might be you or it might be a computer, located at the equipment or connected over the Internet.

>> **Logging and recordkeeping:** Sometimes on paper but most commonly using a computer.

In this chapter, I start by asking questions about the kind of operating you'd like to try and what bands you want to use. Then I review the different types of station equipment to help you choose among different styles of equipment, select an appropriate antenna, and connect everything. This chapter also helps you figure out what kind of computer and interface accessories work best for operating and recordkeeping or logging.

At first, these choices can seem impossibly complicated. Rest assured that by thinking things through a little bit and taking one step at a time, you can make it a lot easier on yourself. Don't forget — this is supposed to be fun! You can change your mind! You can try different things! Let's just get started, shall we?

Planning Your Station

Don't tell anybody, but you're about to embark on a journey called *system design*. You may think that making decisions is impossible, but all you have to do is a little thinking up front.

Deciding what you want to do

You can find many activities in ham radio, which I cover in Part 3. You can use the same equipment to participate in most of them. Before you start acquiring equipment, think about these questions:

>> What got you interested in ham radio?

>> Can you pick two or three activities, styles, or modes that really pique your interest?

>> If you know and admire a ham, does he or she do something that you want to do?

>> What sounds most intriguing: Using the digital modes, chatting by voice, mastering Morse code, exchanging images?

>> Are you interested in a specialized type of operating, such as satellite or DX or meteor scatter?

>> Will your station be set up at home, in a vehicle (mobile), or as portable equipment (or all three)?

>> Will you set up your station to be permanent or temporary?

>> Do you intend to participate mainly for enjoyment or for a specific purpose, such as public service or travel communications?

All these considerations affect your choice of equipment.

Knowing your ham radio resources is also important. Answer these questions:

>> What's your budget for getting on the air?

>> How much space do you have available for your station?

>> How much space do you have for antennas?

>> Do you have restrictions in your property deed or rental/lease agreement on transmitting or putting up antennas?

The following sections help you determine what your options are.

Deciding how to operate

The first decision to make is where you expect to operate your station most of the time — at home, in a car, or out of a backpack, for example. This choice determines the size and weight of the equipment, what kind of power source it needs, and the type of antennas you'll be using. All those characteristics have a big effect on what features and accessories you'll want and need.

Home operation

A home station is usually a semi-permanent fixed installation. (In CB and commercial systems, this is a "base" station.) Along with the radio equipment, you need a little furniture and space to put it in. Choose a location for your station that minimizes the effect on other family members. A basement station shouldn't be right below a bedroom, for example. All in all, a spare bedroom or dry basement is about the best place because it won't be wet, hot, or cold.

Using voice modes means speaking out loud and probably listening on a speaker. The tones of digital mode or CW signals can bother non-hams, as well as pets. That mean a good, comfortable pair of headphones is a must. You'll find it much easier to understand voice and CW signals by using headphones, too.

Hams sometimes refer to headphones as "cans." This is an old term from the days in which headphones did look like little cans and were just about as comfortable to wear!

Because most hams operate with external antennas, plan appropriate ways of getting feed lines to them. What's going to hold the antennas up? Larger structures, such as masts or towers, will probably need building permits or approvals. Even small external installations are prohibited under some real estate agreements. You may decide to operate with indoor antennas and those have their own issues. (I cover antennas later in this chapter.)

A big part of the amateur service is being available in emergencies. Because you may lose power when you need it most, consider how you might operate your station without AC power. A radio that runs on 12 volts DC can run from a car battery for a while. All your computing gear and accessories also need power. If you have a generator, consider how you can use it to safely power your station, if necessary. An auxiliary power source can also charge smartphones and tablets.

Most "12-volt" radios are really designed to work best from 13.8 V, which is the typical voltage of a vehicle's electrical system with the engine running. Operating at voltages below 11 V can result in erratic operation or not working at all! Make sure your "12-volt" power supply can supply 13.8 V at full current.

Handheld radio operation

Regardless of what other pursuits you choose in ham radio, you will find it useful to have a handheld VHF/UHF radio. It's just so darn handy! A handheld radio keeps you in touch with local hams and is very useful on club and personal outings. Many handheld radios also feature an extended receive range that includes commercial FM stations, weather stations, or police and fire department bands.

If you're buying your first handheld radio, get a simple, single- or dual-band model. You can make a more-informed decision later, when you upgrade to a model with all the bells and whistles. Simple radios are also easy to operate. No new radio you can buy today will be missing any significant feature.

If your club or public service team uses a repeater that supports one of the digital voice modes (see Chapters 8 and 9), make sure your handheld radio is compatible with that mode.

Accessories can extend the life and usefulness of a portable radio. Here are some of the most popular:

>> The flexible rubber duck antenna supplied with handheld radios is great for portable use but isn't as efficient as a full-size whip antenna. An external base-station antenna greatly extends the range of a handheld radio while you're at home.

>> Use a high-quality, low-loss feed line for cables more than a couple dozen feet long. (See "Feed line and connectors," later in this chapter, for more information.) You'll need an adapter from the radio's RF connector to the feed line, too.

>> A speaker–mike combination allows you to operate without holding the radio up to your face. There are handheld combinations that look like a microphone and *boom sets* that are headphones with an extension to hold the microphone near your mouth.

>> A case or jacket (like a smartphone sleeve) protects the radio against the rough-and-tumble nature of portable use.

>> Spare batteries are musts. If you have a rechargeable battery pack, be sure to have a spare, and keep it charged. A drop-in charger works faster than the supplied wall-transformer model. If the manufacturer offers one, a battery pack that accepts ordinary AA cells is good to have, especially in emergencies, when you may not be able to use a charger.

>> A vehicle power cable that connects to the auxiliary power ("cigarette lighter") is also handy. Make sure your radio accepts vehicle power directly or get an automotive adapter to convert vehicle power to the right voltage.

Regardless of what kind of radio you have, be sure to keep a record of model and serial numbers. Engrave your name and call sign, if you like, in an out-of-the-way location on the case, such as under the battery. Even larger radios sometimes get taken outside sometimes. Protect your investment against theft and loss! Check your homeowner's and auto insurance for coverage of radio equipment. If you're an ARRL member, you can use the ARRL equipment insurance program (www.arrlinsurance.com).

TECHNICAL
STUFF

When choosing a power supply, add up all of the current required by your station equipment that will be connected, including the transmitter operating at full power output. (User manuals have this information.) Then add at least 10 percent to provide some margin. You will probably add more equipment in the future so consider buying an even larger supply if your budget supports it. The exam asks about choosing a power supply.

Choosing a Radio

Now comes the fun part: shopping and choosing! Many new hams find this first adventure of buying a radio intimidating, so I wrote an article about it for the ARRL — "Choosing Your First Radio." You can download it at www.arrl.org/buying-your-first-radio, along with a checklist for comparisons, other articles about buying and evaluating radios, and a primer on ham jargon. To get an idea of what products are available, check the advertisements of the latest models in recent copies of QST, On the Air (OTA), and CQ magazines. If you have a license, no doubt you've received a copy of a catalog from Ham Radio Outlet (www.hamradio.com), DX Engineering (www.dxengineering.com), or Universal Radio (www.universal-radio.com). (If not, call them and ask for one!) Perhaps MFJ Enterprises (www.mfjenterprises.com) sent you a catalog of its extensive line of accessories and antennas. If you have a local radio store, pay a visit to browse through the catalogs and product brochures. Inquire about upcoming sales or promotions. Manufacturers often exhibit their new gear at larger hamfests and conventions. Many of the vendors advertise on the popular ham radio websites, too. Your job at this point is to gather a wide variety of information.

TIP

New gear comes with a warranty, but used equipment is often an excellent bargain, and hams do love a bargain. You can find nearly any imaginable piece of gear with a little searching on online swap sites, including eBay. (Look for these sites through the portals and reflectors listed in Chapter 3.) I like to buy and sell through the ham radio websites such as the Classified pages on the eHam.net portal (www.eham.net/classifieds), and QRZ.com also has for-sale forums. Enter *ham trade* or *ham swap* in an Internet search engine to find more. Your local club is also a good source of used gear. As with shopping at hamfests, get help from an experienced friend before buying.

Dozens of handheld and mobile radios are for sale, so use a checklist of features to help you decide on a model. Note the capabilities you want as well as the ones that fall into the nice-to-have category.

TIP

If you plan on doing a particular type of operating, such as Parks On the Air or satellites, join a related mailing list or group and ask what type of gear works best.

Radios called *all-mode* or *multimode* include SSB, AM, CW, and FM modes. This distinguishes them from single-mode (usually FM-only) mobile radios. They may also include special DATA configurations for modes like RTTY and FSK digital modes. An *all-band* radio usually covers the 1.8 MHz through at least 50 MHz bands and usually 144 and 430 MHz, too. (Because 222 MHz operation is U.S.-only, it is usually not offered by world-wide manufacturers.)

After you have started to narrow your list of choices, you can follow up with reviews of specific equipment. The ARRL publishes Product Review columns covering all sorts of gear. The articles are all archived and available to members online. The popular eham.net and QRZ.com portals have product review forums that you can search by manufacturer and model.

TIP

Each radio manufacturer has online users groups that are also good sources of information about specific models and accessories.

As you can see in Table 12-1, regardless of what kind of station you plan to assemble, a new radio consumes at least half of your budget. This is appropriate because the radio is the fundamental piece of equipment in ham radio. New hams interact with the radio more than any other equipment.

Allocating your resources

When you start assembling a station, you have a range of items to obtain — not only the radio itself, but also antennas, accessories, cables, and power sources. Table 12-1 shows some comparisons of relative costs based on the type of station you're setting up. If you pick a radio first, the remaining four columns give you a rough idea of how much you should plan on spending to complete the station. These figures are approximate but can get you started. I assume that all the gear is purchased new.

TABLE 12-1 **Relative Cost Comparisons**

	Total Cost Relative to Basic HF Base	Radio and Power Supply or Batteries	Antennas	Accessories	Cables and Connectors
Handheld VHF/UHF	0.1–0.3	75%	Included	25%	Included
Mobile VHF/UHF	0.2–0.5	75%	20%	Not required	5%
All-mode VHF/UHF	1.0	50%	30%	5%	15%
Portable-Mobile HF	0.5-0.7	60-75%	10-25%	10%	5%
Basic HF base	1.0	50%	25%	15%	10%
Full-featured HF	2.0–5.0	75%	15%	10%	5%

Software defined radios

The software-defined radio (SDR) in which most of a radio's functions are implemented by digital signal processing (DSP) software is one of the most significant changes in ham radio technology since the transistor. The SDR has become the primary technology used for commercial radios. Analog radios are still around and useful but most new radios are SDRs.

TECHNICAL STUFF

Digital signal processing (DSP) refers to a microprocessor in the radio running special software that operates on, or *processes,* incoming signals. An SDR radio is built around DSP hardware and firmware to perform most of the radio's function. (SDR is discussed more in Chapters 8 and 9.)

The most advanced SDR converts signals at radio frequencies to digital data. This is called *direct sampling.* In general, the higher the frequency at which the conversion from analog RF to digital data occurs, the better the SDR will perform. (Look in the radio's operating manual or product specification sheet for more information.)

In an SDR, except for the receiver input and transmitter output circuits, the signal-handling part of the "radio" is a program. You can change the program any time you want to make the SDR do something else. A manufacturer can upgrade your equipment with new software just like a new operating system version on your phone.

After some filtering and conditioning, an SDR converts the incoming radio signal to digital data. From there, the signal is all data until it's converted back to audio for the operator or to data characters for a computer. Going the other way, the outbound signal only needs to be amplified before being applied to the antenna and radiated into space.

As introduced in Chapter 8, a typical SDR displays signals across a frequency range called a *spectrum display.* This is also called a *panadapter* view after an older piece of equipment that was used for this purpose. The other primary view is the *waterfall* display, with colors that show which signals are strongest over time. If a PC is used as the SDR's "front panel," a mouse is used to select signals and click operating controls.

The main benefit of SDRs over analog radios is flexibility. SDR transceivers also have excellent receivers that are less susceptible to overload by strong signals than most superheterodyne radios. Because all of the signal processing is done by software, SDRs don't require a special circuit for each function, such as noise blanking or filtering. Software functions like filtering are more adjustable than those performed by analog electronics. The SDR still requires an analog amplifier to reach useful transmit signal power levels, though.

Some SDRs are packaged as stand-alone radios with everything done in a single unit. These are good choices for small stations and for portable or mobile operating. SDRs that divide functions between an RF package and a PC to perform display and control functions are best suited for permanent and home stations. SDR transceivers may have 100 watts of RF output but some require an external power amplifier to get to the usual 100-watt output.

As of late 2020, several self-contained SDR ham radio transceivers are available, and hams are finding the latest models to be strong performers. FlexRadio's FLEX-6000 series (www.flexradio.com) with the Maestro control console, shown in Figure 12-1, is the latest in the company's SDR line of equipment. The popular IC-7300 is a completely stand-alone SDR packaged like a traditional transceiver but with all the SDR flexibility and functions. Kenwood, Yaesu, and Elecraft all offer SDR transceivers, as well. New manufacturers like Apache Labs (apache-labs.com), ELAD (ecom.eladit.com), Expert Electronics (eesdr.com/en), and Xiegu (www.xiegu.eu) also build SDR transceivers.

FIGURE 12-1:
(a) A FlexRadio Maestro control console; (b) Icom IC-7300 stand-alone SDR.

(a)

(b)

Photo courtesy American Radio Relay League

TECHNICAL STUFF

There are many receive-only SDRs. The RTL-SDR (www.rtl-sdr.com) is a popular inexpensive VHF/UHF receiver in an USB thumb-drive "dongle." It doesn't offer top performance but it is a great way to try SDR functions without spending a lot of money. More capable SDRs with better receiver performance and wider frequency ranges include the HackRF, Airspy, Kiwi, Lime, and the versatile Red Pitaya which is also capable of being used as a test instrument. (Online searches will turn up these and many more.) Some can receive HF while others are VHF/UHF only and require an *up-converter* to receive lower frequencies.

Radios for VHF and UHF operating

There are two basic styles of operating on the VHF and UHF bands. The most common is repeater-based voice operation, which uses FM or digital voice modes on evenly spaced channels. Direct point-to-point communication between stations uses both analog and digital modes. Radios are available for each style, and some can do both equally well. Here I start with the radios designed for repeater operation — your most likely first ham radio! Then I'll introduce the "all-mode" radio and some auxiliary equipment.

Mobile FM radios

(The material in this section also applies to VHF/UHF digital voice radios. Most can operate on both analog FM and digital voice, but hams tend to refer to them as "FM radios.")

Many hams use FM on the VHF and UHF bands regardless of their favorite operating style or mode. A newly minted Technician licensee is likely to use an FM mobile or handheld radio as a first radio. FM is available on the all-mode radios, but because of the mode's popularity and utility, FM-only radios are very popular.

FM radios come in two basic styles: *mobile* and *handheld,* as shown in Figure 12-2. Although they are intended for use in vehicles, you can use them as base stations at home, too. Both mobile and handheld radios offer a wide set of features, including loads of memory channels to store all your region's repeater information, powerful scanning modes, and several types of squelch systems.

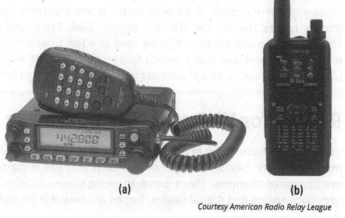

FIGURE 12-2: FM radios: (a) A Yaesu FT-7900A mobile and (b) a Kenwood TH-D74A handheld.

(a) (b)

Courtesy American Radio Relay League

I suggest that you start with a mobile radio shared between your vehicle and your home. Assuming you live in an area with average or better repeater coverage, you can simply put a magnet-mount (*mag-mount*) antenna on top of the refrigerator and you're in business. (If you live in a rural area, you probably need an outdoor antenna.) The stronger signal from the mobile allows you to operate successfully over a wider range, which is important at first. When you know more about what type of FM operating you want to do, you can buy a handheld radio with the right features and save money.

The more-powerful mobile transceivers used with an external antenna extend your range dramatically. Receivers in mobile radios often have better performance than those in handheld radios; they're capable of rejecting the strong signals from commercial transmitters on nearby frequencies. Information about how receivers perform in such conditions is available in product reviews. Your own club members may have valuable experience to share, because they operate in the same places as you.

TIP

If you are part of a public service team, ask what type of radios are popular with the group members. That model would be a good first choice because they can help you learn how to use it and share their programming choices. They may even have equipment you can borrow.

You can use mobile radios for digital data operation on the VHF/UHF bands, such as packet radio and APRS as discussed in Chapter 11. The most common setup was shown in Chapter 2 — the radio connects to an external data interface which is connected to a computer. For packet (which includes APRS and Winlink system connections), this interface is called a *terminal node controller,* or *TNC.* Radios that support D-STAR and other digital voice modes also have a digital interface that connects directly to the computer.

TIP

A radio that's APRS-enabled can connect to (or may include) a GPS receiver and transmits your location, as described in Chapter 11. These radios are also useful for navigation and geocaching.

You can expect the radio to include as standard features encoding and decoding of CTCSS subaudible tones (tones used to restrict access to analog repeaters), variable repeater offsets, dozens of memory channels, and a numeric keypad. Digital voice radios will include the majority of the digital system features.

TIP

Extended-coverage receiving is a useful feature. I find it very useful to listen to commercial FM broadcast and the National Weather Service (NWS) weather alert stations at 162 MHz (www.weather.gov/nwr).

Programming a radio with dozens (if not hundreds) of memory channels can be a chore if you do it all with the front-panel controls. Most manufacturers offer free or low-cost programming software that connects to the radio with an optional cable. The software package *CHIRP* (chirp.danplanet.com/projects/chirp/wiki/Home) is a free, open-source programming tool, and RT Systems (www.rtsystemsinc.com) offers comprehensive packages including cables — both support numerous radios.

Ask around in your club or public service team to see if someone has programming software and a cable to connect your radio to a computer; you'll be really glad you did. The software enables to you to quickly set up your radio for the local repeaters and simplex channels, including alphanumeric labels for each channel, if your radio supports that feature.

TECHNICAL STUFF

DMR (Digital Mobile Radio) system radios are programmed by loading them with a set of data called a *code plug*. (The term originated back when an actual special plug was required to program a radio.) Data in the code plug is transferred from a computer over a USB or proprietary interface. Be sure to get the necessary cable with your radio.

TIP

Most radios can also be *cloned*, meaning that you can transfer the contents of one radio's memories to another radio of the same model by using a cloning cable. This method can save a lot of time if you buy a new radio and a friend has the same model already programmed.

Handheld FM radios

Handheld radios come in single-band, dual-band, and multiband models. With the multiband radios covering 50–1296 MHz, why choose a lesser model? Expense, for one thing. The single-band models, particularly for 2 meters, cost less than half the price of a multiband model. You'll do the lion's share of your operating on the 2 meter (VHF 144–148 MHz) and 70 cm (UHF 420–440 MHz) bands, so the extra bands may not get much use.

WARNING

Inexpensive handhelds made in China are very popular due to their low price. Before you buy one, check out its ARRL Product Review. These radios are often adaptations of models made for Land Mobile Radio or LMR use (a commercial/public-safety service, often referred to as *Part 90* for the section of the FCC rules that apply). Some of these radios have been found to have poor output signal quality. Be sure the radios meet the specifications for amateur radio use. Your club or public-service team may have the latest information.

A rechargeable battery and simple charger come with the radio. Be sure to get a spare battery and the base charger, if your budget allows. Other useful accessories include a full-size whip antenna, adapters for connecting external antenna cables, battery packs that accept standard AA or AAA batteries, speaker-mikes or boom sets, and earphones.

All-mode radios

All-mode VHF/UHF radios operate using single sideband (SSB), Morse code (CW), FM, and AM as opposed to FM-only radios. Along with 50, 144, and 440 MHz operation, the Kenwood TS-2000 and Icom IC-9700 can go all the way to 1200 MHz. (None support 222 MHz operation as of late-2020.) Note that the all-mode radios often do not include digital voice modes.

TIP

The small all-band, all-mode radios available today provide HF, VHF, and UHF operation from either a home station or even the smallest vehicle. Although an all-mode radio is very capable, it's common to have a mobile VHF/UHF FM radio, as well, for using the local repeaters. The FM-only radio provides an emergency backup to your primary radio and you can use it in your vehicle, too.

Some VHF/UHF all-mode radios have special features, such as full duplex operation and automatic compensation for transponder offsets, that make using amateur satellites easier. (I introduced amateur satellite operation in Chapter 11.) Satellite operations require special considerations because of the need for cross-band operation and the fact that satellites are moving, which results in a Doppler shift on the received signal.

An all-mode radio can also form the basis for operating on the amateur microwave bands. Commercial radios aren't available for these bands (900 MHz; 2.3, 3.4, 5.6, 10, and 24 GHz; and up), so you can use a transverter instead. A *transverter* converts a signal received on the microwave bands to the 28, 144, or 440 MHz band, where the radio treats it just like any other signal. Similarly, a transverter converts a low-power (100 milliwatts or so) output from the radio back up to the higher band. Bringing the output signal up to 10 watts or more requires an external amplifier.

VHF/UHF amplifiers

Increasing the transmitted power from a handheld or low-power mobile radio is common. Amplifiers can turn a few watts of input into more than 100 watts of output. Solid-state commercial units are known as *bricks* because they're about the size of large bricks, with heat-sinking fins on the top. A small amp and external antenna can improve the performance of a handheld radio to nearly that of a mobile radio.

TIP

The exam includes several questions about using amplifiers properly. Amplifiers are either FM-only or SSB/FM models. Amplifiers just for FM use cause severe distortion of an SSB signal. An amplifier designed for SSB use is called a *linear amplifier*, and SSB/FM models have a switch that changes between the modes. You can amplify Morse code signals in either mode, with more gain available in FM mode.

RF safety issues are much more pronounced above 30 MHz because the human body absorbs energy more readily at those frequencies. An amplifier outputs enough power to pose a hazard, particularly if you use a beam antenna. Don't use an amplifier at 50 MHz or above if the antenna is close to people. Revisit your RF safety evaluation if you plan on adding a VHF/UHF amplifier to your mobile or home station. See www.arrl.org/rf-exposure for more information.

VHF/UHF preamplifiers

You may find that you need more sensitivity at VHF and higher frequencies. This could be due to being far away from other stations or you could have a long run of coaxial cable that soaks up some of the signal due to *cable loss*. Either way, adding a *preamplifier* or just *preamp* is a common solution.

Preamps are sensitive, low-power amplifiers that increase the level of a received signal. To be most effective, they must be installed *at the antenna*. The preamp receives power through feed line connecting it to the station and has a switching circuit that detects when you are transmitting to route the powerful transmitted signal around the sensitive preamp circuits. (Not all preamps have automatic switching — be sure you understand how the preamp works.) There are a couple of exam questions about using and installing preamps.

Preamps and receivers for the VHF/UHF/microwave bands are evaluated on the basis of *gain* (how much larger they can make the signal) and *noise figure*. Gain is usually 10, 20, or 30 dB. Remember that the higher the gain the more susceptible the preamp is to overload from strong signals. Only buy as much gain as you need. Noise figure is a measure of how much noise the preamp adds along with the signal. Lower values of noise figure mean less noise is added and the preamp is more effective.

Radios for HF operating

All modern radios have perfectly usable receive and transmit performance. The differences involve performance in several key areas, such as capability to receive in the presence of strong signals, signal filtering, coverage of one or more VHF/UHF bands, operating amenities such as sub-receivers, and whether a built-in antenna tuner is available.

HF radios for the home station fall into these categories:

>> **Basic:** This radio has a simplified set of controls with basic receiver filters and signal adjustments. Controls may be fixed-value, with on and off settings. The radio has limited displays and metering, connects to a single antenna, and has

minimal support for external accessories. A basic radio is good for a beginning ham and makes a great second or portable radio later.

>> **Portable/mobile and QRP:** Smaller radios (some are all-band/all-mode) intended for mobile or portable use. Most have a complete set of operating features although few front-panel controls. QRP radios are small enough to be carried on a hike and may operate on a few or even a single band with minimal controls.

>> **Standard:** This radio includes all the necessary receive and transmit adjustments. It has front-panel controls for *incremental tuning* (fine tuning) plus filter selection and adjustment. Most SDR transceivers will provide a waterfall and spectrum or panadapter display. It has an expanded set of memory and metering functions. You can find models that have additional bands and direct connections for digital operation. Internal antenna tuners are common. Many radios have "accessory" connectors for external equipment, such as data interfaces and band-switching equipment.

>> **High-performance:** This radio has state of the art receive capabilities and extensive receive and transmit controls on the front panel, including all the standard features, plus menu settings and options. Multi-channel receiving is included, along with complete interfaces for digital modes. Internal antenna tuners are standard, and some antenna switching usually is provided.

Nearly all commercial radios are supported by control software from the manufacture or third-party vendor that enables remote operation over the Internet or within your home network.

Digital data on HF

More and more HF radios provide a connector or two with a USB data interface built in so that it's easy to connect a personal computer and operate on the digital modes, such as FT8 or to access the Winlink system. A few even have a built-in decoder for some types of digital mode signals. The key features to look for are accessory sockets on the radio carrying some of the following signals:

>> **FSK (frequency-shift keying):** A digital signal at this connector pin causes the transmitter to output the two tones for frequency-shift keying, a method of transmitting using two frequencies, usually used for radioteletype (RTTY).

>> **Data in/out:** If a radio has an internal data modem, you can connect these digital data inputs and outputs to a computer. You may need an RS-232 (a type of serial communication) or USB converter interface.

>> **Line in/out:** Audio inputs and outputs compatible with the signal levels of a computer's sound card, this input is used for digital data when a computer sound card is used as the data modem.

>> **PTT:** This input (the same as the push-to-talk feature on a microphone) allows a computer or other external equipment to turn the transmitter on and off.

A USB *audio codec* is built-in firmware that transfers audio signals between the radio and computer as streams of digital data. Use the operating system audio device control functions to route signals to and from your data software via the audio codec. This eliminates the need for a stand-alone data interface or audio cables from the sound card to the radio.

To find out how to configure a radio to support digital data, look on the manufacturer's website, or ask the dealer for the radio manual. Connection diagrams for AFSK and direct FSK operation should be included. If the manual doesn't provide an answer, contact the manufacturer to ask how to hook up the radio.

Searching online for the combination of your radio model and the digital mode you want to use may turn up interface information for your radio.

HF amplifiers

I recommend that you refrain from obtaining an amplifier for HF operations until you have some experience on the air. Most HF radios output 100 watts or more, which is sufficient to do a lot of operating in any part of the hobby.

You need extra experience to add an amplifier and then deal with the issues of power, feed lines, RF safety, and interference. Also, the stronger signal you put out when using an amplifier affects more hams if the signal is misadjusted or used inappropriately. Get the basic techniques right first.

When do you need output of 1500 watts, or even 500 to 800 watts? In many circumstances, the extra punch of an amplified signal is required. DXers need amplifiers to make contact over long paths on difficult bands. A net control station's amplifier gets switched on when a band is crowded or noisy so that everybody can hear. In emergencies, an amplifier may get the signal through to a station that has a poor or damaged antenna or in noisy conditions.

HF amplifiers come in two varieties: vacuum-tube and solid-state. Tubes are well suited to the high power levels involved. Solid-state amplifiers, on the other hand, tend to be complex but require no tuning or warm-up; just turn them on and go. Tube amplifiers are less expensive per watt of output power than solid-state amps, but they're larger, and the tubes can be expensive.

Before you buy a used tube amplifier, be sure replacement tubes are available at a reasonable cost! Many older types of tubes are no longer in production and can be hard to replace at any cost.

Don't attempt to use CB "footlocker"-type amplifiers. These amps are not only illegal, but they won't work with most ham radios and often have serious design deficiencies that result in poor signal quality.

Filtering and noise

After *sensitivity*, the ability to pick up a weak signal, the most important capability for a receiver is distinguishing one signal from another. This is called *selectivity*. Receivers use filters to remove unwanted signals from interfering with the one you want to receive while *attenuating* (reducing the strength of) unwanted signals just a few hundred hertz away. The range of frequencies that are passed is the filter's *bandwidth*, measured in hertz or kilohertz. Filters that remove a range of frequencies are called *band-reject* filters. If the range of rejected frequencies is very narrow, the filter is a *notch filter*.

Older radios use *crystal filters* made of quartz crystals. Crystal filters have a fixed bandwidth, and most can't be adjusted.

Current mid-level and high-end radios use DSP technology to perform filtering in software. The bandwidth of DSP filters is adjustable, which allows you to use just the right amount of filtering. DSP can also perform filtering functions to remove off-frequency signals, reduce or eliminate several kinds of noise, or automatically detect and remove interfering tones.

Radios often use fixed-width *roofing filters* that help the receiver reject extremely strong signals often encountered on busy bands. If roofing filters are available for your radio, adding one is a good way to get the best performance from the receiver. *Preselectors* are tunable filters that reject very strong out-of-band signals, such as broadcast stations.

The standard filter bandwidth for HF SSB operation is 2.4 to 2.8 kHz. FM and AM operation use 6 to 10 kHz filter bandwidths. Filters with bandwidths of 1.5 to 2.0 kHz are used for operating under crowded conditions with some loss of fidelity. For Morse code and digital modes, the standard filter is 500 Hz wide. Narrower filters, down to and below 250 Hz, are sometimes used under extremely crowded conditions. (With practice, your ears are pretty good filters, too!)

Buying additional receiving filters for an older radio is a whole lot cheaper than buying a new radio. Inrad filters (available at www.vibroplex.com) are available for most recently manufactured radios.

Another kind of filtering removes noise that is received along with the desired signal. This is called *noise reduction* and is usually turned on and off with a switch labeled NR. The radio may just have one general purpose NR function or different ones for voice and CW. NR often doesn't work well with digital modes. You can try each setting and decide which you like best or whether you want to use them at all.

TIP

Digital modes are generally tolerant of noise that bothers human hearing. Audio distortion caused by noise reduction or blanking can actually make digital mode reception worse so don't use those features when using digital modes.

TIP

The exam will ask about *noise blankers*. (The controls for noise blankers are labeled NB.) Noise blankers are used to remove short noise bursts called *impulse noise*. Noise blankers work by detecting the noise pulse and cutting off or "blanking" the receiver's output during the pulse. Most radios have a general purpose NB feature, and some higher-end models have more than one. Noise blanking is intended primarily for on ignition noise (from gas engines) and power-line noise.

WARNING

Noise blanking, noise reduction, and preamps can all make a receiver susceptible to overloading by strong signals. When this happens, it can sound like the strong signals themselves are generating the spurious signals and noise. Before assuming that a signal is causing interference, turn off these features to be sure your own receiver isn't at fault!

Choosing an Antenna

Antennas are at least as important as the radio, if not more important. A good antenna can make a weak radio sound better than the other way around. You need to give antenna selection at least as much consideration you do to radio selection. This section touches on several types of useful and popular antennas.

Hams use two basic types of antennas: *directional* and *omnidirectional*. Directional antennas focus signals in a preferred direction, both for transmitting and receiving. This focusing is called *gain* because it appears the signal is stronger, as if it was amplified. Directional antennas can also reject unwanted signals from other directions. Omnidirectional antennas transmit and receive equally well in all directions and are used when signals may come from any direction at any time. They, too, can have gain but it comes from focusing toward the horizon not in any horizontal direction.

TIP

If you want to know more about antennas and want to try building a few yourself, you need more information. I can think of no better source than *The ARRL Antenna Book*. It's a good ham resource, and many professional antenna designers have a copy, too. If you would like to get a general introduction to antennas, try the ARRL book *Basic Antennas*.

TECHNICAL STUFF

Vertical or *horizontal polarization* describes how an antenna and radio waves are oriented with respect to the ground. If an incoming radio wave and the antenna are oriented differently, called *cross-polarization*, the antenna won't receive the radio wave very effectively. Polarization is most important on the VHF and UHF bands, where signals usually arrive with their polarization largely intact. On the HF bands, travel through the ionosphere scrambles the polarization so that it's much less important.

Beam antennas

TIP

The most common type of directional or *beam* antenna is called a *Yagi*, after Japanese scientists Yagi and Uda, who invented the antenna in the 1920s. A Yagi has several straight rods or tubes (called *elements*) mounted on a long supporting tube called a *boom*. *Log-periodics* similar to large TV antennas with lots of elements are another type of beam antenna used by hams.

The most common HF Yagi today is a three-element design (a reflector, a driven element, and a director). The antenna works on three popular ham bands (20, 15, and 10 meters) and so is called a *tri-bander*. Figure 12-3a shows a three-element Yagi beam with a lowest operating frequency of 14 MHz. The antenna is on a 55-foot telescoping mast.

On VHF and UHF, beams are very popular because they give signals a big boost inexpensively and without requiring large supports. On 144 MHz and higher frequencies, beams are about the same size as large TV antennas. Figure 12-3b shows a dual-band Yagi for 144 and 430 MHz, intended for portable satellite operating but it can be installed permanently, as well.

Log periodics are also used, with popular models available that cover 20 through 10 meters on HF and others that cover VHF and UHF frequencies. These antennas have a similar appearance to Yagis but have more elements, and all of them are connected to the feed line. TV antennas are usually log periodics.

TECHNICAL STUFF

Beams can be made from square or triangular loops. They work on the same principle as the Yagi, but use loops of wire instead of straight elements made from rod or tubing. Square-loop beams are called *quads*, and the triangles are called *delta loops*.

A beam antenna, such as those shown in Figure 12-4, can be rotated, which allows you to concentrate your signal or reject an interfering signal in a certain direction. You can place small HF beams on inexpensive masts or rooftop tripods, although they are too big for most structures designed for TV antennas. You also need a *rotator* that mounts on the fixed support and turns the beam. You can control the rotator from inside the station with a meter to indicate direction (see "Supporting Your Antenna," later in this chapter).

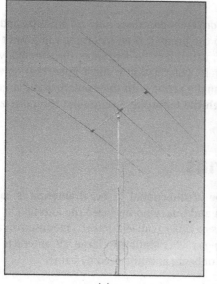

FIGURE 12-3: An HF beam antenna for 20 through 10 meters (a) and a dual-band VHF/UHF beam for portable satellite operating (b).

(a)   (b)

VHF/UHF antennas

Most antennas used above 50 MHz at fixed stations are either short *whips* (thin steel or aluminum rods) or beams. Whips mounted so that they're vertical are used for local FM operation, whereas beams are used to access faraway repeaters and for VHF/UHF DXing on SSB and CW.

FM operating is done with vertically polarized antennas because vertical antennas on vehicles are omnidirectional This important for mobile operation — the most widespread use of FM. To prevent cross-polarization, base and repeater antennas are vertical, too. This convention is universal.

You can mount mobile antennas as removable or permanent fixtures. The most easily removed antennas are the *mag-mount* models, which use a magnetic base to hold themselves to a metal surface. Mag-mount antennas are available from HF through UHF. Lip-mounts, such as the Diamond K412 (www.diamondantenna. net/k412snmo.html), can be attached to a trunk, hood, or hatchback. The drawback of these mounts is that the installation isn't as clean as that of a permanently mounted antenna.

Drilling a hole in your car for a permanent mount, such as for an NMO-style base, looks best of all. Trunk-mount antennas for VHF and UHF look good and perform well. Roof-mount antennas perform the best but are most exposed to damage and have the biggest visual impact. All three options are fairly close in performance.

WARNING

Watch out for internal wiring when drilling holes in a car. It's very common to run wiring inside hollow panels or hidden trays. Check a service manual or ask a dealer service department to find out where it's safe to drill holes.

Whichever method you choose, be sure that you can remove the antenna from the mount to deter theft and for clearance, such as in a car wash or parking garage. You can generally route antenna cables under trim, carpet, and seats.

TIP

A popular and inexpensive vertical antenna is the *quarter-wave whip,* or *ground-plane,* antenna. You can build a simple ground-plane antenna from wire and a coax connector as a first antenna project (www.arrl.org/files/file/Technology/tis/info/pdf/ab18-16.pdf).

Another type of vertical antenna for base or home stations consists of several lengths of transmission line connected end to end to act as an *array.* The array is enclosed in a fiberglass tube for weatherproofing. The array focuses the signal toward the horizon for a stronger signal. The Comet GP-series of antennas (www.cometantenna.com/amateur-radio/base-antennas/ba-dual-band) is typical of this design and several variations are available.

Operators chasing long-distance VHF and UHF contacts use beam antennas that are horizontally polarized. Many of the long-distance VHF and UHF propagation mechanisms respond best to horizontally polarized waves. If you have an all-mode radio and want to use it for both FM and SSB/CW/digital operating, you'll need both vertically and horizontally polarized antennas.

TIP

For operating on CW and SSB where horizontal polarization is used, you can use horizontal loops called *halos.* These antennas are practical for in-motion use from 50 MHz through the 432 MHz bands. (You can see pictures of these loops in the popular catalogs.)

If you choose to use a beam on VHF and UHF bands, it's a good idea to use 3 to 5 elements on 6 meters and 5 to 12 elements at higher frequencies. These antennas are small enough to mount and turn with heavy-duty TV antenna rotators.

HF antennas

Wires, verticals, and beams are the three basic HF antennas used by hams all over the world. You can build all these antennas with common tools or purchase them from the many ham radio equipment vendors.

At HF, antennas can be fairly large. An effective antenna is usually at least ¼ wavelength in some dimension. On 40 meters, for example, a ¼-wavelength vertical antenna is a metal tube or wire 33 feet long. At the higher HF frequencies, antenna sizes drop to 8–16 feet but are still larger than even a big TV antenna. Your physical circumstances (and any regulations or restrictions) have a great effect on what antenna you can put up. Rest assured that there are a large variety of designs to get you on the air.

Wire antennas

Figure 12-4 shows the *full-wavelength loop* and the *half-wavelength dipole*, two popular types of simple wire antennas you can build and install in your backyard or use for portable operation. (Wavelength and frequency are discussed in Chapter 2.) These cost only a few dollars and can be built and installed in an afternoon. There's a special sense of satisfaction in making contacts via antennas that you built yourself.

The simplest wire antenna is a *dipole*, which is a piece of wire cut in the middle and attached to a feed line, as shown in Figure 12-4. The dipole gives much better performance than you might expect from such a simple antenna. To construct a dipole, use 10- to 18-gauge copper wire. It can be stranded or solid, bare or insulated.

TIP

Wire antennas are great as *stealth antennas* for use where a bigger antenna might not be possible or allowed. They can be nearly invisible outside when made out of thin wire. They can be installed inside in attics or along ceilings or dangled from windows or balconies.

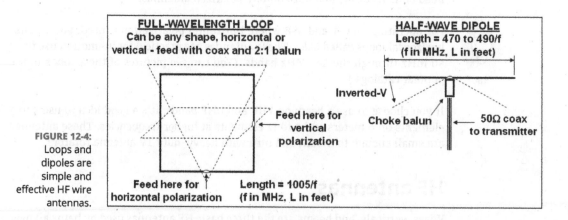

FIGURE 12-4:
Loops and dipoles are simple and effective HF wire antennas.

Although you often see the formula for dipole length given as 468/f, the building process is a lot more reliable if you begin with a bit longer antenna and make one or two tuning adjustments. Start with an antenna length of

Length in feet = 490 / frequency of use in MHz

Allow an extra 6 inches on each end for attaching to the end insulators and tuning and another 12 inches (6 inches × 2) for attaching to the center insulator. If you're building a 10 meter dipole, you should start with a length of 490 / 28.3 = 17.3' + 6" + 6" + 12" = 19.3 feet. (The length of wire in a full-wavelength loop is twice as much as in a half-wavelength dipole.)

To assemble the dipole, follow these steps:

1. **Cut the wire exactly in the middle, and attach one piece to each end insulator, just twisting it back on itself for the initial check.**

2. **Attach the other end to the center insulator in the same way.**

3. **Attach the feed line at the center insulator, and solder each connection.**

4. **Attach some ropes, and hoist the antenna in the air.**

5. **Check the dipole.**

   Make some short, low-power transmissions to measure the standing wave ratio (SWR, a measure of RF energy reflected back to the transmitter by the antenna — see the explanation following this procedure), as explained in your radio's operating manual, or use an antenna analyzer (see Chapter 15). The SWR should be somewhat less than 2:1 on the frequencies you want to use.

6. **Adjust the antenna's length.**

   Find the frequency at which SWR is lowest. If that frequency is too low, shorten the antenna by a few inches on each end. If that frequency is too high, you may have to add some more wire. Repeat the process until the frequency of minimum SWR is where you want it to be.

7. **Make a secure wrap of the wire at the end insulators, solder the twist if you like or use a clamp, and trim the excess.**

   You've made a dipole!

TIP

You can follow the same steps for most simple wire antennas; adjusting the lengths as necessary.

TECHNICAL STUFF

SWR stands for *standing wave ratio*, and it indicates how well the antenna is accepting power from your feed line. An SWR of 1:1 means no power is coming back to your transmitter from the antenna. Higher values indicate that some of the power is bouncing back and forth in the feed line before being radiated away by the antenna. There are analyzers and meters to let you measure SWR, and I discuss them in Chapter 15.

You can connect the dipole directly to the transmitter with coaxial cable and use the dipole on the band at which it's ¼ wavelength long or any odd number of ½ wavelengths. A dipole tuned for 7 MHz, for example, works reasonably well on the 21 MHz band, too, where it's approximately 3½ wavelengths long.

You can install a loop to be either vertical or horizontal. If installed vertically, you can control the signal polarization by attaching the feed line, as shown in Figure 12-4. The loop's minimum SWR will be about 2 to 2.5 to 1 so you may have to use an antenna tuner if that causes your transmitter to automatically reduce output power. Trim the antenna length in the same manner as for the dipole.

Other common and simple wire antenna designs include

>> **Inverted-V:** This dipole is supported at its midpoint, with the ends angling down at up to 45 degrees. This antenna requires only one tall, central support and gives good results in nearly all directions.

>> **End-fed half-wave (EFHW) or Off-Center-Fed Dipole (OCFD):** These are just regular half-wave dipoles with the feed line attached at the end or part way along the antenna. This allows the antenna to be used on several bands. Commercial versions are popular since a matching network is required.

>> **Non-resonant doublet:** The wires of this antenna are fed at the center with *open-wire* or *ladder-line* feed line and used with an antenna tuner to cover several bands. These antennas usually aren't ½ wavelength long on any band, so they're called *doublets* to distinguish them from the ½-wavelength dipoles.

>> **Trap dipole:** This antenna uses some appropriately placed components to isolate portions of the antenna at different frequencies so that the dipole acts like a simple ½-wavelength dipole on two or more bands (www.arrl.org/hf-trap-antennas).

For more information on these and many other antennas you can build, check out the ARRL's antennas page at www.arrl.org/antennas.

If you don't have the perfect backyard to construct the antenna of your dreams, don't be afraid to experiment. Get an antenna tuner (or use the one in your radio), and put up whatever you can. You can even bend wires or arrange them at strange angles. Antennas want to work!

Vertical antennas

Vertical antennas are nearly as popular as wire antennas on the HF bands. Verticals don't require tall supports, keep a low visual profile, and are easy to move or carry. Verticals radiate fairly equally in all horizontal directions, so they're considered to be *omnidirectional* antennas.

The ¼-wavelength design is a lot like a ½-wavelength dipole cut in half and turned on end. The missing part of the dipole is supplied by an electrical mirror of sorts, called a *ground screen* or *ground plane*. A ground screen is made up of a dozen or more wires stretched out radially from the base of the antenna and laid on top of the ground. The feed line connects to the vertical tube (it can also be a wire) and to the radials, which are all connected together. The ¼-wavelength verticals are fairly easy to construct and, like dipoles, work on odd multiples of their lowest design frequency.

Ground-independent verticals are up to twice as long as their ¼-wavelength counterparts but don't require a ground screen. The lack of a ground screen means that you can mount them on masts or structures above the ground. A special impedance matching circuit is used to work with low-impedance coaxial feed lines. Several commercial manufacturers offer ground-independent verticals, and many hams with limited space or opportunities for traditional antennas make good use of them.

Both types of verticals can work on several bands. Commercial multiple-band verticals are available that work on up to nine HF bands.

TIP

A dipole for 14 MHz or a higher frequency is short enough to be rotated like a beam. This antenna can be surprisingly effective. Rotatable dipoles are available from several manufacturers that work on several of the HF bands.

Most hams start on HF with a wire or vertical antenna. After you operate for a while, the signals you hear on the air give you a good idea of which antennas are effective. Visit other ham stations to get more ideas for antennas suitable for you. After you have some on-the-air experience, you can decide whether you need a beam antenna.

TIP

If you start building and repairing antennas, you'll want to locate sources of metal tubing and rod that sell small quantities to individual buyers. Surplus metal outlets are good to have in your list of contacts, too.

Feed line and connectors

Gee, how tough can picking a feed line be? It's just wire, right? Not at all, as Figure 12-5 shows. As I mention in Chapter 2, there are two basic types: *coaxial cable* and *open-wire*.

TECHNICAL STUFF

Feed lines are *transmission lines*, just like high-voltage power lines. They're basically the same thing but for different types of signals.

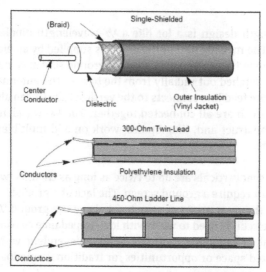

(Braid)   Single-Shielded

Center
Conductor
         Dielectric   Outer Insulation
                            (Vinyl Jacket)

300-Ohm Twin-Lead

Conductors   Polyethylene Insulation

450-Ohm Ladder Line

Conductors

FIGURE 12-5:
Different types of
feed lines.

TIP

Coaxial cable or "coax" (pronounced "co-ax") has a *center conductor* surrounded by a plastic insulator (also called the *dielectric*), which, in turn, is surrounded by a metal sleeve called the *shield*. The shield is covered by a protective plastic *jacket*. The signal is contained entirely inside the cable, as the figure shows. Coax can be bundled with other cables and laid directly on metal surfaces without affecting the signals inside. Most amateur feed lines are coax.

Open-wire feed line also goes by several other names: parallel-conductor, window line, ladder line, and twin lead. It is just two wires held apart by plastic insulation (window line) or by individual plastic or ceramic spacers (ladder line). The signal flows equally on both wires. Open-wire line has to be kept away from other materials by at least twice its width so it is harder to run through or mount on walls. It requires standoffs to run along pipes or masts.

TIP

Because in open-wire feed line the wires are not surrounded by any shielding material, this type of feed line has to be separated from other feed lines and any conductive surfaces.

Each type of feed line has a different *characteristic impedance* (the symbol is Z_o), which describes how the signal inside divides itself between voltage and current. Most coaxial cable has a characteristic impedance of 50 ohms. Common feed line impedances are 75–, 300, 450–, or 600–ohms. This value is important for choosing the right feed line for an antenna. The antenna manufacturer will recommend the right feed line's characteristic impedance.

TECHNICAL
STUFF

TIP

You may be surprised that feed line can turn some of your signal into heat, both on transmit and receive. This is called *feed line loss*. Some types of feed lines are lossier than others and there are some questions about feed line on the exam.

When I started hamming, I used 100 feet of RG-58 with a 66-foot dipole that I tuned on all bands. I didn't realize that on the higher bands, I was losing more than half of my transmitter output and received signals in the coax. Making up feed line loss with an amplifier costs hundreds of dollars. Changing antennas to a beam with 3 dB of gain costs at least that much, not counting the mast and rotator. That makes the extra cost of lower-loss RG-213 cable look like a pretty good bargain.

TECHNICAL STUFF

The dB or *decibel* is a way expressing ratios using powers of ten. To find out more about the "dee-bee" read my tutorial at www.arrl.org/files/file/A%20 Tutorial%20on%20the%20Decibel%20-%20Version%202_1%20-%20Formatted.pdf.

Table 12-2 compares several popular feed lines in terms of their relative cost (based on RG-58) and the loss for a 100-foot section at 30 MHz (10 meters) and 150 MHz (just above 2 meters). The loss is shown in dB and in S units on a typical receiver, assuming that one S unit is equivalent to 6 dB.

TABLE 12-2   **Relative Cost and Loss of Popular Feed Lines**

Type of Line and Characteristic Impedance	Outside Diameter (Inches)	Cost Per Foot Relative to RG-58	Loss of 100' at 30 MHz in dB and S Units	Loss of 100' at 150 MHz in dB and S Units
RG-174A/U (50 ohms)	0.100	2/3	6.4 dB/1 S unit	>12 dB/>2 S units
RG-58C/U (50 ohms)	0.195	1	2.6 / 0.5	6.7 / 1.1
RG-8X (50 ohms)	0.242	1¼	2.0 / 0.3	4.6 / 0.7
RG-213/U (50 ohms)	0.405	3	1.2 / 0.2	3.1 / 0.5
1" open-wire line	1" width	1 to 2	0.1 / <0.1	0.4 / <0.1

A good approach is to use the feed line with the lowest loss you can afford. Open-wire feed line is a special case because you must add an impedance transformer or tuner to convert its higher characteristic impedance to the 50 ohms needed by transmitters. This adds some extra expense.

TIP

To save money on feed line, buy it in 500-foot spools from a distributor. If you can't afford to buy the entire spool, share the spool with a friend or two. Splitting the expense is an excellent club buy and can save more than 50 percent compared with buying cable 50 or 100 feet at a time. Do the same for coaxial connectors, buying them in quantity and sharing the cost.

WARNING

Beware of used cable unless the seller is completely trustworthy. Old or used cable isn't always bad but can be lossy if water has gotten in at the end or from cracks or splits in the cable jacket. If the cable is sharply bent for a long period, the center conductor can migrate through the insulation to develop a short or change the cable properties. (Migration is a particular problem with foam-insulation cables.)

Before buying used or reusing cable, examine the cable closely. The jacket should be smooth and shiny, with no obvious nicks, dents, scrapes, cracks, or deposits of adhesive or tar (from being on a roof or outside a building). Slit the jacket at each end for a few inches to expose the braid, which should be shiny and show no signs of corrosion or discoloration whatsoever. Slip the braid back. The center insulator should be clean and clear (if solid) or white if foam or Teflon synthetic. If the cable has a connector on the end, checking the cable condition may be difficult. Unless the connector is newly installed, you should replace it anyway, so ask if you can cut the connector off to check the cable. If you can't cut it off, you probably shouldn't take a chance on the cable.

The standard RF connectors used by hams are SMA, BNC, UHF, and N-type connectors. SMA connectors are miniature, low-power connectors used at VHF and higher frequencies and are often found on small pieces of equipment such as handheld transceivers. BNC connectors are used for low power (up to 100 watts) at frequencies through 440 MHz. UHF connectors are used up to 2 meters and can handle full legal power. N connectors are used through 1200 MHz, can handle full legal power, and are waterproof when properly installed. (Photos or drawings of connectors are shown in vendor catalogs, *The ARRL Handbook*, and WA1MBA has created a handy online chart of connector types at www.wa1mba.org/rfconn.htm.

Good-quality connectors are available at low prices, so don't scrimp on these important components. A cheap connector works loose, lets water seep in, physically breaks, or corrodes, eating up your valuable signals. By far the most common connector you'll work with is the PL-259, the plug that goes on the end of coaxial cables. The Amphenol 83-1SP model is the standard PL-259 connector. By buying in quantity (another good opportunity for a group purchase), you can get these high-quality connectors at a steep discount compared with purchasing them individually.

TIP

Installing and waterproofing connectors is an important part of station building and maintenance. It's important to have the right tools and materials to do the job properly, so I discuss the subject in Chapter 16.

Supporting Your Antenna

Antennas come in all shapes and sizes, from the size of a finger to behemoths that weigh hundreds of pounds. All antennas, however, need to be clear of obstacles.

Antennas and trees

Although Marconi used a kite for his early experiments, a handy tree is probably the most common antenna support. A tree often holds up wire antennas, which tend to be horizontal or use horizontal support ropes. The larger rotatable antennas and masts are rarely installed on trees (not even on tall, straight conifers) because of the mechanical complexity, likelihood of damage to the tree, and mechanical interference between the antenna and tree.

Nevertheless, for the right kind of antennas, a tree is sturdy, nice to look at, and free. The goal is to get a pulley and *halyard* (the lifting rope) into the tree at the maximum height. If you're a climber (or can find someone to climb for you), you can place the pulley by hand. Otherwise, you have to figure out some way of getting a line through the tree so you can haul up a pulley. You may be able just to throw the antenna support line over a branch. Bear in mind that when a line is pressing against the bark of a tree, the tree can rub and chafe against the line until the line breaks (or squirrels can gnaw through the rope). (This catastrophe always happens at night, in a storm, or right before an important contact.) If the line stays intact, the tree tries to grow around the line, creating a wound that makes raising or lowering the antenna impossible.

ANTENNA AND TOWER SAFETY — DO IT RIGHT!

Before you start installing your antenna, take a minute to review some elementary safety information for working with antennas and their supports. The article "Antenna and Tower Safety" at www.arrl.org/antennas points out a few common pitfalls in raising masts and towers. To work with trees, aside from using common sense about climbing, you may want to consult an arborist.

If you plan on putting up or working on a tower or mast, two books should be on your reading list: *Antenna Towers for Radio Amateurs,* by Don Daso (K4ZA), and *Up the Tower,* by Steve Morris (K7LXC). Both are available from the ARRL and other booksellers. The *ARRL Antenna Book* has a lot of information on tower and antenna safety, too.

TIP

Most simple antennas can be supported with ropes over a branch. If you want to avoid damage to the tree, bringing in an arborist or a tree service professional to do the job right, using sturdy, adequately rated materials, may be a good idea. Radio Works has a good introduction to antennas in trees at www.radioworks. com — see the Jim's Notebook section.

Masts and tripods

A wooden or metal mast is an inexpensive way to support an antenna up to 30 feet above ground. If you're handy with tools, making a homebrew mast is a good project; numerous articles about their construction are in ham magazines. Masts are good candidates to hold up wire HF antennas and VHF/UHF antennas, such as verticals and small beams. If you're just supporting a VHF or UHF vertical, you won't need a heavy support and probably can make a self-supporting mast that doesn't need guy wires. If you area has high winds or if the mast is subjected to a side load (such as for a wire antenna), however, it needs to be guyed.

One commercially available option is a telescoping *push-up mast,* designed to hold small TV antennas and often installed on rooftops. Push-ups come in sizes up to 40 feet, with guying points attached. You can also construct masts by stacking short sections of metal TV antenna mast, but you have to add your own guying points. You can't climb either telescoping or sectional masts, so mounting the antenna and then erecting the whole assembly is up to you. You can also mount a section or two of stacking mast on a chimney to support a small vertical. Push-up and TV masts are available (along with all the necessary mounting and guying materials) online — just search for *tv antenna push-up mast* and you'll get lots of choices.

One step beyond the mast is the roof-mount tripod. The lighter tripods are used for TV antennas and can hold small amateur antennas. Larger tripods can handle midsize HF beams. Tripods are good solutions in urban areas and in subdivisions that may not allow ground-mounted towers. Tripods ranging from lightweight to heavy-duty are available from several tower and antenna manufacturers.

In recent years, multiple-section telescoping fiberglass masts have become available at low cost, such as those from Spiderbeam (distributed by Vibroplex in the U.S. at www.vibroplex.com). These masts can't hold a lot of weight but are perfect for small wire antennas and verticals. They make great portable antenna supports, too.

Towers

By far the sturdiest antenna support is the tower. Towers are available as self-supporting (unguyed), multisection crank-up, tilt-over, and guyed structures 30 feet tall and higher. Towers are capable of handling the largest antennas at the highest heights, but they're substantial construction projects, usually requiring a permit to erect. Table 12-3 lists several manufacturers of towers.

TABLE 12-3	Tower Manufacturers			
Antenna	Website	Lattice	Crank-Up	Self-Supporting
U.S. Towers	www.ustower.com/product_cat/ham-radio-fixed-towers/		X	
Rohn Industries	www.rohnnet.com	X		X
Trylon	www.trylon.com			X
Aluma Tower	www.alumatower.com	X	X	X
Heights Tower	www.heightstowers.com	X	X	X
Universal Towers	www.universaltowers.com/products.html			X

The most common ham tower is a *welded lattice tower.* This tower is built from 10-foot sections of galvanized steel tubing and welded braces. You must guy or attach it to a supporting structure, such as a house, at heights of 30 feet or more. A modest concrete base of several cubic feet is required to provide a footing. Lattice towers for amateur use are 12 to 24 inches on a side and can be used to construct towers well over 100 feet high. Lattice towers are sufficiently strong to hold several large HF beam antennas, if properly guyed. *Tilt-over towers* are lattice towers hinged in the middle so that you can pivot the top sections toward the ground by using a winch. Because of mechanical considerations, most amateur tilt-overs are limited to less than 100 feet in height.

Crank-up towers are constructed from telescoping tubing or lattice sections. A hand-operated or motorized winch raises and lowers the tower with a cable and pulley arrangement. A fully nested crank-up is usually 20 to 25 feet high, reducing visual effect on the neighborhood, and when fully nested and blocked for safety you can climb it to work on the antennas. Crank-ups also usually have a tilting base that aids in transporting and erecting the tower. Crank-ups don't use guy wires, so they depend on a large concrete foundation of several cubic yards.

This keeps their center of gravity below ground level to prevent tipping over when fully extended. You can install crank-ups in small areas where guying isn't possible; they're available in heights of up to 90 feet.

WARNING

Crank-up towers are very dangerous to climb since their sections can move and cause serious injury. Never climb a crank-up tower unless the tower is completely lowered and nested, with sturdy blocking inserted through the tower to prevent any section from slipping and falling.

Self-supporting towers rely on a large concrete base for center-of-gravity control. They're simpler and less expensive than crank-ups. Available at up to 100 feet, they have carrying capacity similar to that of fixed lattice towers. Mounting antennas along the length of a self-supporting tower is more difficult than for a fixed lattice tower with vertical supporting legs.

WARNING

Be extremely careful when buying a used tower or mast. Unless the item has been in storage, exposure to the elements can cause corrosion, weakening welds and supporting members. If it's disassembled improperly, the tower can be damaged in subtle ways that are difficult to detect in separate tower sections. A tower or mast that has fallen is often warped, cracked, or otherwise unsafe. Have an expert accompany you to evaluate the material before you buy.

WARNING

In almost any urban or suburban area, you need a building permit to put up a tower. There may also be restrictions included in your property deed or homeowners association (HOA) rules. Your homeowners insurance also has to be updated to include the tower as part of the house or as an auxiliary structure. Check these out first to avoid an expensive mistake. Check with club members and other local hams who may have experience with local building requirements.

Regardless of what you decide to use to hold up your antennas, hams have a wealth of experience to share in forums such as the TowerTalk email list. You can sign up for TowerTalk at www.contesting.com. The topics discussed range from mounting verticals on a rooftop to which rope is best to giant HF beams, to how to locate true north. The list's members include experienced hams who have dealt with some difficult questions.

Rotators

A *rotator* is a motorized gadget that sits on a tower or mast and points antennas in different directions. Rotators are rated in terms of wind load, which is measured in the number square feet of antenna surface it can control in strong winds. If you decide on a rotatable antenna, you need to figure out its wind load to determine the size rotator it requires. Wind load ratings for antennas are available from antenna manufacturers. The most popular rotators are made by Hy-Gain (www.mfjenterprises.com/collections/hy-gain) and Yaesu (www.yaesu.com).

TIP

Although the words "rotor" and "rotator" are used interchangeably, they don't mean the same thing. The rotator is the device that causes something to rotate, like an antenna mast. The rotor is something that turns. So, the correct term for what points your antenna is "rotator."

You need to be sure that you can mount the rotator on your tower or mast; some structures may need an adapter. Antennas mount on a tubular mast that sits in a mast clamp on top of the rotator. If you mount the rotator on a mast, as shown on the left side of Figure 12-6, you must mount the antenna right at the top of the rotator to minimize side loading. This type of installation reduces the maximum allowable antenna wind loading by about half.

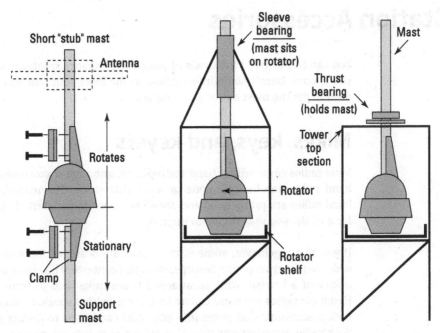

FIGURE 12-6:
Mounting a rotator on a mast and on a tower.

Inside a tower, the rotator is attached to a *rotator plate* (a shelf inside the tower the rotator is attached to), and the mast extends through a *sleeve* or *thrust bearing* (a tube or collar that hold the mast centered above the rotator), as shown in the center and on the right side of Figure 12-6. Because using a bearing in this way prevents any side loading of the rotator, you can mount antennas well above the tower top if the mast is sufficiently strong.

An indicator assembly called a *control box*, which you install in the station, controls the rotator. The connection to the rotator is made with a multiple-conductor cable. Install the feed lines to the antennas in such a way so that they can accommodate the rotation of the antennas and mast. If you have a rotator but no control box, replacements are probably available from the original manufacturer. The Green Heron controllers are configurable to operate a wide variety of rotators and can be computer-controlled, as well.

WARNING

Used rotators can be risky purchases. They're always installed in exposed locations and wear out in ways that aren't visible externally. Even if the rotator turns properly on the bench, it may jam, stall, or slip under a heavy load. Buy a new rotator or get a used one from a trusted source for your first installation. Norm's Rotor Service (www.rotorservice.com) and C.A.T.S. (www.rotor-doc.com) both sell refurbished rotators.

Station Accessories

You can buy or build hundreds of gadgets to enhance whatever style or specialty you choose. Here's some information on the most common accessories that you need to get the most out of your station.

Mikes, keys, and keyers

Most radios come with a hand microphone, although if you buy a used radio, the hand mike may be long gone or somewhat worn. The manufacturer-supplied hand mikes are pretty good and are all you need to get started. After you operate for a while, you may decide to upgrade.

If you're a ragchewer, some microphones are designed for audio fidelity with a wide frequency response. Net operators and contesters like the hands-free convenience of a headset with an attached boom mike held in front of your mouth. Handheld radios are more convenient to use, with a speaker–microphone combination accessory that plugs into your radio and clips to a shirt pocket or collar. Your radio manufacturer may also offer a premium microphone as an option or accessory for your radio.

Heil Sound (www.heilsound.com) and Vibroplex (www.vibroplex.com) manufacture a wide range of top-quality microphones. Several manufacturers offer boom sets, which are great for extended operating.

TIP

The frequency response of a microphone can make a big difference on the air. If you operate under crowded conditions, the audio from a microphone whose response emphasizes the midrange and higher frequencies is more likely to cut through the noise. Some microphones have selectable frequency responses so that you can have a natural-sounding voice during a casual contact and then switch to the brighter response for some DXing. If you're not sure which is best, ask the folks you contact, or do an over-the-air check with a friend who knows your voice.

Antique microphones may look great but if their elements are worn out or damaged their fidelity will be terrible. Old microphones can also have wiring problems or being clogged with cigarette smoke or "dried breath." Be prepared for a cleaning and rebuilding project if you want to use them.

Morse code enthusiasts have thousands of keys to choose among, spanning more than a century of history (see Chapter 11). Beginners often start with a straight key and then graduate to an electronic keyer and a paddle. If you think you'll use CW a lot, I recommend going the keyer/paddle route right away.

Most rigs now include a keyer as a standard option. You can plug the paddle into the radio, and you're on your way! CW operators tend to find paddle choice very personal, so definitely try one out before you buy. A hamfest often has one or more key-bug-paddle collectors, and you can try many styles. The ham behind the table is likely to be full of good information as well.

Programmable memories are very handy for storing commonly sent information, such as your call sign or a CQ message. Sometimes, I put my keyer in beacon mode to send a stored CQ message repeatedly to see whether anyone is listening on a dead band. (If everybody listens and nobody transmits, the band sounds dead but may be open to somewhere surprising.)

If you decide on an external keyer, you can choose kits or finished models, such as the popular Winkeyer-based keyers made by K1EL (k1el.tripod.com). Several computer programs send code from the keyboard. Browse to www.ac6v.com/morseprograms.php for an extensive list of software.

A *voice keyer* is a device that can store short voice messages and play them back into your radio as though you were speaking. Some keyers are stand-alone units, and others use a PC sound card. Voice keyers are handy for contesting, DXing, calling CQ, and so on. Some models also store both CW and voice messages, such as the MJF Contest Keyer (www.mfjenterprises.com). Contest logging software such as N1MM Logger+ (www.n1mm.com) and Writelog (www.writelog.com) can create a voice keyer by using the computer's sound card.

Antenna system gadgets

Antenna tuners

Antenna tuners don't really "tune" your antenna, but they allow your transmitter to operate at maximum efficiency no matter what impedance appears at the station end of your feed line. Tuners are explained in the article "Do You Need an Antenna Tuner?" at www.arrl.org/transmatch-antenna-tuner.

TIP

Although your new radio may be equipped with an antenna tuner, in some situations you may need an external tuner installed between your radio and the antenna system. Antennas being used far from their optimum frequency often present an impedance that the rig's internal tuner can't handle. External tuners often include *baluns* (an abbreviation of "balanced-to-unbalanced") for connecting open-wire feed lines to coaxial cable.

Tuners are available in sizes from tiny, QRP-size units to humongous, full-power boxes larger than many radios. Table 12-4 lists a few of the manufacturers offering an assortment of tuners. If you decide to purchase a tuner, choose one that's rated comfortably in excess of the maximum power you expect to use. I highly recommend getting one with the option to use balanced feed lines. The ability to switch between different feed lines and an SWR meter (which measures reflected RF power) is a nice-to-have feature.

TABLE 12-4 **Antenna Tuner Manufacturers**

Manufacturer	Website	Balanced Feed Line	High-Power (>300 Watts)	Automatic Tuning
MFJ Enterprises	www.mfjenterprises.com	Yes	Yes	Yes
Ameritron	www.ameritron.com	Yes	Yes	No
Vectronics	www.vectronics.com	Yes	Yes	No
LDG Electronics	www.ldgelectronics.com	External balun adapter	Yes	Yes
Palstar	www.palstar.com/en/tuners	Yes	Yes	Yes
SGC	www.sgcworld.com	Yes	Yes	Yes
Inrad	www.vibroplex.com	No	No	Yes

Power and SWR meters

TIP

Although most tuners and radios have an SWR meter built in (see Chapter 18 for more about SWR), a stand-alone meter is a good piece of equipment to have available and there are exam questions about them. A *directional wattmeter* will measure power in your feed line flowing toward the antenna (forward power) and power that has been reflected from the antenna (reflected power). You can compute SWR from those two values but most meters have special scales to show you both power and SWR measurements. Daiwa Industry (www.daiwa-industry. co.jp/EN/products_hammtr.html) meters are typical designs. Inexpensive SWR-only meters don't show power but can serve as an indicator that everything is okay (or not) in your antenna system. The Bird wattmeter shown in Figure 16-6 is an excellent meter, although it doesn't show SWR.

Dummy load

TIP

Along with the tuner, you need a *dummy load*, a large resistor that can dissipate the full power of your transmitter. The dummy load keeps your transmitted signals from causing interference during tuneup. The MFJ-260C can dissipate 300 watts, which is adequate for HF transceivers. High-power loads, such as the MFJ-250, immerse the resistor in cooling oil. (These are paint cans filled with oil, sometimes called *cantennas*, the name of an old Heathkit product.) HF dummy loads may not be suitable for use at VHF or UHF, so check the frequency coverage specification before you buy. The exam asks at least one question about using a dummy load.

Digital mode interfaces

Operation on the digital modes, except for PACTOR, uses a sound card to send and receive data. With a simple data and radio control interface, your computer and radio form a powerful data system. Chapter 2 shows a typical station configuration for operating this way. Use of PACTOR 3 or later protocol versions require the use of an SCS modem (www.scs-ptc.com/en/Modems.html).

TIP

To connect your computer to your radio there are several options. Your radio might have a USB audio codec built in so all interaction is digital. If you use your PC's sound card, an analog data interface such as made by MFJ (www. mfjenterprises.com) or the West Mountain Radio line of Rigblasters (www. westmountainradio.com) handles receive and transmit audio as well as PTT (push-to-talk) and your regular microphone. You'll need to know what a data interface is on the exam.

An external sound card like a Signalink by Tigertronics (www.tigertronics.com/ slusbmain.htm) takes care of the interfacing, provides high-quality sound card functions, and connects to the PC via USB.

If you choose to use an external multiple-mode controller for the digital modes, such as the Timewave PK-232 or DSP-599zx (www.timewave.com), Kantronics KAM (www.kantronics.com), or MFJ-1278B, you need only a terminal program such as Hyperterm (Windows), xterm (Linux), or Terminal (Mac). Terminal Node Controllers (TNCs) for packet radio are available from all three companies, as well.

Remote Control Stations

More and more hams are setting up stations and operating them by remote control. Why? There are several reasons: The most common is that they can't put up effective antennas where they live. If you rent or lease, the property owner might not allow you to install antennas. (Asking nicely and promising not to cause interference, eyesores, or safety hazards sometimes gets you a temporary tryout and a possible waiver of the rule.) Without antennas, ham radio is a challenge!

Another reason is noise. Not the noise the station makes but the noise made by home electronics, appliances, computer equipment, power lines, and other electric-power devices. The old wisdom, "You can't work 'em if you can't hear 'em" is true. You can use special receiving antennas and noise-cancellers, but a lot of noise can be just as bad as having ineffective antennas.

And there is always the possibility of your transmitted signal causing interference at home and to the neighbors. Turning your transmit power down can solve the problem, but you might not make as many contacts, either. What's a ham to do?

The answer is often to build or share a station somewhere that you can put up antennas and that is relatively free of noise. Then, use the Internet or some other kind of data connection to access the station. With the technology we have today, this can work amazingly well! It's pretty much like operating a regular station but with a *really* long microphone cord.

Remote control rules

Before you start getting that gleam in your eye, imagining big towers holding antennas high in the air, remember there are a few rules you must follow:

>> **License authority:** You have to be licensed at the location of the transmitted signal. Some U.S. hams have set up stations in other countries or in other areas of the world, such as the Caribbean. This requires either a local license or a reciprocal operating permit. You can find out more about either of those at the ARRL's web page for International Operating. Some types of operating permission require you to be physically present in the licensing country, so be sure to read the fine print!

>> **Permission:** Trite but true, you have to have permission to use a transmitter. Whether you have a license to operate at the site or not, you still have to have permission to use the equipment.

>> **Identification:** When you send your call sign, be sure it indicates the location from which you are transmitting. If you are a U.S. ham and the station is in the U.S., then no problem — just send your regular call sign. If you are connecting to the station across a national border, you'll have to use a call sign from the station's country. And some awards and operating events require hams to use call signs that indicate where they are operating from.

>> **Control:** Regardless of how well you design the station, you have to be able to turn off the transmitter no matter what. Some security controllers accessible by phone can turn a relay on and off to control AC power to the station. You can even buy power strips with a web interface. Having a method that uses a different method of access than the usual control link is a good idea.

Remote operation is a lot of fun and can make ham radio accessible to you wherever you may be. For example, one well-known contest operator frequently fires up his home station in Ohio while sitting in a hotel room in Tokyo, Japan! He loves that he can get on the air in his favorite events even when his business takes him out of the country. But follow the rules and don't abuse the privilege — it wouldn't take more than a few bad apples to spoil the remote operating barrel.

Accessing a remote control station

A good overview of the state of remote control is available in *Remote Operating for Amateur Radio,* by Steve Ford, WB8IMY. Even though the software and network options are continually changing, the basics are sound. The book is available from the ARRL and other ham radio booksellers. *The ARRL Handbook*'s section on remote stations is updated regularly, including site evaluation and alternative power.

The most common setup is to use the Internet with a PC at each location running a station control program. The remote station is typically configured something like Figure 12-7. (The additional DTMF controller connected to the phone line is needed in case the computer loses control of the radio and you need to shut everything down and start over.)

If you set up your station at home but operate it away from home, the controlling software can run on a laptop so you can fire up the rig from a hotel room or coffee shop. There's no reason why you have to lug around a laptop, either. A smartphone or tablet computer can be quite enough. The remote-control software developed by Pignology (www.pignology.net) runs on an iPhone and even includes logging software so that everything fits in the palm of your hand. The RigPi Server (www.rigpi.net) uses a Raspberry Pi microcomputer to perform all of the necessary control functions and can be accessed without special software — just a web browser!

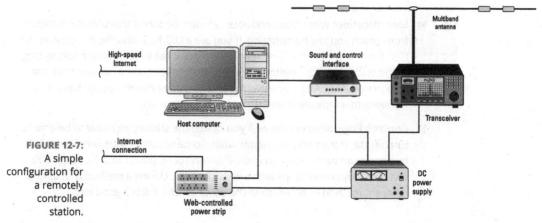

FIGURE 12-7:
A simple
configuration for
a remotely
controlled
station.

Courtesy American Radio Relay League

TIP

If you think you might like to try remote operating, start by checking out one of the many online remote receivers at WebSDR (www.websdr.org). There are dozens of receivers available for you to use located all over the world. Try listening to the same station from different receivers to get an idea of how propagation varies!

Why go to the trouble of building the station yourself? "Time-share" remote stations are maintained by Remote Ham Radio (www.remotehamradio.com). The stations are available for a per-minute fee (with a paid membership). Be careful though, because using these capable stations will spoil you!

Remote operating is becoming easier and more common. More radios support it, more software is available, and high-speed Internet is available in more places than ever. The technology to make it work is available, radios themselves are easier to control over the Internet, and commercial products are appearing that provide plug-and-play operation. You'll have the option of building a traditional home station and be able to operate it from wherever you are. Some clubs and informal groups build and share a remote station. As I say at the beginning of the chapter, this is an exciting time!

Upgrading Your Station

Soon enough, usually about five minutes after your first QSO, you start thinking about upgrading your station. Keep in mind the following tips when the urge to upgrade overcomes you. Remember the adage "You can't work 'em if you can't hear 'em!"

>> The least expensive way to improve your transmit and receive capabilities is to use better antennas. Dollar for dollar, you get the most improvement from an antenna upgrade. Raise antennas before making them larger.

>> Consider increasing transmit power only after you improve your antennas and eliminate local noise sources. Improve your hearing before extending the range at which people can hear you. An amplifier doesn't help you hear better.

>> The easiest piece of equipment to upgrade in the station is the multiple-mode processor between your ears. Before deciding that you need a new radio, be sure you know how to operate your old one to the best of your abilities. Improving your know-how is the cheapest and most effective improvement you can make.

By taking the improvement process one step at a time and by making sure that you improve your own capabilities and understanding, you can achieve your operating goals quicker and get much more enjoyment out of every ham radio dollar.

Chapter 13

Organizing a Home Station

A well-organized home station provides many benefits for occasional and serious ham enthusiasts alike. You will spend many hours in the station, so why not make the effort to make your experience as enjoyable as possible? This chapter explains how to take care of the two most important inhabitants of the station: the gear and you. The order of priority is up to you. I cover mobile stations in your vehicle and how to set up temporary stations for portable operating in Chapter 15.

Designing Your Station

One thing you can count on is that your first station layout will need to be changed. It's guaranteed! Don't bolt everything down right away. Plan to change the layout several times as your operating style and preferences change along with your interests and skill level. By thinking ahead, you can avoid some common pitfalls and save money, too.

Keeping a station notebook

Before you unpack a single box or put up one shelf in your station, you should start a station notebook to record how you put your station together and to help you keep your station operating. The notebook can be a paper, spiral-bound notebook or a three-ring binder. Your notebook can be electronic files, too — take notes in text files and pictures with your phone or camera. The important thing is to have a record of additions, changes, and repairs. Sooner or later, you'll be thankful you have a notebook, no matter what it's made of!

You might not want to have your notebook out "in the field" when you're working on antennas or operating a portable station. Use a note-taking app or voice recorder on your phone or make a pocket notebook and pencil part of your tool kit so you can take notes whenever you need to.

A three-ring binder full of report cover pockets is a great place to keep equipment manuals and warranties. Instruction sheets can be taped into a notebook or kept in the binder. You will accumulate a surprising number of these, so give them a home!

Here are some typical things that go in my notebook:

>> The plans for a circuit or antenna before I start building it along with the final as-built information and a description of how it performs.

>> SWR versus frequency for antennas that are working properly so I can tell if something has changed later.

>> Connector wiring details, such as what is wired to each pin and wire colors.

>> Software configuration and settings, file names, passwords (not of my bank account!), and screen shots.

>> How things fit together so you can get them together again: This is a great reason to keep a camera or camera app handy.

All these methods save you tons of time and frustration by recording important details that are hard to remember. Don't rely on memory! Taking a few minutes to record information saves tenfold the time later.

Keep a list of parts and pieces you run low on or use the last of. The next time you visit the store, hamfest, or website, you won't forget to stock up!

Create a list or spreadsheet with the make, model, and serial number of your station equipment along with the date of purchase, amount paid, whether it was used, and who it was purchased from. This information is important in filing an insurance claim.

Building in ergonomics

Spending hours in front of a radio or workbench is common, so you need to have the same concerns about ergonomics in your radio station that you do at work. You want to avoid awkward positions, too-low or too-high furniture, and harsh lighting, to name just a few. By thinking about these things in advance, you can prevent any number of personal irritations.

The focal point

Remember your main goals for the station. Whatever you plan to do, you'll probably use one piece of equipment more than half the time. That equipment ought to be the focal point of your station. The focal point can be the radio, a computer keyboard and monitor, or even a microphone or Morse code paddle. Paying attention to how you use that specific item pays dividends in operating comfort.

The computer monitor

You may be building a radio station, but in many cases, you'll be looking at your computer monitor more than the radio. (I talk about the types of computers in Chapter 14.) Certainly, monitors are the largest pieces of equipment at your operating position. Follow the guidelines for comfortable computer use. Position the desktop at the right height for extended periods of typing, or a keyboard tray might work.

Buy a high-quality monitor (larger is better), and place it at a height and distance for relaxed viewing. Used or recycled PC stores have many excellent monitors available at deep discounts. Most have adjustable tilt, but try to find one with adjustable height and with the ability to rotate for additional adjustment options. Figure 13-1 shows a few ways of integrating a monitor with radio gear.

Tablets and laptops are very popular and often support external monitors along with the built-in display. Putting these devices front-and-center with the monitor above and behind them often works well.

Monitors mounted too far above the desk give you a sore neck. If you place the monitor too far away, your eyes hurt; if you place it too far left or right, your back hurts. Now is the time to apply computer ergonomics and be sure that you don't build in aches and pains as the reward for the long hours you spend at the radio.

If you wear eyeglasses, get a special pair that focuses at the distance to the monitor. Experiment with your software to display different color combinations and font types or sizes until you are comfortable. Your eyes will thank you!

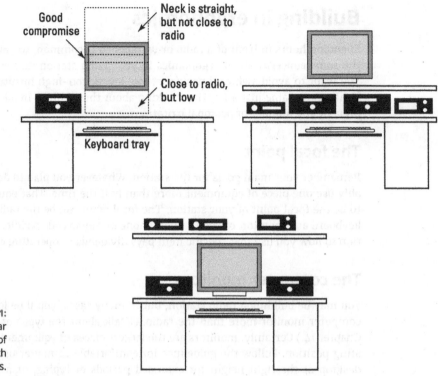

Good compromise

Neck is straight, but not close to radio

Close to radio, but low

Keyboard tray

The radio

Radios (and operators) come in all shapes and sizes, which makes giving hard-and-fast rules difficult. HF operators tend to do a lot of tuning, so placement of the radio is very important. If the tuning knob is at an uncomfortable height or if there's not enough desktop on which to rest your arm, a case of sore "tuning elbow" can result. VHF-FM operators do less tuning, so the radio doesn't have to be as close to the operator. Placing your most-used radio on one side of the keyboard or monitor is probably the most comfortable arrangement.

TIP

If you're right-handed, have your main radio on your left and your mouse or trackball on your right. If you're left-handed, do the opposite. PC-based software-defined radio gear often uses the mouse for tuning, which means you'll be using it a lot! Keep the mouse or trackball in a place where it's comfortable to use for long periods of time.

Adjust the front of the radio to a comfortable viewing angle. You should be able to see all the controls and displays without having to move your head up or down. Try to keep your primary radio and the computer monitor together in your field of view so you only have to move your eyes between them without moving your head.

The operating chair

A key piece of support equipment, so to speak, is the operator's chair. I'm surprised to visit state-of-the-art radio stations and find cheap, wobbly, garage-sale chairs at the operating positions. After spending all that money on electronics, don't compromise performance because of your chair!

Your body is in contact with your chair longer than with any other piece of equipment.

Get a roll-around office-style chair with good lower-back support and plenty of padding in an adjustable seat. Choose a chair that lets you sit in several comfortable positions at the desk without leaning on your arms or stressing your lower back. You may find that chairs with arms make sitting close to the operating desk difficult. It is usually possible to remove the arms. Avoid leaning forward or backward while operating.

Have a couple of different chair pads or cushions for a different "feel" now and then. Make a short foot rest to lift your feet off the floor and your thighs off the chair seat, as in Figure 13-2.

The desk and shelves

The top surface of your operating desk is the second-most-contacted piece of equipment. As is the case with chairs, many choices are available for desks suitable for a ham station. Consider height and depth when looking at desks.

Before choosing a desk, you need to decide whether the radio will sit on the desk or on a shelf above it. Figure 13-2 illustrates the basic concerns. Do you like having your keyboard on the desk? You need to be comfortable sitting at your desk with your forearms resting comfortably on it. If you tune a radio a lot, such as for most HF stations, avoid arrangements that cause your arm to rest on the elbow or on the desk's edge for prolonged periods. Nothing is more painful! Make sure you have enough room for wrist support if using the keyboard is your main activity.

The most common height for desks is 28 to 30 inches from the floor. A depth of 30 inches is about the minimum if a typical transceiver that requires frequent tuning or adjustment is sitting directly on it. You need at least 12 inches between the front edge of the desk and the tuning knob. With your hand on the radio control you use most frequently, your entire forearm needs to be on the desktop. If the radio is sitting on a shelf above the desk, be sure that it's close enough to you that tuning is comfortable and doesn't require a long reach. Be sure you can see the controls clearly.

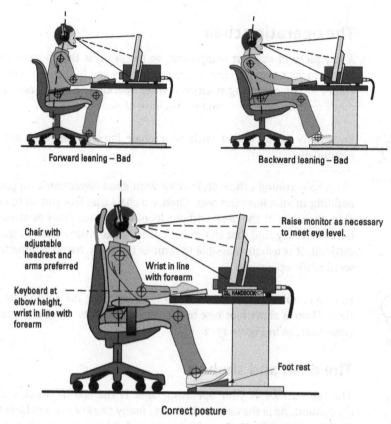

Forward leaning – Bad

Backward leaning – Bad

Raise monitor as necessary to meet eye level.

Chair with adjustable headrest and arms preferred

Wrist in line with forearm

Keyboard at elbow height, wrist in line with forearm

HANDBOOK

Foot rest

Correct posture

FIGURE 13-2:
Station desk and chair ergonomics should be arranged for operator comfort.

Graphic courtesy American Radio Relay League

TIP

For small spaces, a desk-and-shelf computer workstation may be a good solution. You'll probably have to add some shelves, but the main structure has all the right pieces and may be adjustable to boot. Some have wheels so they can be moved for access behind the station.

TIP

Plan for expansion, if at all possible! You may start with just a few pieces of gear, but rest assured that more gadgets and accessory items will appear. Adding shelves is nearly always a good way to make more space.

Viewing some example ham stations

Because every station location and use is going to be different, the most helpful thing I can do is provide examples. Then you can decide what works for you. The goal of this section is to get you thinking about what works for you, not to suggest that you duplicate these stations exactly.

Almost every ham's home station starts on a desktop or table. Shown in Figure 13-3, Mike Adams, N1EN, built a station on HF and VHF/UHF bands just like that! Along with compact transceivers, he uses a laptop and tablet with a full-size keyboard for typing convenience. He has added a small shelf and storage drawer that hold monitors at the right height. Mike's station features:

>> Monitor and table easy to see and position

>> Plenty of writing and arm support surface

>> Transceivers within easy reach

Another option is to use a computer workstation with vertical shelves. Some models sit on wheels, allowing you move the whole station out of the way (or work on the back!) when it's not being used. A fold-down desk gives you more room for the keyboard, mouse, and any other operating accessories. Internet searches for *apartment ham radio station* will turn up many inventive solutions to getting on the air without taking up a lot of space.

FIGURE 13-3:
A tabletop station built around laptop and tablet computers.

Photo courtesy of the American Radio Relay League

TIP

Many hams like soft light in the station because it's not distracting and makes reading the indicators and displays on the radios and accessories easy.

Ash, KF5EYY, operates in a lot of contests, such as in Figure 13-4 at the 4U1ITU club station in Geneva, Switzerland. This simple station can make a lot of contacts with its clean layout, monitor placed for neck and eye comfort, and uncluttered layout:

» The computer is on the floor so it doesn't occupy desktop space.

» There is lots of arm support on the desktop.

» Putting the monitor above the radio minimizes head and eye movement.

» Ear buds can be more restful than headphones during long contests.

Photo courtesy of the American Radio Relay League

FIGURE 13-4:
A simple, uncluttered layout is effective for long periods of operating.

TIP

Put paper labels on the front-panel tuning controls of an amplifier and antenna tuners to make changing frequencies easy without extended tuning. This method reduces adjustment time and stress on the amplifier and minimizes interference to other stations.

Figure 13-5 shows that you don't have to have gear stacked to the ceiling to have an effective station. This station, shared by Portuguese contesters Ana, CR7ADN, Jose CT1CJJ, and Felipe, CT1ILT, often finishes with top scores in worldwide contests. The radios are conveniently placed and the CW paddle is just to the left of the keyboard so Ana doesn't have to move her hand and arm very far. Station controllers are just below the monitor so she can see in a glance how the equipment is configured.

FIGURE 13-5:
A station with two main transceivers for high-performance contest operation.

The station in Figure 13-6 is another example of a simple layout that makes good use of the equipment without complicated clutter. Bryant, KG5HVO, is working on his CW skills from this clean station. Bryant uses a boom-set that combines the headphones and microphone. Above the center monitor are some maps and charts to let him know where to point his beam. A second monitor to the right shows information from the spotting networks and scoreboards during contests.

FIGURE 13-6:
Adding a second monitor provides more information to the operator.

Building in RF and Electrical Safety

Whatever type of station you choose to assemble, you must keep basic safety principles in mind. Extensive literature is available for hams (see the sidebar "Sources of RF and electrical safety information," later in this chapter).

WARNING

Don't think that you can ignore safety in the ham station. Sooner or later, the equipment is likely to get damaged or someone gets hurt. Take a little time to review and follow the safety fundamentals.

Electrical safety

Electrical safety isn't particularly complicated. It consists of following the few simple rules that prevent shock and fire hazards. Electricians and engineers have developed these rules over many decades. Take advantage of what they have learned to keep your station safe.

>> **Know and follow the fundamental wiring rules for AC power.** The National Electrical Code (NEC) contains the rules and tables that help you do a safe

wiring job. *The NEC Handbook,* as well as numerous how-to and training references, is available in your local library or at home-improvement centers. *The Complete Guide to Wiring* (published by Cold Spring Press, 7th Edition) is an excellent source of instructions and guidelines for safe and correct wiring in your station that is "up to code" as of 2020. If you're unsure of your skills, hire an electrician to do the work or inspect yours.

>> **Deal with DC power carefully, especially in a car, to prevent short circuits and poor connections.** Treat vehicle batteries and connections to them with respect and caution. A short circuit results in high currents that can cause expensive fires. Poor connections result in your radio operating erratically. Be sure to install fuses and protect the wiring from abrasion or cuts. As with AC power, follow the safety literature or hire a professional installer to do the job right.

>> **Think of your own personal and family safety when constructing your station.** Don't leave any kind of electrical circuit exposed where someone can touch or damage it accidentally. Use a *safety lockout* (a device that prevents a circuit breaker from being closed, energizing a circuit) on circuit breakers when you're working on AC wiring or AC-powered equipment. Have fire extinguishers handy and in good working order. Show your family how to remove power from the ham station safely.

RF exposure

TIP

The signals your transmitter generates can also be hazardous if you are too close to the antenna. The human body absorbs radio frequency (RF) energy, turning it into heat. The biological effect of RF energy varies with frequency, being most hazardous in the VHF and UHF regions. A microwave oven operates at the high end of the UHF frequencies, for example. One whole section of the exam is dedicated to questions about RF Safety.

REMEMBER

Don't let the word *radiation* frighten you! It applies to both radio waves (RF) and the particles emitted by radioactive substances. Only particles have enough energy to cause genetic damage. Radio waves are many, many times weaker than particles — all they can cause is heat. You need to be careful with transmitted RF, but it is much less of a concern than radioactivity.

Amateur signals are usually well below the threshold of any harmful effects but can be harmful when antennas focus the signal in such a way that you're exposed for a long period of time. High-power VHF and UHF amplifiers can definitely be hazardous if you don't handle them with caution.

TIP

To reduce the possibility of exceeding an RF exposure guideline, place your antennas where people can't get close to them. If you have a directional antenna (such as a beam), don't aim it at people (or yourself). Both of these tips will reduce safety hazards.

A comprehensive set of RF safety guidelines is available in *The ARRL Handbook* and at `www.arrl.org/rf-exposure`. As you construct your own station, do a station evaluation to make sure you're not causing any hazards due to your transmissions.

First aid

As in any other hobby that involves the potential for injury, having some elementary skills in first aid is important. Have a first-aid kit in your home or station, and be sure your family members know where it is and how to use it. Training in first aid and CPR is always a good idea for you and your family, regardless of your hobby.

SOURCES OF RF AND ELECTRICAL SAFETY INFORMATION

Be responsible and check out these inexpensive safety references:

- The American Radio Relay League promotes safety for all manner of amateur radio activities. You can find excellent discussions of ham station hazards and how to deal with them in *The ARRL Handbook* and *The ARRL Antenna Book*. Several articles are available in PDF format at `www.arrl.org/safety`.

- For safety issues relating to power circuits, the Electrical Safety Forum (`www.electrical-contractor.net/ESF/Electrical_Safety_Forum.htm`) is a good source of information.

- Brush up on lightning and grounding issues in a series of engineering notes on the Protection Group website (`www.transtector.com/resources/white-papers`). Look in the *Resources* area for *white papers*.

- The ARRL publications *RF Exposure and You*, *The ARRL Handbook,* and *The ARRL Antenna Book* all discuss RF exposure safety issues.

Grounding and Bonding

Grounding and bonding are two techniques for insuring electrical safety and managing the RF currents flowing around your antenna system and station equipment. *Grounding* means a connection to your power system's voltage reference point, usually connected to the Earth. *Bonding* means connecting equipment together so there can be no voltage difference between them. Both are important for safety and good station design.

The simple guidelines in this section will get you started, but you need to learn more about grounding and bonding practices. To answer your questions and help you perform grounding and bonding correctly, I wrote the book *Grounding and Bonding for the Radio Amateur.* (See Figure 13-7.) It's available from the ARRL, online, and from ham radio distributors. The book goes into detail about AC safety, lightning protection, and RF management, including examples to guide you as your station grows.

FIGURE 13-7:
Grounding and Bonding for the Radio Amateur by the author shows you how to build in safety and control RF.

Courtesy American Radio Relay League

AC and DC power

TIP

AC and DC power systems employ grounding and bonding to prevent shock and fire hazards. Shock hazards are created when there is a path for current to flow from an energized conductor through you. Fire hazards are created by high currents that heat wires enough to melt or ignite flammable materials. Grounding provides safety by connecting exposed conductors, such as equipment cases, directly to the system's voltage reference point, such as a building's metal structure. The ground conductor also provides a path to guide current away from you in the event of a short circuit between the power source and the exposed conductors. Electrical safety is important and the exam will ask questions on the subject.

AC power safety grounding uses a dedicated conductor — the so-called "third wire." Three-wire AC outlets connect the ground pin to the ground connection at the master circuit-breaker box. The ground conductor must be heavy enough to handle any possible fault currents and cause the circuit breaker to trip, removing power from that circuit.

All equipment in your station with exposed metal enclosures or connections, whether powered directly by AC or not, should be bonded together with wire or strap. This keeps everything at the same voltage and prevents a shock when touching two different pieces of equipment at the same time. Bonding all the equipment to the ground system also provides current a safe path if any short-circuits occur.

Because DC systems generally use low voltages (less than 30 volts), grounding is less concerned with preventing shock than with preventing excessive current and poor connections. Both can result in a lot of heat and significant fire hazards. Follow the manufacturer's recommendations for the size of wire to use for power connections. Keep connections tight and clean to prevent erratic operation by your equipment. Be sure to use wire that is heavy enough to carry the expected current without a significant drop in voltage.

Lightning

The power and destructive potential of lightning are awesome. Take the necessary steps to protect your station and home. The ARRL web page on lightning protection (www.arrl.org/lightning-protection) includes several articles by Ron Block (NR2B) on how to create "zones of protection" in your home and station. Whether you decide to simply disconnect your feed lines outside the station or install a more complete protection system, be sure to do the job thoroughly and completely.

The most important part of lightning protection is giving all that energy a direct path to the Earth — where it is going anyway! That's what your ground system should do. To deal with any energy that does visit your station, bonding the equipment together will keep it all at the same voltage, limiting destructive currents. There are plenty of resources and instructions available to help you build a solid ground system right from the start.

WARNING

Even if you have a good grounding and lightning protection system, it is unwise to operate when lightning is in the area. Stay away from station equipment when lightning is within a few miles of your station. You can keep an eye on lightning strikes in your area by using the Blitzortung (map.blitzortung.org) or Lightning Maps (www.lightningmaps.org) websites.

RF management

TIP

The techniques that work for AC and DC power safety often don't work well for the high frequency signals that hams use. For RF, a wire doesn't have to be very long before it starts acting like an antenna or transmission line. At 28 MHz, for example, an 8-foot piece of wire is about ¼ wavelength long. It can have high voltage on one end and very little voltage on the other. The way to control RF voltages and current in your station is to bond all the equipment together. Bonding keeps the equipment at the same RF voltage, not necessarily zero volts. This helps prevent *RF interference,* or *RFI,* which can disrupt radio or computer operation. The term *RF ground* is discouraged to keep from getting the wrong idea about the purpose of bonding. Bonding also reduces the effects of *ground loops* that are formed when multiple cables connect multiple pieces of equipment.

You can bond equipment together at RF by connecting each piece of gear to a copper strap or pipe (called an *RF ground bus*) with a short piece of strap or wire, as shown in Figure 13-8. Ham gear usually has a ground terminal just for this purpose. Then connect the bonding strap or pipe to your station's AC safety ground with a heavy wire.

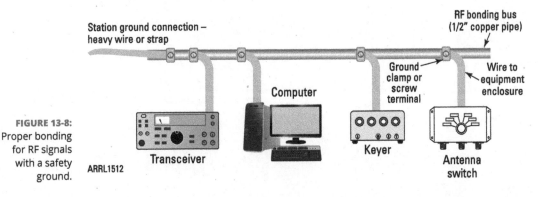

FIGURE 13-8: Proper bonding for RF signals with a safety ground.

TIP

Aluminum or copper roof *flashing* is a good source of strap or sheet for use in your station. 20 gauge is a good thickness to use and easy to work with. Roofing contractors and metal scrap dealers may also have extra flashing you can buy.

Placing a sheet of metal (solid or mesh or screen) under your equipment as shown in Figure 13-9 creates an *RF reference plane* that helps you bond your equipment together. This sheet is connected to your AC safety and lightning protection ground systems. It reduces RF voltage differences in your station that can cause unwanted "hot spots" or current.

FIGURE 13-9:
Metal sheeting under equipment helps control RF picked up from your transmitted signal.

Photo courtesy of the American Radio Relay League

Chapter **14**

Computers in Your Ham Station

O riginally used as a replacement for the paper logbook, the computer in ham radio has evolved nearly to the point of becoming a second operator, controlling radios, sending and receiving CW and digital messages, and linking your station to thousands of others through the Internet. As I discuss in the previous chapter, the computer is often part of today's radio.

Regardless of what operating system you prefer, software tools and programs are available to help you enjoy any type of operating you like. Some software is supplied by commercial businesses, and the amateur community has developed an amazing amount of shareware and freeware. Hams freely contribute their expertise in any number of ways, and developing software is a very popular activity.

What Type of Computers Do Hams Use?

What type of computer you use really depends on the job you want it to do. Let's start with some resources for the main types or *platforms* of ham radio computers.

Windows

Most software available for ham applications runs on the Windows operating system, Windows 7 or later. If you browse to ham radio software listings like www.dxzone.com/catalog/Software, most of the titles will be Windows-based. (Listings on reference sites like DX Zone often include inactive websites.) In addition, hams also develop special applications in spreadsheets like Excel. If you have a Windows computer already, you'll find lots of software available.

TECHNICAL STUFF

Some pre-Windows 7 software is very useful even though it is not actively supported, such as the *TLW* (Transmission Lines for Windows) and *HFTA* (High-Frequency Terrain Analysis) packages that were originally distributed with ARRL reference books. To run software of this vintage, try using Run as Administrator or Run in Compatibility Mode, both of which may work. More information is available at www.howtogeek.com/228689/how-to-make-old-programs-work-on-windows-10 and windowsreport.com/run-windows-xp-games-windows-10.

Linux

Linux has an increasing number of adherents, particularly among digital-mode enthusiasts. Along with desktop and laptop PCs, the miniature Raspberry Pi single-board computers run various versions of Linux, called *distributions* or *distros*. A listing focused exclusively on Linux software titles is available at www.dxzone.com/catalog/Software/Linux.

Macintosh

The Macintosh computing community is also well-represented in ham radio software with programs available for all of the common ham radio uses. A useful website devoted to Macintosh computers and ham radio is www.machamradio.com. The popular QRZ.com web portal has an Apple ham radio software forum, Mac Ham Radio on macOS & iOS, at forums.qrz.com/index.php?forums/mac-ham-radio-on-macos-ios.109. (In Forums, look for the Technical Forums section.)

TIP

The Ham-Mac mailing list (www.mailman.qth.net/mailman/listinfo/ham-mac) is not as active as it once was but the searchable archives have a lot of useful information.

Android and iOS

There are also many apps for Android and Apple tablets and phones with more added every day. Most are free or very inexpensive. (When shopping in the online app stores search both for *amateur radio* and *ham radio* so you find the widest selection.) Most apps play a supporting role to guide your operation. The main limitation is connecting to other devices, such as your radio, to operate or process signals. The most common connection is through the headphone/external microphone jack as if you were plugging in a headset. If your phone or tablet supports Bluetooth, though, an audio adapter will allow you to connect a radio's audio input and output.

TIP

The combination of an inexpensive tablet and a portable radio creates an effective simple station you can take anywhere. For example, combine a PSK31 app like DroidPSK (www.wolphi.com/ham-radio-apps/droidpsk) on your phone or tablet with a QRP radio and — *voilà!* — you've created a portable digital mode station in a backpack as British ham, Carl 2EØEZT, shows in this YouTube video — www.youtube.com/watch?v=NCaRmCbKoCA&ab_channel=2E0EZTamateurradio.

What about integrating an app with a radio? Do any radios run Android or iOS software? Stand-alone radios only run the software that controls their electronics because they must meet stringent FCC type-certification requirements in order to be sold in the U.S. This is changing, however, as radios for commercial use are adapted to amateur radio. For example, the handheld RFinder B1 (rfinder.shop/product/rfinder-b1-dual-band-dmr-4g-lte) combines the Android OS with a VHF/UHF DMR and analog radio. Expect more of these hybrids to become available for all of the ham bands and modes!

Microcontrollers

Along with the Raspberry Pi, the Arduino and other microcontrollers like the STM32 ("Blue Pill"), ESP32, and Teensy 4.0 are widely used for ham projects. The PIC family of processors is also popular. (The ARRL has several books available to help you learn about using microcontrollers at arrl.org/shop — search for "microcontroller.") Typical projects include customized test instruments, switch and accessory controls, and simple receivers and transmitters. There is a lot of processing power available in these small packages, enough for even some of the more sophisticated digital modes like WSPR.

What Do Ham Computers Do?

It would be easier, frankly, to list things that computers *can't* do in the ham station — like amplify kilowatt signals (but they can control the amplifier) or radiate your signal on the air (even though they can control your rotator). Almost everywhere else in the station, though, you can connect and use a computer. Let's list a few of the ways in which computers have made it possible to operate in new ways.

Software-defined radio

Many SDR receivers consist of an RF "front end" and a PC to process the signals. The front end filters the incoming signals and converts them to two channels of digital data called "I" and "Q" because of how they are created. Those signals are then transferred over a high-speed digital interface to the PC where they are processed to recover the transmitted information. (Some front-ends just provide the I and Q channels as audio and the PC sound card does the digital conversion.) Transmitted signals follow the same path in reverse with an amplifier circuit feeding the antenna instead of receiving filters.

Figure 14-1 shows typical SDR software (*PowerSDR*) displaying a 30 meter CW signal. In the middle of the screen, the signal's audio is displayed as if it was connected to an oscilloscope. At the same time, the audio of dots and dashes is played back over the PC speakers. (Some SDR software can also decode the CW into text characters.) The software can also recover audio from SSB, FM, AM, and other types of signals, which can then be decoded by other software for digital mode operation.

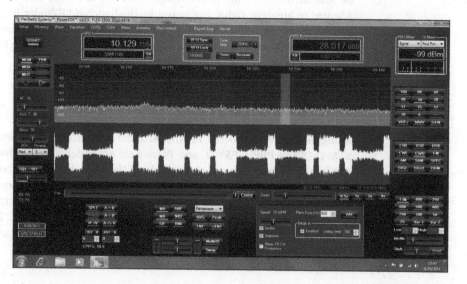

FIGURE 14-1: The *PowerSDR* software's operator interface showing a received CW signal along with all of the operating settings and controls.

Along with the signal, you can also see the settings of all the signal processing and tuning controls. Most SDR software has a similar display, which is configurable to show or hide controls and move them around the screen. You can see other examples of SDR operation in Chapters 8 and 12.

New SDR hardware and software are being introduced every day. To keep track of what's available and learn about new products, websites like Passion-Radio (`www.passion-radio.com/store/sdr-software-10`) are good references to bookmark. Most manufacturers will make new product announcements on the various online forums (`QRZ.com` and eham.net are typical), social media, and user groups, as well.

WSJT-X and fldigi

In Chapter 11, you learned about *WSJT-X* and *fldigi* software that support different digital modes. (The exam has several questions about them.) Most of these modes use AFSK (audio FSK) modulation that "fits" within a standard voice channel. The PC sound card is connected to the radio's microphone input and either headphone or speaker audio output and a serial connection (USB or RS-232/COM) controls the radio's frequency, mode, and keying. All of the magic happens in the PC software which is available for Windows or Linux.

Part of the configuration for these and similar programs is an adjustment for how powerful the PC's processor is. Fast PCs like those used for gaming can run signal processing software at full throttle, which allows the weakest signals to be decoded as quickly as possible. Older PCs can be set to work slower.

Radio and remote control

Most radios have a serial port (COM), USB, or Ethernet interface through which you can monitor and control nearly every radio function. (Icom uses a proprietary interface called CI-V that requires a converter.) Because of that flexibility, some control programs replicate the radio's front panel on the computer's screen. Some radio manufacturers offer radio control software that you can download. Third-party programs such as Ham Radio Deluxe (`www.hrdsoftwarellc.com`) and DXLab Suite (`www.dxlabsuite.com`) integrate radio control with logging software and other features. Computer control of radios is the subject of a couple of exam questions.

In Chapter 12, you were introduced to remote control stations with the radio over "here" and the control operator (that is, you) over "there" connected through the Internet. Although there are stand-alone accessories like the RigPi (`www.rigpi.net`) and RemoteRig (`www.remoterig.com/wp`) to manage the connection, most hams operating a remote control station use a PC at each end. Neither of the PCs needs to be all that powerful, but your Internet connection needs to be stable and at least 1 Mbit/s. The remote-control software developed by Pignology (`www.pignology.net`) runs on an iPhone and even includes logging software so that everything fits in the palm of your hand.

NETWORKING HAM RADIO STUFF

With just about everything connected to the Internet in one way or another, it is a reasonable question to ask why more ham radio "stuff" doesn't have a network-style interface. There are two main reasons: cost and RF interference. Until recently, adding a full networking "stack" of software and the necessary hardware just wasn't cost-effective for ham radio hobbyist-budget gear. The second reason is that networking equipment and the strong RF fields around ham radio antennas don't get along well, plus network equipment tends to leak RF signals of its own that interfere with received signals. Hamation (www.hamation.com) has pioneered station networking using RS-485 serial data technology and careful interface design. Their ShackLAN4 system works well under these challenging conditions.

The networking situation is changing, though, with the introduction of extremely inexpensive microcontrollers and single-board computers like the Arduino and Raspberry Pi that have network and WiFi interfaces built in. The addition of WiFi to low-cost consumer stuff will drive the cost down even more, along with the use of inexpensive tablets and the ever-present smartphone. WiFi solves the two hardware problems because it doesn't need any cables (except at the access point) to pick up RF and it operates at 2.4 GHz, well above most amateur bands. The growing use of WiFi and inexpensive host hardware is encouraging hams to incorporate more network capabilities. For example, the Hamlib project (github.com/Hamlib/Hamlib/wiki) is developing a common radio and rotator control interface (an API to you coders out there). My guess is they are just getting started and will extend that model to other common equipment such as tuners, amplifiers, switches, and test instruments. This is an exciting beginning!

TIP

You can use your computer to send CW, too! There are many simple CW interfaces that use a COM port's control signals to drive a transistor connected to the radio's KEY input. Electronic keyers like the Winkeyer (k1el.tripod.com/WhatisWK.html) accept text messages via USB and convert them to Morse code to key the radio.

Hardware considerations

If you don't do complex antenna modeling or high-performance software defined radio, you don't need to own the latest and greatest speed-demon computer. If you're thinking about upgrading a home computer, a computer that's a couple of years old does just fine in the ham station, even for digital mode processing.

If you decide to purchase a new computer for the station, many ham radios and accessories use the RS-232 serial COM port, which has been phased out on PCs in favor of USB. (COM ports are now referred to as *legacy* ports.) Integrating a

USB-only PC into the ham station means that you either have to purchase a serial port expansion card or use USB-to-RS-232 converters. More and more radios and accessories are converting to USB interfaces, however.

WARNING

USB to Serial converters are usually based on one of two brands of converter ICs: Prolifics or FTDI. Converters based on FTDI chips seem to be the most compatible with ham software and equipment.

TIP

With old computers so inexpensive (check out your local PC recycler for great deals) it's easy to dedicate an old PC to a single task or two, such as displaying an APRS map or monitoring propagation.

Keeping a Log of Your Contacts

As you get ready for contacts with your new station, you'll need a way to keep track of them. Maintaining a detailed logbook — a *log* — is no longer an FCC requirement, but there are a lot of good reasons to record what you do on the ham bands. The main log is a nice complement to your station notebook.

Paper logging

Your station log can be paper log sheets in a notebook or binder with handwritten entries for every contact. Some hams use paper logs when operating mobile or portable, transferring the contacts to a computer log at home. Figure 14-2 shows a typical format for a paper logbook.

Be sure to record the basics:

>> **Time:** Hams keep time in UTC (or World Time) for everything but local contacts.

>> **Frequency and mode:** Just recording the band in either MHz or wavelength is sufficient (20 meters or 14 MHz, for example). An abbreviation for the mode (SSB, CW, FM, PSK, etc.) is sufficient.

>> **Call sign:** Identify each station you contact.

Those three pieces of information are enough to establish the who, when, and where of ham radio. Beyond the basics, you probably want to include the power you used, the signal reports you gave and received, and any personal information about the other operator, like name and location. Most people don't keep a log of casual local contacts made via repeaters, but you may want to log your participation in nets or training exercises. If so, it's okay to use local time.

Record year somewhere on each sheet

Watts assumed   Use UTC!   RST optional   QSL sent & received

DATE	FREQ	MODE	POWER	TIME	STATION WORKED	REPORT SENT	REPORT REC'D	TIME OFF	COMMENTS QTH NAME OSL VIA	QSL S	R
3 Nov	20 m	USB	100	1431	G3SXW	58	57	1437	Roger, funny chap	x	
				1438	OK1RI	57	58		Jiri	x	
				1442	DL6FBL	59	57		nr Munich, Ben	x	
				1515	V51AS	59	59		Namibia, Ralph direct	x	
				1538	A61AJ	59	59		UAE Ali K2UO	x	
4 Nov	146.82	FM	25	0200	N7UK	-	-	0224	ARES net		
				0225	N7FL	-	-		Debbie, needs battery		
8 Nov	40 m	LSB	100	0330	K8CC	57	55	Lots of QRN	MI Dave		
	3.985	LSB	100	0405	W7EMD	59	57		WA Emergency Net		
9 Nov	10.124	PSK	10	0445	N7FSP	gud	OK		nice copy tonight		

Band or frequency OK   Can be used for any information about the QSO

FIGURE 14-2: A typical paper log sheet showing basic information.

UTC, also known as Coordinated Universal Time (www.timeanddate.com/time/aboututc.html) is the time at the Prime Meridian (0° longitude, London). Your "UTC Offset" is the number of hours your local time is ahead or behind UTC. UTC doesn't change through the year, so your UTC offset will change twice a year in areas that implement daylight saving time.

Don't limit yourself to just exchanging the information recorded in your logbook or logging program. Another person is on the other end of the QSO with lots of interesting things to say!

Computer logging

If you're an active ham, I highly recommend keeping your log on a computer. This is probably the most common function for computers in the ham station — keeping track of your contacts and other activities. The logging software makes it easy to look up previous contacts with a station or operator, apply for some of the operating awards, or participate in radiosport contests like you learned about in Chapter 11. Just do an Internet search for *ham radio logging software*, and you'll be surprised at the variety available.

You can use your logging program like a blog as a day-to-day radio diary to keep track of local weather, solar and ionospheric conditions, equipment performance, and behavior. (In fact, the word *blog* is a shortened version of *web log*, meaning "a log on the web.")

Figure 14-3 shows the window for logging contacts in DXLab Suite mentioned previously. Not only can you enter all the information but the program automatically interfaces to your radio's computer control port and fills in the frequency and mode for you. The software can also upload your log to contact confirmation services that you learn about later in this chapter. Programs like Ham Radio Deluxe (www.ham-radio-deluxe.com) and DX4WIN (www.dx4win.com) are also available.

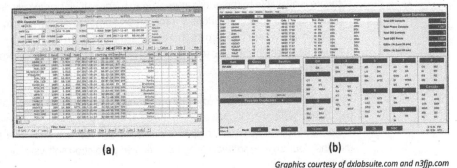

FIGURE 14-3:
(a) *DXLab Suite* is a general purpose logging program; (b) N3FJP's *Field Day Log* is a contest-style logger.

(a)   (b)

Graphics courtesy of dxlabsuite.com and n3fjp.com

TIP

Online logging has become more and more popular, and the most used service as of 2020 is Club Log (clublog.org/about.php) designed by Michael, G7VJR. Not only can it save your QSO information, but it also provides analysis, support for award programs, and helpful operating tools. Look for more of these services in coming years, although Michael is doing an awfully good job right now!

Contest operation is a little more demanding so the software is focused on what you need to participate efficiently. There are several "contest loggers" available, but I often recommend N3FJP's software (n3fjp.com) as easy-to-learn and very inexpensive. Figure 14-3 shows the version of N3FJP's logger you might encounter at Field Day. Other excellent choices include N1MM Logger+ (n1mm.hamdocs.com) and WriteLog (writelog.com).

Logging programs can also export data in standard file formats such as ADIF (Amateur Data Interchange Format) and submit your contact information to electronic confirmation services like eQSL.net and the ARRL's Logbook of the World (www.arrl.org/lotw). For contest logs, your software will generate a log in Cabrillo format (www.arrl.org/cabrillo-format-tutorial).

For mobile or portable operation, portable logging apps such as *HamLog* (pignology.net/hamlog.html) or *CQ/X* (www.no5w.com) will run on a smartphone or tablet. Specially designed for mobile operation in VHF+ contests (called *roving*), the program *RoverLog* (roverlog.2ub.org) helps you maximize your score.

PREDICTING PROPAGATION

As you start to make HF contacts, you will fill in more and more of the needed locations to qualify for, say, Worked All States, or even DX Century Club. You'll start to notice patterns of when and where you can make contact — can it be predicted? Yes, and there is some excellent software around to help. One of the best is free, and it produces beautiful graphics. Called *Voice of America Coverage Analysis Program,* or *VOACAP,* this excellent software is available to the public at www.voacap.com. It was originally developed by the National Telecommunications & Information Administration (NTIA) to help the VOA plan when and where to beam its shortwave signals. A team of amateurs led by Finnish ham Jari Perkiömäki (OH6BG) took the software one step further and combined it with maps and graphs.

Figure (a) below is a VOACAP *coverage map* showing where my 30 meter signal might be heard from Jefferson City, Missouri, in the early evening hours of mid-November. There are other maps and charts to help with point-to-point contacts and other types of predictions. Your logging software may have its own prediction software, or you can use other packages

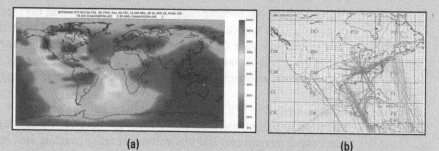

(a)   (b)

VOACAP graphics courtesy Jari Perkiömäki (OH6BG), James Watson (HZ1JW)
and Juho Juopperi (OH8GLV); 6 meter activity map courtesy DXmaps.com

After you've made the prediction, how can you tell if it's accurate? The best way, of course, is to get on the air and make contacts! Many of your fellow hams are doing exactly the same thing and posting what they hear online on spotting network websites like DXsummit.fi and DX Heat (dxheat.com/dxc). All that information can be hard to manage, but EA6VQ has written software to collect it and display it on his DXmaps.com website. Figure (b) here shows a 6 meter activity map from DXmaps.com when the 50 MHz band opened all across the eastern U.S. and across the equator deep into South America. By listening to your radio and watching the online maps, you will "be there" when the conditions are right! There are lots of propagation tools for you to use, as you will find, from newsletters, websites, and other hams.

Submitting a contest log

Perhaps you heard hams exchanging rapid-fire contacts in a contest (see Chapter 11), made a few QSOs, and discovered it was a lot of fun! It won't take you long to have quite a few contacts in your log — why not send your log to the contest sponsors? This helps them cross-check the contacts; you might even get a certificate!

It's pretty easy — here's how to send in your log. Start by generating a Cabrillo-formatted log file. Your logging software does this automatically. Check the software's Help services to find out how. Name the file "*yourcallsign*.log." Use a plain text editor like *Notepad* to view the file and make sure the information is correct. Don't change the file to a word processor format; keep it in plain-text format.

Next, visit the sponsor's web page to find out whether you should email the log or use a web-based upload page. (Most sponsors accept but don't really want paper printouts of your log.) If email is preferred, attach the log file to the email with your call sign in the Subject line. Otherwise, use the log upload web page and follow its instructions.

After you have submitted a log, you may receive a confirmation message from the sponsor. Save that message in case there are any problems in processing your log. The sponsor may have a claimed-score or logs-received web page as well. If your call is listed there, the log has been received and will be checked and scored. All you have to do is wait for the results!

TIP

It is a good idea to create a folder on your computer just for contest logs. Save the Cabrillo log in the folder and rename it something like "*Yourcall – Date – Contest*," such as "NØAX–Feb 2020–CW Sprint." If you get a confirmation message, save it in the same folder for future reference until the contest results have been published.

THERE'S AN APP FOR MAPS

Hams *love* maps — if you have even a passing interest in geography, you will find ham radio to be a very friendly hobby. There are quite a few reasons for this high level of interest. First, you need to know where signals are coming from. Second, you need to know where *your* signals are going to! Maps are important to help you orient antennas properly so that you are heard where you need to be heard.

The *azimuthal-equidistant* map in the figure below may look unfamiliar, but it's common in ham radio. Centered on your location, this kind of map squashes the whole world

(continued)

onto a disk the outer edge of which is halfway around the planet — your *antipode*. A straight line starting at your location shows the path a radio wave will likely take. For example, from the central point of the map in the figure (near Chicago, Illinois) pointing your antenna straight east sends your signal not to Europe (as you might expect from a rectangular Mercator projection map) but across Africa! To a radio wave, Congo in the heart of Africa is straight east of the U.S. If propagation is good, that signal might keep going across the Indian Ocean and reemerge on the opposite edge to reach Australia and then the Pacific Ocean. Who knew?

Each of the labels in the map in the list on the left is the call-sign prefix for hams in that country. In addition, you can see a symbol for the Sun in the South Pacific (near VP6 or Pitcairn Island, home to the descendants of the Mutiny on the Bounty mutineers). The map also shows where the world is in daylight or darkness, also important for what happens to your signals. This useful map is just one of the features of *DX Atlas* software by VE3NEA (dxatlas.com). Another great map package, *Mapability*, is produced by EI8IC (www.mapability.com/ei8ic). You'll find lots of ham radio mapping software to enjoy as you chase awards, engage in contests, or simply listen to the world turning.

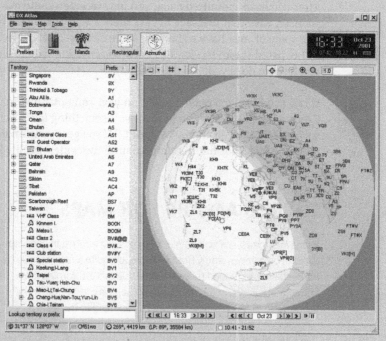

Courtesy Afreet Software, Inc.

Confirming Your Contacts

After you've logged the QSO, you may want to get a formal confirmation of the contact for an award. It's easy to do this electronically using your computer logging program. Or you might want to exchange the ham radio equivalent of business cards, the *QSL card*.

QSL cards

QSL cards, which are the size of standard postcards, range from simple to ornate. DXpeditions (see Chapter 11) often creates a multi-panel folding cards with lots of information and pictures from the rare location. QSLs are primarily exchanged for HF contacts and are used to qualify for operating awards. Exchanging QSL cards — called *QSLing* — is one of my favorite activities, and I've built a lifetime collection of several thousand from all over the world.

TIP

With all of the online contact confirmation services, you don't need to exchange QSLs for every QSO. I like to get a paper QSL for a first contact but then rely on the electronic systems after that.

The usual format for QSLs is a photo or graphic with the station's call sign on the front (see Figure 14-4). Information about a contact is written on the front or back, depending on the design. The card is then mailed directly to the other station or through an intermediary.

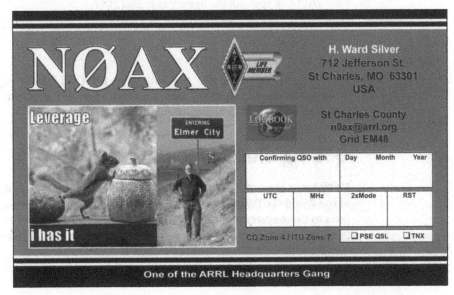

FIGURE 14-4:
One of my QSL cards.

You can find many varieties of QSLs, but three basic rules can help make exchanging them as quick and error-free as possible:

>> Have your call sign and QSO information on one side so that the receiver doesn't have to look for it. The front of my card has all the information.

>> Print your call sign and all contact information in a clear, easy-to-read typeface or in capital letters. Artistic lettering can be hard to read!

>> Beyond your mailing address, make sure that the QSL shows the physical location of the station, including county (for U.S. stations) and four- or six-character grid square (see Chapter 11).

You can find advertisements for QSL printers in ham magazines and on websites such as ac6v.com/qslcards.php. You can design your own cards and have them printed at a local print or copy shop. Using a photo-quality printer, you can even print each one at home. If you make a LOT of contacts, there are even QSL services that will take your log and send a printed card to each station!

Follow these suggestions for accuracy:

>> **Double-check the dates and times of your contacts.** Date and time are frequent sources of error. Start by making sure that your own clock is set properly. Use UTC or World Time for every QSL except those for local contacts.

>> **Use an unambiguous format for date.** Does 5/7/17 mean May 7 or July 5? The date is crystal-clear if you show the month with a Roman numeral, as in 7/V/17, or spell out the month, as in 5 Jul 2017.

>> **Use heavy black or blue ink** that won't fade over time. Never use pencil that can be erased or altered.

TIP

Most logging software has the ability to print all of the necessary contact information directly on a QSL card. Some packages will even print the entire card with a nice image and the log information in one operation.

QSLing electronically

Many hams are confirming their contacts on two sites: eQSL and ARRL's Logbook of the World (LoTW). Your logging software may even be able to upload your contacts to these systems automatically as you make them. With these systems, there is no need to exchange paper cards although many hams send a card for a first contact with a station for their collections.

eQSL (www.eqsl.net) was the first electronic QSL system and is extremely easy to use. Its site has a tutorial slideshow that explains just how eQSL works and how to use it. eQSL offers its own operating awards, as well, verified by contacts uploaded to the eQSL system.

The ARRL's LoTW (www.arrl.org/logbook-of-the-world) is more complicated to use. You're required to authenticate your identity and license, and all submitted contacts are digitally signed for complete trustworthiness. LOTW provides electronic verification of QSOs for award purposes. It currently supports the ARRL awards and CQ's WPX and WAZ award programs.

TIP

DXpeditions often uses an online QSLing system such as Club Log's OQRS system (secure.clublog.org). You can support the expeditioners with a donation and request your QSL at the same time. It's speedy, secure, and highly recommended.

Direct QSLing

If you want to send a paper card, the quickest (and most expensive) option is *direct*, meaning directly to other hams at their published addresses. You can find many ham addresses on QRZ.com. This method ensures your card will get to recipients as fast as possible and usually results in the shortest turnaround time. Include the return postage and maybe even a self-addressed, stamped envelope. Direct QSLing costs more than electronic QSLing but makes it as easy as possible for you to get a return card on its way from the other ham — many times, with a colorful stamp.

TIP

The "Forever" stamps at the post office are guaranteed to be worth one unit of postage, even if the cost goes up in the future. That's a good deal!

WARNING

Postal theft can be a problem in poorer countries. An active station can make hundreds of contacts per week, attracting unwelcome attention when many envelopes start showing up with those funny number–letter call signs on them. Don't put any station call signs on the envelope if you have any question about the reliability of the postal service. Make your envelope as ordinary and as thin as possible. If the station gives QSL instructions online or during the contact, be sure to follow them!

Using QSL managers

To avoid poor postal systems and cut postage expenses, many DX stations and DXpeditions use a QSL manager. The manager is located in a country with reliable, secure postal service. This method results in a nearly 100 percent return rate. QSLing via a manager is just like direct QSLing. If you don't include return postage and an envelope to a manager for a DX station, you'll likely get your card back via the QSL bureau (see the next section), which takes a few months at minimum. You can locate managers on websites such as QRZ.com's QSL Corner, which is free to members (`www.qrz.com/page/qsl-corner.html`). If the station has a website or has posted information on the QRZ.com page, a manager will usually be listed there. DX newsletters like the Daily DX (`www.dailydx.com`) and the OPDX Bulletin (`www.papays.com/opdx.html`) are also good sources of information.

If you send your QSL outside the U.S., be sure to do the following:

>> Use the correct global airmail letter rate from the U.S. Postal Service website (`postcalc.usps.com`).

>> Ensure airmail service by using an Air Mail sticker (free at the post office), an airmail envelope, or an Air Mail/Par Avion stamp on the envelope.

>> Include a self-addressed envelope (include USA after your state and Zip code).

>> Include return postage by enclosing a couple of dollars with your card. Be sure the currency isn't visible — place it inside the self-addressed envelope.

Greenstamp is slang for a $1 bill.

Bureaus and QSL services

All that postage can mount up pretty quickly. A much cheaper (and much slower) option exists: the QSL bureau system. You should use this method when the DX station says "QSL via the bureau" or on CW and digital modes, "QSL VIA BURO." The QSL bureau system operates as a sort of ham radio post office, allowing hams to exchange QSLs at a fraction of the cost of direct mail.

If you are an ARRL member, you can bundle up all your DX QSLs (you still have to send domestic cards directly) and send them to the outgoing QSL bureau, where the QSLs are sorted and sent in bulk to incoming QSL bureaus around the world. The cards are then sorted and distributed to individual stations. The recipients send their reply cards back in the other direction. Go to `www.arrl.org/qsl-service` for more details. To get your cards, you must keep postage and envelopes in stock at your incoming QSL bureau. (Anyone can use the incoming QSL bureaus.) Then, when you least expect it, a fat package of cards arrives in the mail. What fun!

You can watch a nice video about the ARRL Outgoing QSL Bureau by its current manager, Rose-Anne Lawrence, KB1DMW, at `https://www.youtube.com/watch?v=Vxgcn3sMVtU&ab_channel=ARRLHQ`.

Applying for awards

Each award program has its own method for submitting QSL cards to qualify for an award. All of them have a few things in common, though. There is a form to fill out listing each contact individually. For more than a few contacts, you'll need to enter the information in alphabetical order by prefix. For example, a contact from KX9X will be listed before N1EUZ before WBØGQP. (For DX prefix order, use the ARRL's most-current DXCC List at www.arrl.org/country-lists-prefixes). Print clearly so the award manager does not misread your information. Pay the award fee, if any, with a check or money order or electronically if that option is available. (Don't send cash unless it is necessary.)

Next, sort the cards into the same order as on the form. Orient them with the contact information facing up, even if it is on the back of the card. Bundle the cards together so that the top card is the first on the application form. You then mail or ship the cards to the award manager as directed by the award's sponsor. If you are sending a lot of cards or if the cards are particularly rare, send the cards by certified mail or with a signature-required service.

Don't forget to include return postage or shipping costs in your award fee. It is also a good idea to include a self-addressed postcard with the application that the sponsor can return so you know the package was received. Assuming all your information checks out, you'll receive your certificate and QSL cards!

TIP

Cards for the ARRL DXCC Award can be checked by a local or regional "field checker" (www.arrl.org/dxcc-field-checking). You can make arrangements to attend a club meeting and submit your cards to him or her directly. Card checkers often have a booth or table at the larger hamfests. Other award programs may allow you to submit a list of QSOs according to General Certification Rules (GCR — www.ka5wss.com/posts/gcr-general-certification-rules) and not have to send in the actual cards.

Chapter **15**

Operating Away from Home

The equipment available to today's hams for operating away from home is the best ever; it's small, it's lightweight; it's efficient. This makes it easier than ever to have fun with ham radio from your car, campsite, or even while taking a hike! If you like being outdoors, visiting new places, and taking the occasional road trip, you can mix in a little ham radio! This chapter introduces you to the first steps and points you at resources to take you the rest of the way.

Mobile Stations

"Goin' mo-byle" has returned to ham radio in a big way — hams are activating rare counties, parks, and grid squares from mobile stations as good as those at home! State QSO parties and OTA ("on the air") programs have created a great demand for the mobile-er. Whether you have a station at home or not, operating on the road offers challenges and opportunities all its own.

HF mobile radios

Recognizing the rapid growth in mobile and portable operation, most manufacturers offer small, rugged radios. Each year, more bands and better features are crammed into these amazing radios. These radios are quite capable as base or fixed stations if you want to use the radio at home, too. Many radios include coverage of VHF and UHF bands on the weak-signal modes (SSB and CW) as well as FM.

TIP

An excellent website for mobile operators is Alan Applegate (KØBG)'s website for mobile operators (www.k0bg.com). The site covers everything from powering your mobile rig to using it effectively.

Because they're so small, these radios must make some compromises to save space. The operator interface is constructed as a series of menus. This makes some adjustments a little harder to get to, although the most-used controls remain on the front panel. These smaller rigs don't include internal antenna tuners at the 100-watt output level, as many larger rigs do.

As when evaluating these radios' larger base-station siblings, you have to consider features and accessories. Because these radios have a minimal set of controls and are menu-driven, I recommend that you try one before you buy it. If you don't have a friend who owns one, you can often find an owner online, maybe even with a video to share, who will answer basic questions.

What you're looking for in a mobile/portable station is one of the small all-band radios introduced in Chapter 11. The typical model produces 100 watts output on all amateur bands between 1.8 and 50 MHz. If VHF/UHF is covered, you can expect around 50 watts on 144 MHz and 25 watts on 432 MHz. The radio has a detachable *control head* or *faceplate* that can be securely mounted near the operator. The radio itself is typically installed under a seat or in the trunk. Figure 15-1 shows the Yaesu FT-891 HF+6 meter transceiver; you can also try the Icom IC-7100, both currently available.

TIP

Connect the radio headphone output to your vehicle audio system's AUX input with a short stereo audio cable. You'll find high-fidelity mobile hamming to be easy on the ears.

Mobile operation in an RV works just as well as from a car. You can set up the gear in the driving area as a regular mobile station or back in the living area like a small home station. Try searching online for *rv ham radio station* for a wide range of examples.

FIGURE 15-1:
The FT-891 transceiver is designed for mobile and portable operation on HF and 6 meters.

WARNING

The extra space in an RV can lead you to set up your equipment as if it were in a stationary home. Don't let a "shakedown cruise" become a literal description! All of your gear and accessories must be safely stowed or secured whenever the RV is moving, both for your safety and to protect the equipment.

If you have an RV or camping trailer and adding ham radio sounds like fun, you should put the annual Quartzfest week-long ham radio camping hamfest on your to-do list (www.quartzfest.org). It combines living outdoors in the Arizona desert — in January — with a week of ham radio fun. There are lots of seminars and demonstrations so you'll learn a lot about ham radio on the road!

Maybe you'd like something less tied to an automobile or RV. Well, how about a bicycle mobile station? The bicycle mobile station shown in Figure 15-2 was built by Christian Bravo, W4ALF. He built the station on his 16-inch folding bike (it even fits in his trunk!) with an Elecraft KX2 QRP transceiver and Buddipole antenna coils and whips mounted on the luggage rack. The counterpoise extending behind the rear wheel is a fiberglass fishing rod with 20 feet of #18 AWG wire coiled along it. Christian uses the *HamLog* app on his phone as his logging software. He's active in the Summits On the Air (SOTA) program with this lightweight and mobile station.

Mobile installations

Driving your station creates its own set of unique considerations. Because vehicles come in so many styles, every installation is a custom installation. Leave some of your radio budget for automotive fixtures and wiring. You may find it prudent to spend a few dollars for a professional shop to make recommendations about power wiring, how to route wires and cables in your vehicle, and safe installation methods.

FIGURE 15-2:
This folding
bicycle-mobile
station was built
by Christian
Bravo, W4ALF, to
operate on the
HF bands.

Courtesy American Radio Relay League

WARNING

Lease agreements may prohibit modifications to the vehicle or hold you responsible for repair costs. If you share the vehicle, get agreement on where to place the radio before drilling any holes.

Where can you fit a radio in your vehicle or boat? If you have an RV or a yacht, you may not have a problem, but in a compact car or an 18-foot runabout, the space issue can be a challenge. Luckily, many radios designed for mobile use, such as the IC-7000 shown in Figure 15-3, have detachable *control heads* or *faceplates*. A cup-holder flex-mount designed to hold phones or tablets can also hold a small control head.

WARNING

Poorly secured radio gear can be *lethal* in an accident. Please don't drive without making sure your equipment is securely mounted. Anything free to move inside the vehicle is a hazard to you and your passengers.

As you can see, this arrangement provides lots of functionality in a small space. This mobile station can operate on the HF bands and several VHF/UHF bands with all controls right at the driver's fingertips. (The small unit below the control head is a controller for an adjustable antenna.)

FIGURE 15-3:
Most mobile radios have detachable lightweight control heads or faceplates that are easy to secure on flexible or dashboard mounts.

Courtesy American Radio Relay League

TIP

If you can borrow a mobile radio before buying it, make sure that the control head can be mounted in your vehicle where the display is easy to see and the controls are easy to operate. Don't build in a driving distraction! Use hook-and-loop adhesive tape to try out the mounting locations without drilling any holes.

A detachable control head allows you to put the body of the radio under the dash or a seat or on a bulkhead. The radio is attached to the control head with a special cable available from the manufacturer in various lengths. The combination of cable and mounting hardware for the radio and control head is called a *separation kit*. The separation kits can be expensive purchased separately but may be available at reduced cost if purchased with the radio.

TIP

Make sure that the separation kit cable is long enough to connect the control head to the main radio unit, wherever you decide to mount them. Include extra length to run the cable under mats and around obstacles.

WARNING

Don't install a mobile radio control head (or anything else) on or near a panel that conceals an airbag! If the air bag deploys, the control head will become a projectile. A heavy radio or improper mounting might prevent the airbag from deploying properly, leading to injuries inside the vehicle. Take care to keep cables and microphones away from airbag deployment areas as well. Check the vehicle's service manual for the locations of *all* airbags, which may be on the door panels and other unexpected locations. Take caution when working around the airbag systems!

TIP

Multimode radios are very popular for the special type of mobile contesting called *roving*. One of these small rigs and a *transverter* or two for other bands makes for a lot of fun as you rove from grid square to grid square, racking up points.

It can be "interesting" trying to find a solid connection to vehicle power that can supply enough current for your radio at full power. A 100-watt rig can draw as much as 25 amps at peak output. This is far too much current to use the typical auxiliary power jack, also known as a cigarette or cigar lighter. Those are usually rated at less than 10 amps — check the vehicle's manual. You will have to find a connection rated for the load.

The most robust power connection is directly to the battery. (Newer cars and trucks have a battery charge control system — see your owner's manual or consult the dealer for the proper point at which to connect your radio.) To prevent a short circuit from finding a path through your radio, be sure to include a fuse in *both* the positive *and* negative leads. Make sure your connections are secure so they won't come loose in the car's harsh mechanical environment. Use anti-corrosion compound on the terminals because the connection will be exposed to corrosive chemicals, salt, and water.

TECHNICAL STUFF

Vehicles sold in the past few years usually have a *battery management system* (BMS) that monitors the charge of the battery using a sensor in the battery's negative lead. So you don't affect BMS operation, connect the radio's negative lead to the right place in your vehicle's wiring. This is usually where the battery's ground strap connects to the engine block but there may be more convenient points. Check with the dealer first.

Each vehicle is different in the way cables have to be routed. If you can find protected access holes and cable trays, usually under removable panels or covers, use them because they keep the cables from being abraded or pinched, which can lead to damaging short circuits. Fight the temptation to make a hasty or unsecured connection to vehicle power. Speaking from personal experience, a loose connection or wire can cause a lot of trouble in a big hurry. Be safe and do things right.

TIP

Cars and pickup trucks usually have extra holes in the firewall between the engine and passenger compartments. The holes are usually covered by a rubber plug and may be behind carpet or protective panel.

If you are unsure of how to mount your radio properly, an automotive sound shop can usually do the job. They might even be able to do the whole radio system installation at a reasonable cost.

Mobile antennas

Obviously, you won't be able to drive, boat, or cycle around with a full-size HF antenna for most of the bands. Quarter-wave whips for 10, 12, and 15 meters are 8.3, 9.4, and 11 feet long. Mounting the antenna on the vehicle adds at least a foot or two to the overall height. Because maximum allowed vehicle height is 13 feet, 6 inches, longer antennas will exceed that limit and become increasingly impractical.

For mobile operation on the HF bands, many hams use Hamstick-type antennas attached to a mag-mount or a permanent mount attached directly to the vehicle, as shown in Figure 15-4. These antennas consist of wire wound on fiberglass tubes about 3 feet long, with a stainless-steel whip or "stinger" attached at the top. The antennas work on a single band and are sufficiently inexpensive that you can carry a whole set in the car. (A mailing tube or fishing rod carrier works well to carry several antennas.) You will have to change the antenna to use a different band. Another design uses resonators attached to a permanent base section to operate on different bands. The resonators and fiberglass whip antennas use a standard 3/8-24 threaded mount.

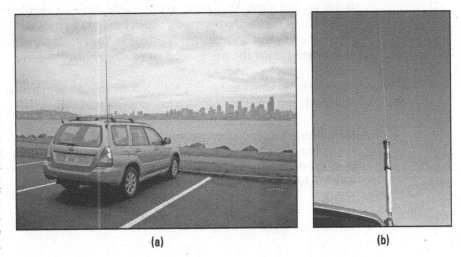

FIGURE 15-4: (a) A Hamstick-type mobile antenna mounted on the NØAX mobile along with a small VHF/UHF dual-band whip; (b) a screwdriver-style adjustable antenna.

(a)

(b)

An adjustable design that has become popular in recent years allows the antenna to tune over nearly any HF frequency. (Antennas of this type are known as *screwdriver* antennas because they were first made with DC motors similar to those in battery-powered electric screwdrivers.) The antenna is similar to a Hamstick — its base section is a coil with a whip attached at the top. The coil is adjustable, however, with sliding contacts that are controlled from inside the vehicle. The coil is adjusted for minimum SWR.

The antenna also has a small switch in its base that closes temporarily once with each turn of the motor. Screwdriver antenna controllers count the number of switch closures as the coil contacts move up and down. This way you can store *pre-sets* to position the antenna automatically. Advanced controllers can use the radio's computer interface to read the operating frequency and set the antenna accordingly. This is more convenient than swapping an antenna since you can adjust the antenna without stopping or getting out of the vehicle.

Hamstick-type antennas are least expensive, and screwdriver antennas are most expensive. Performance varies dramatically, depending on mounting and installation, so predicting which gives the best results is difficult. An advantage of the screwdriver antennas is that they can be adjusted to suit the circumstances even if something is close to the antenna or wind is bending the whip section.

Mobile signals are generally stronger at 20 meters and up. The need to keep antennas small makes operating on the "low bands" (10 MHz and below) more of a challenge.

The type of antenna mount you select depends on where it can be mounted and how your vehicle is constructed. HF antennas are pretty large, and it takes a three-magnet design for a mag-mount to be secure at highway speeds. Adjustable trunk-lip and hatch-lip mounts are available, such as the Diamond K400 (www.diamondantenna.net/k400.html), which can hold a good-sized mobile antenna, even one of the smaller screwdriver antennas. You should try to mount the antenna so it extends above the trunk, hood, or roof.

To avoid scratching paint with a mag-mount you can buy circular magnet pads or in a pinch, use a plastic sandwich bag beneath the magnet.

No matter what kind of mount you use, there should be a solid connection from the mount to the body of the car. The connection may be made by the mount itself, such as by set-screws. If a mag-mount is being used, the connection can be made with a heavy wire attached to a body screw somewhere nearby, such as in a door frame.

WARNING

Mobile antennas depend on the vehicle's metal body to provide a "ground" or counterpoise for proper operation. Many vehicles are made with plastic or composite panels, hoods, trunk lids, and so forth. These look just like metal, but a magnet-mount antenna will not "stick" to them. Aluminum panels have the same problem. Clamp-on or drilled-hole mounts depend on being attached to a metal surface to work. A body shop can help you find a metallic support such as a roof support column or door frame.

TIP

To get some good ideas for mounting antennas, go to a hamfest and take a walk through the parking lot looking for mobile antennas. See how they are installed and with what type of mount. If you can find a vehicle like yours, even better! Take photos of good mounting ideas for reference later on.

When running the feed line between the mount and the radio, be sure it will not be pinched or worn by moving seats or doors. If the mount is on a hatchback or swinging door, leave a loop of cable to allow for the motion. Running the feed line beneath weather stripping helps protect it, as well.

WARNING

Be very careful when drilling into a vehicle panel or pillar. It is common to run cables and wires inside hollow body parts. Drilling into a wire can create a short circuit or break a control or signal cable. Limit how far the drill bit can extend into the hole and inspect with a flashlight to see if there are wires present before enlarging and using the hole. Similarly, use the shortest screw length to avoid touching a hidden wire or cable.

If adjusting your antenna doesn't produce a low-enough minimum SWR, you will probably have to try adding a matching component or antenna tuner. The MFJ-909 or -910 (www.mfjenterprises.com) may be all you need. If that doesn't do the job, you will need an adjustable antenna tuner, either manual or automatic, for the transmitter to output full power.

Portable Operating

Portable operation is getting more popular every week. The "self-contained" style involves carrying or packing the entire radio package, including a power source, to the location where you plan to operate. You can hike, bike, or paddle to your station site, combining outdoor action with ham radio! Even if hauling the gear all by yourself isn't your cup of tea, setting up a small station from a scenic location is a great way to enjoy ham radio, as you can see from Paul, W6PNG's setup by the Golden Gate bridge in Figure 15-5.

FIGURE 15-5:
W6PNG operated
this portable
station from the
Golden Gate
National
Recreational
Area.

Courtesy American Radio Relay League

TIP

To help you set up an operate from portable station, the ARRL publishes "Portable Operating for Amateur Radio" by KB1HQS (www.arrl.org/shop/Portable-Operating-for-Amateur-Radio). It's a great book that answers a lot of questions and gives you pointers to be successful.

Planning is important for portable operation to be a success. If you are going to be vehicle-based, you can haul considerably more gear. Will your power source be a battery, the vehicle battery, commercial AC, AC from an inverter connected to a battery, or an AC generator? How about operating tables or tent? Fabric camp chairs can be awfully uncomfortable for operating a radio on a table — a plastic folding chair will be more comfortable than a fabric chair for operating. Before setting out, set up in your backyard and take the station for a spin so you can see what works and what doesn't. The author operated from Emerald Island, North Carolina during a July contest (see Figure 15-6) and managed to bring a fan along, too!

If you're going to be hauling your gear, treat it like a backpacking trip. Start by assigning yourself a total weight budget. Get creative with antennas and accessories to maximize your options for the radio and power. Some amazingly small and lightweight radios are available. These radios aren't always the easiest to operate, however. If you're just starting, you may want to pass up a minimal radio in favor of one that's easier to operate and has more features until you know more about operating. When you have more experience, you'll know what features you can do without. Practice setting up and using the gear at home so you're not trying to learn how it works while swatting mosquitos!

FIGURE 15-6:
The author operating from a campground on Emerald Island, NC.

When you are just starting, concentrate on one or two bands. On HF, the 14, 17, and 21 MHz bands are favorites with low-power and portable operators. These bands are active for a large portion of the day, and the antennas are small enough to carry easily. 7 MHz and 10 MHz open up in the mid- to late afternoon and are active all evening. Picking one band from each group is a good beginning strategy.

On VHF, 50 MHz and 144 MHz operation from high elevations is common. Plenty of operators are available, particularly during weekend contests, and those bands often feature interesting propagation.

LET'S GET OTA!

OTA? It means "On the Air." You will find a lot of OTA activities these days. You can be the station OTA (called "activating") with other stations trying to contact you! That's a great way to have some fun almost any day of the week. Or maybe you prefer calling the OTA stations to get those elusive contacts and qualify for an award? Let's see what's out there, OTA-wise:

One of the fastest-growing activities is the Summits on the Air (SOTA; www.sota.org.uk) award program. It's a favorite with backpackers and hikers like Steve Galchutt (WGØAT), who clamber to the top of hills and mountains, put flea-power stations on the air for a little while, then head back down. WGØAT is famous for activating SOTA summits using a QRP station carried by his pack goats, Boo and Peanut. (See figure.)

An original, the Islands On the Air program (IOTA: www.iota-world.org) keeps track of contacts with salt-water islands and island groups all around the world. If you are a geography or map buff, this is an especially great award program. There are literally thousands of these islands, and some are activated every week. The U.S. Islands program (USI: usislands.org) includes freshwater and coastal saltwater islands.

You don't have to go far to participate in the Parks On the Air (POTA) program (parksontheair.com and www.facebook.com/groups/parksontheair). You can operate from parks, historical trails, conservation areas, monuments, and the list goes on. This fast-growing program finds portable stations ("activators") handing out contacts every day on the HF bands.

If you're a ham in your teens or twenties, check out the world-wide Youngsters On the Air program (YOTA: ham-yota.com). This very active group sponsors conferences in the summer each year and several special events with -YOTA call signs. Japan is home to the newest YOTA group, and similar groups are forming here in the Western Hemisphere, too. Start your own YOTA group and join the fun.

Scouts have their own OTA program, too. Listen for Jamboree On the Air (JOTA: www.scouting.org/international/jota-joti/jota) each year on the third weekend of October as thousands of scouts around the world sample ham radio.

And let's not forget GOTA or Get On the Air — which is part of ARRL Field Day every year. GOTA stations are "free" and encourage new hams (maybe like you?) to get on the air and make some contacts. If you're a GOTA veteran, why not help them out next year?

Courtesy Steve Galchutt

Portable antennas

At VHF and UHF, portable antennas are very small, lightweight, and easy to pack and carry. Even for the 6 meter band, a three-element Yagi or 2-element quad can be quickly disassembled and carried in a gym bag. On higher-frequency bands, longer, higher-gain antennas are practical, as well. Two or three sections of a painter pole or telescoping tubing are sufficient to hold the antenna with a minimum of guying. Three light-duty ropes and some tent pegs are sufficient to hold up most antennas and masts.

At HF, however, the larger antennas are more difficult to deal with. You can try a lightweight wire antenna if you can find a way to support it well above the ground. Trees or lightweight fiberglass masts are your best choices. An *end-fed half-wave* (*EFHW*) only needs one support and can be used on several bands. Vertical antennas need a sturdy base, often a set of guy ropes, and usually a set of wires to make a ground screen.

A more convenient choice may be the portable antennas designed exactly for this type of operation. The Buddipole designs (Figure 15-7a; www.buddipole.com) are well-known and easy to set up. Each of the horizontal sections is adjustable so you can create a dipole for the 40 through 2 meters bands. Turned vertically, the TW vertical dipole (Figure 15-7b: www.dxengineering.com/parts/dxe-tw-20101-p) can be adjusted to cover several bands. The *magnetic loop* antenna is another popular choice. These antennas are small and can be supported by a tripod which makes them easy to put up, even in the wind.

FIGURE 15-7:
(a) Chuck Greeno, WA7BRL, using a Buddipole portable dipole antenna to activate a lighthouse; (b) A TW portable vertical dipole at a scenic overlook — a great radio location.

(a)　　　　　　　　　(b)

Courtesy Chuck Greeno

TIP

Small, portable antennas are a compromise, trading performance for convenience. If you can put up a higher, longer antenna, do so. You'll find the performance to be much better.

Portable power

There are several things to consider when choosing a power source with size, weight, and operating time (or capacity) usually being the most important. Whether you are driving, biking, or walking to the station site makes a lot of difference! Here are the most common options:

>> **Lead-acid storage:** Large and heavy but these batteries can supply a lot of power. When fully charged, they can run a QRP transceiver for an entire weekend.

>> **Lead-acid gel-cells:** Available in a wide range of sizes, these are still heavy but smaller than car or tractor batteries; can be trickle-charged.

>> **Li-ion:** Multi-cell battery packs are fairly lightweight and can be charged quickly; good for QRP operations; must have a Li-ion charger.

>> **Li-iron-phosphate (LiPO):** Better performance than Li-ion and constant voltage output, but expensive; must have an LiPO charger.

>> **Solar panels:** Bulky at the size needed to run a radio directly; can be combined with a battery pack and charger; combination panel-charger-battery units available.

>> **Generator:** Small generators like the Honda EU-series are very popular. Quiet and efficient, they can run a 100-watt station all weekend on less than 5 gallons of gasoline. The 1000-watt EU-1000i weighs about 30 pounds (not counting gas and cords) so it is suitable if you're driving.

TECHNICAL STUFF

When you're not using an AC-powered supply, you need to use equipment that can operate from a wide range of voltages. Batteries discharge or a cloud can shade a solar panel. Most 100-watt transceivers need to be supplied with 13.8 V plus-or-minus a couple of volts. Low-power QRP radios are more tolerant but you still need to keep the input voltage above the specified minimum. When power supply voltage drops to the minimum, the radio may begin operating erratically or transmit a poor-quality signal.

Field Day

The annual ARRL Field Day (www.arrl.org/field-day) is one of ham radio's largest events. (See Figure 15-8.) Held on the fourth full weekend of June, more than 50,000 hams from around North America are involved in some years. Clubs, informal groups, and individuals are all involved in this annual exercise. See the Field Day web page for special rule changes during the pandemic.

Field Day operating frequently involves disassembling and moving equipment from a fixed station to a temporary location and operating for the weekend. Assuming you can drive to the site, this kind of portable operation can be approached much like building a fixed station, but with the expectation that it will be disassembled after a couple of days.

Much of a successful Field Day revolves around antenna selection and installation. It's a challenge to erect effective HF and VHF/UHF antennas using temporary supports. You have to choose between quick-but-low and difficult-but-high antenna systems. Instead of heavy towers, you'll have to make do with masts. Can you really get a line over that high branch in a tree, and will it stay there for the weekend? Remember that a compromise antenna that can be put up and put on the air makes more contacts than a complicated "Field Day Special" that turns out to be a little too special to stay up!

FIGURE 15-8:
Members of the St Charles Amateur Radio Club (KOØA) during the Field Day night shift.

Field Day is a great opportunity to see if you are really prepared to respond following a disaster or to provide effective public service. Is all the gear ready to go? Can you find all the accessories? Does it all work when away from home or a club station? These are good questions to answer *before* you need to respond to a real emergency.

TIP

Many public service teams treat Field Day as an "incident response," including organizing the event according to the IMS plan that a real served-agency would require. (See Chapter 10.) An emergency operations center (EOC) or communications trailer station can be put to the test during Field Day — there is even a special category for them in the competition.

Field Day "gotchas"

If you're going to be operating with more than one radio being active at a time, take along some *band-pass filters* to keep Radio A from interfering with Radio B. Suffering through local interference all weekend is stressful and makes it darned hard to have a good time. Try to keep the antennas for each radio far apart and don't point them at each other.

Whether just using 100-watt transceivers or adding amplifiers, having "RF in the shack" can be a real problem. Make sure you bring some ferrite RFI suppressors, follow good cabling practices, and pay attention to bonding of the equipment and computer.

TIP

A sheet of aluminum foil on the operating table under the equipment can serve as an RF bonding bus. Just use clip leads to connect the enclosures to the foil. This helps prevent RF "hot spots" and RFI from the transmitted signal.

Assuming you're using AC power, don't scrimp on AC safety. Be sure the generator is protected by a GFCI (Ground-fault Circuit Interrupter) outlet or build your own (kits for outdoor GFCI outlets are available in the electrical department of home improvement stores). Check the ground and neutral wiring of extension cords and power strips.

Because Field Day tends to attract larger antennas, don't get sloppy about putting them up. "Walking up" a tower or mast with an HF antenna attached can be perilous — keep the center of gravity between the lifters and the base of the tower, which should be securely held down. Don't let anyone climb a poorly guyed tower! Watch out for power lines and other hazards — remember that you're not familiar with the area.

Protect yourself and visitors by clearly marking feed lines and power cables, antenna wires, guy wires and stakes, fuel cans, batteries, and other safety hazards. Assign one member of the team to be a "safety captain." Keep vehicles away from the stations and antenna systems. And watch the weather — rain, wind, and lightning can appear quickly.

TIP

Bring a weather-alert radio and tune in the nearest NWS weather station. If severe weather is on the way, you'll get an alert. A good second choice is to use a VHF handheld to monitor the NWS broadcasts.

IN THIS CHAPTER

» **Acquiring tools and components**

» **Maintaining your station**

» **Troubleshooting**

» **Repairing your equipment**

» **Building equipment yourself**

Chapter **16**

Hands-On Radio

H am radio is a lot more fun if you know how your station equipment works. You don't have to be an electrical engineer or a whiz-bang programmer, but by keeping things running smoothly and dealing with the inevitable hiccups, you will learn a variety of simple skills. As you tackle problems, you'll find that you're having fewer of them, getting on the air more, and making more contacts. Trying new modes or bands will also be much easier for you.

To help you get comfortable with the hands-on part of ham radio, this chapter provides some guidance on the three parts of keeping a ham radio station on the air: making sure your equipment doesn't break often, figuring out what's wrong when it does break, and fixing what's broken.

TIP

Before exploring the insides of your equipment, please take a minute to visit the section on RF and electrical safety in Chapter 13. Ham radio is a hobby, but electricity doesn't know that. I'd like to keep all my readers for a long, long time, so follow one of ham radio's oldest rules: Safety first!

TIP

The chapters in *The ARRL Handbook* on construction techniques and on troubleshooting are great references. The chapter on component data and references is full of useful tables and other data you'll need at the workbench. You might also consider my book *Circuitbuilding Do-It-Yourself For Dummies* (Wiley) — it provides a lot of information about working with electronics and common tools.

Acquiring Tools and Components

To take care of your radio station, you need some basic tools. The job doesn't take a chest of exotic tools and racks of parts; in fact, you probably have most of the tools already. How many you need is really a question of how deeply you plan on delving into the electronics of the hobby. You have the opportunity to do two levels of work: maintenance and repair or building.

TIP

A local electronics parts store is a valuable resource. If you are lucky enough to have a local ham radio dealer, even better! When you're tucked in the middle of an antenna or construction project, you don't want to have to stop and wait for materials to be delivered! My advice: Balance purchasing online with supporting a local business — it's worth it.

Maintenance tools

Maintenance involves taking care of all your equipment, as well as fabricating any necessary cables or fixtures to put it together. Figure 16-1 shows a good set of maintenance tools.

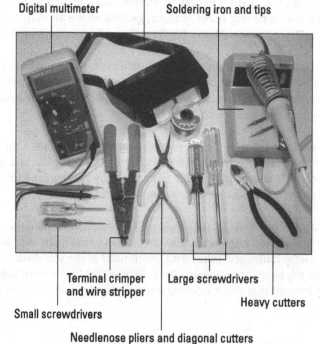

Headmount magnifier

Digital multimeter

Soldering iron and tips

FIGURE 16-1:
A set of tools needed for routine ham station maintenance.

Terminal crimper and wire stripper

Large screwdrivers

Small screwdrivers

Heavy cutters

Needlenose pliers and diagonal cutters

Having these tools on hand allows you to perform almost any electronics maintenance task:

>> **Wire cutters:** Use a heavy-duty pair to handle big wires and cables, and a very sharp pair of diagonal cutters, or *dikes,* with pointed ends to handle the small jobs.

>> **Soldering iron and gun:** You need a small soldering station with adjustable temperature and interchangeable tips. Delicate connectors and printed-circuit boards need a low-temperature, fine-point tip and the station should be certified as static or ESD-safe. (See the note about ESD below.) Heavier wiring jobs take more heat and a bigger tip. A soldering gun should have at least 100 watts of power for antenna and cable soldering. Don't try to use a soldering gun on small jobs or circuit boards.

>> **Terminal crimpers:** Use a real crimper, not pliers — they are not expensive. There are lots of YouTube videos showing how to install crimp terminals (see Figure 16-2) and do the job right the first time. Also use the right terminal size for the wire.

>> **Wire strippers:** Get a good pair of wire strippers with sharp blades for #12 through #24 AWG to handle most jobs. Stripper/crimper combinations are popular.

>> **Head-mounted magnifier:** Electronic components are getting smaller by the hour, so do your eyes a favor. Magnifiers are often available at craft stores. You can also find clamp-mounted, swing-arm magnifier/light combinations.

>> **Digital multimeter (DMM):** Even inexpensive models include diode and transistor checking, a continuity tester, and maybe a capacitance and inductance checker. Some models also include a frequency counter, which can come in handy.

TIP

Electronic kit vendors offer inexpensive "learn to solder" kits. If you haven't soldered before, these are a great way to learn. Most come with a simple soldering iron, too, but you should upgrade to a soldering station as soon as you begin regularly working on electronics.

WARNING

Electronic components can be damaged by static electricity, such as when a spark jumps from your finger to a doorknob — a phenomenon called ESD (for *electrostatic discharge*). Inexpensive mats and ground straps for controlling ESD at your workplace and draining the static from your skin are available from electronic kit and parts vendors.

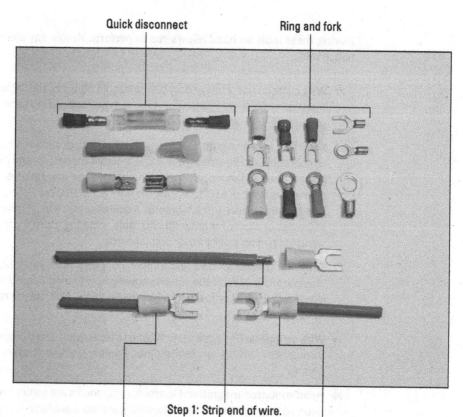

Quick disconnect   Ring and fork

FIGURE 16-2:
The common
types of crimp
terminals.

Step 1: Strip end of wire.

Step 2: Insert wire into terminal.   Step 3: Crimp the terminal on the wire.

You also need to have spare parts on hand. Start by having a spare for all your equipment's connectors. Look over each piece of gear and note what type of connector is required. When you're done, head down to the local electronics emporium and pick up one or two of each type. To make up coaxial cables, you need to have a few RF connectors of the common types shown in Figure 16-3. UHF, BNC, and N. SMA connectors, common on the newer handheld radios, take special tools to install. You'll purchase cables with SMA connectors already installed or adapters, as described next.

TIP

"Reverse SMA" connectors have become common on smaller handheld radios. Be sure you purchase the right type of SMA adapter or cable!

Connectors

UHF-UHF   UHF   F   BNC   SMA   N   N-N

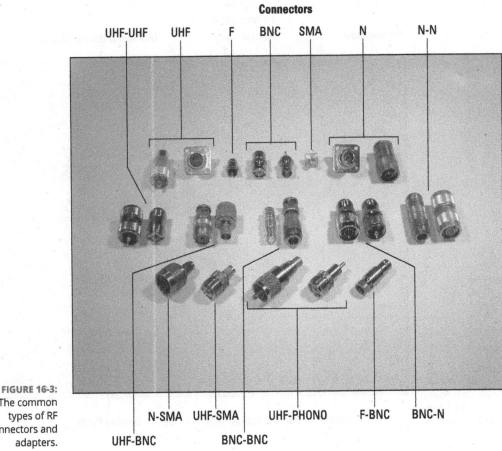

N-SMA   UHF-SMA   UHF-PHONO   F-BNC   BNC-N

UHF-BNC   BNC-BNC

FIGURE 16-3:
The common
types of RF
connectors and
adapters.

Figure 16-4 shows many of the common power connectors used for radios and accessories. Power connectors have large pins and sockets or surfaces to carry the necessary current with low resistance. The audio and data connectors in Figure 16-5 are much smaller. They don't need to carry large currents so the contacts are smaller and more closely spaced.

TIP

Powerpole connectors have become pretty much standard for 12-volt DC power. Get a bag of at least 25 pairs with a simple crimper.

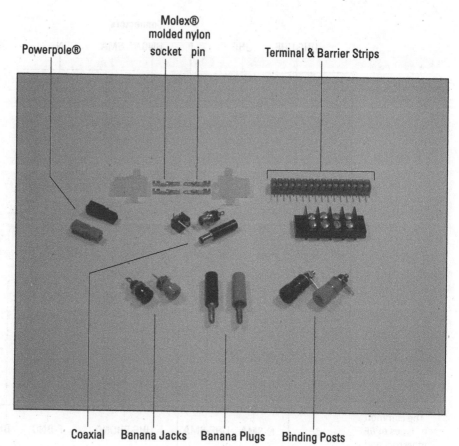

Powerpole®

Molex®
molded nylon
socket   pin

Terminal & Barrier Strips

Coaxial   Banana Jacks   Banana Plugs   Binding Posts

You often need adapters when you don't have just the right cable or a new accessory has a different type of connector. Table 16-1 shows the most common adapter types. You don't have to get them all at once, but this list is good to take to a hamfest or to use when you need to order an extra part or two to make up a minimum order.

TECHNICAL STUFF

A *plug* is the connector that goes on the end of a cable. A *jack* or *receptacle* is the connector that's mounted on equipment. A *male connector* is one in which the signal contacts are exposed pins (disregard the outer shroud or shell). A *female connector* has recessed sockets that accepts male connector pins. A *TRS* connector refers to *tip, ring,* and *sleeve* for stereo phone connectors. A *TRRS* connector has two ring contacts between the tip and sleeve.

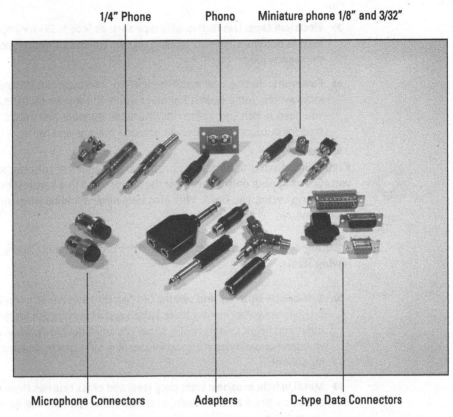

1/4" Phone   Phono   Miniature phone 1/8" and 3/32"

FIGURE 16-5:
Audio and data
connectors.

Microphone Connectors   Adapters   D-type Data Connectors

TABLE 16-1   **Common Adapters**

Adapter Use	Common Types
Audio	Mono to stereo phone plug (¼ inch and ⅛ inch), ¼ inch to ⅛ inch phone plug, right-angle phone plug, phone plug to RCA (phono) jack and vice versa, RCA double female for splices
Data	9-pin to 25-pin D-type, DIN-to-D cables, null modem cables and adapters, 9-pin and 25-pin double male/female (gender benders)
RF	Double-female (barrel) adapters for all four types of connectors, BNC plug to UHF jack (SO-239) and vice versa, N plug to UHF jack and vice versa, male and female SMA to UHF adapter or jumper cable

Along with adapters and spare parts, you should have on hand some common consumable parts:

>> **Fuses:** Have spares for all the fuse sizes and styles your equipment uses. Never replace a fuse with a higher-current or lower-voltage fuse. You'll need glass-cartridge– (3AG) and automotive-blade–type fuses.

>> **Electrical tape:** Use high-quality tape such as Scotch 33+ for important jobs, such as outdoor connector sealing.Use the cheap stuff for temporary or throwaway jobs.

>> **Fasteners:** Purchase an assortment of No. 4 through No. 10 screws, nuts, and lockwashers. Some equipment may require the smaller metric-size fasteners. You need ¼-inch and ⁵⁄₁₆-inch hardware for antennas and masts. Get stainless steel for outdoor jobs, especially antenna building and repair.

TIP

Keep a list of what materials and components that are running low so that when you start shopping online, head for the store, or go to a hamfest or flea market you won't forget what you need. This also helps you avoid buying unnecessary duplicate items.

Cleaning equipment is an important part of maintenance, and you need the following items:

>> **Soft-bristle brushes and swabs:** Old paintbrushes (small ones) and toothbrushes are great cleaning tools. I also keep a round brush for getting inside tubes and holes. Cotton swabs, either the small double-tip version or long, wooden-handle styles are good for cleaning tight spaces and applying solvents.

>> **Metal bristle brushes:** Light-duty steel and brass brushes clean up oxide and corrosion. Brass brushes don't scratch metal connectors but do damage plastic knobs or displays. Don't forget to clean corrosion or grease off a brush after the job.

>> **Solvents and sprays:** Have bottles or cans of lighter fluid, isopropyl alcohol, contact cleaner, and compressed air. Lighter fluid cleans panels and cabinets gently and quickly, and also removes old adhesive and tape. Always test a solvent on a hidden part of a plastic piece before applying a larger quantity.

Repair and building tools

Figure 16-6 shows additional hand tools that you need when you begin doing your own repair work or building equipment. (The figure doesn't show larger tools such as a drill and bench vise.)

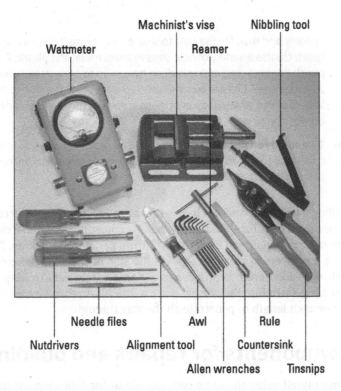

Machinist's vise   Nibbling tool

Wattmeter   Reamer

FIGURE 16-6:
Use these tools
for building or
repairing
electronic
equipment.

Needle files   Awl   Rule

Nutdrivers   Alignment tool   Countersink

Allen wrenches   Tinsnips

Repairing and building go beyond maintenance in that you work with metal and plastic materials. You also need some additional specialty tools and instruments for making adjustments and measurements:

>> **Wattmeter or SWR meter:** When troubleshooting a transmitter, you need an independent power-measurement device. Many inexpensive models work fairly well (stay away from those in the CB shops; they often aren't calibrated properly when not used on CB frequencies). If you do a lot of testing, the Bird Model 43 is the standard in ham radio. Different elements, or *slugs*, are used at different power levels and frequencies. Both used meters and elements are available online and at hamfests.

>> **RF and audio generators and oscilloscope:** Although you can do a lot with a voltmeter, radio is mostly about RF and audio waveforms so you need a way to generate and view them. If you're serious about getting started in electronics, go to www.arrl.org/servicing-equipment to read up on techniques and test equipment for working on radio equipment.

>> **Nibbling tool, stepped drill bit, and chassis punch:** Starting with a round hole, the nibbler is a hand-operated punch that bites out a small rectangle of sheet metal or plastic. Use a nibbler to make a large rectangular or irregular

opening and then file the hole to shape. The stepped drill bit (for smaller holes) creates a variety of hole sizes in sheet metal and plastic. A chassis punch (for big holes) makes a clean hole in up to 1/8-inch aluminum or 20-gauge steel. Chassis punches aren't cheap, but if you plan to build regularly, they can save you an enormous amount of time and greatly increase the quality of your work.

>> **T-handled reamer and countersink:** The reamer allows you to enlarge a small hole to a precise fit. A countersink quickly smooths a drilled hole's edges and removes burrs.

TIP

Drilling a hole in a panel or chassis that already has wiring or electronics mounted on or near it calls for special measures. You must prevent metal chips from falling into the equipment and keep the drill from penetrating too far. To control chips, put a few layers of masking tape on the side of the panel where you're drilling, with the outer layers kept loose to act as a safety net. To keep a drill bit from punching down into the wiring, place a drill bit *depth collar* over the bit, exposing just enough length to penetrate all the way through.

Components for repairs and building

I find myself using the same components in the following list for most building and repair projects. Stock up on these items (assortments are available from component sellers), and you'll always have what you need:

>> **Resistors:** Various values of 5 percent metal or carbon film, 1/4- and 1/2-watt fixed-value resistors; 100, 500, 1k, 5k, 10k, and 100k ohm variable resistors and controls

>> **Capacitors:** 0.001, 0.01, and 0.1 µF ceramic; 1, 10, and 100 µF tantalum or electrolytic; 1000 and 10000 µF electrolytic; miscellaneous values between 220 pF and 0.01 µF film or ceramic

>> **Inductors:** 100 and 500 µH, and 1, 10, and 100 mH chokes

>> **Semiconductors:** 1N4148, 1N4001, 1N4007, and full-wave bridge rectifiers; 2N2222, 2N3904, and 2N3906 switching transistors; 2N7000 and IRF510 FETs; red and green LEDs

>> **ICs:** 7805, 7812, 78L05, and 78L12 voltage regulators; LM741, LM358, and LM324 op-amps; LM555 timer; LM386 audio amplifier

TIP

ICs and transistors should be kept in aluminum foil or static-dissipating bags to prevent damage from static electricity, called *ESD* for *electro-static discharge*.

TECHNICAL STUFF

There are two basic types of components: *through-hole* and *surface-mount*. Through-hole parts have wire or stamped metal leads that go through holes (thus the name) in a circuit board and are soldered to *pads* on the circuit-board. Surface-mount devices (SMD), or surface-mount technology (SMT) components, have metal contacts on their surface. The component is placed on a pad with some solder paste and the whole assembly heated until the solder melts, attaching the component to the pad. Through-hole components are a lot easier for human hands (and eyes) to work with, but SMD components are becoming the norm for electronics. Learning how to work with both is a good idea.

Although having a completely stocked shop is nice, you'll find that building up the kinds of components you need takes time. Rather than give you a huge shopping list, I give you some guidelines to follow:

>> **When you buy or order components for a project, order extras.** The smallest components — such as resistors, capacitors, transistors, and diodes — are often cheaper if you buy in quantities of ten or more. After a few projects, you'll have a nice collection.

>> **Hamfests are excellent sources of parts and component bargains.** Switches and other complex parts are particularly good deals. Parts drawers and cabinets often come with parts in them, and you can use both.

>> **Broken appliances and entertainment devices around the home are worth stripping before throwing out.** Power cords and transformers, headphone and speaker jacks, switches, and lots and lots of other interesting hardware items end up in the dump. Also, seeing how these items are made is interesting.

>> **Build up a hardware junk box by tossing in any loose screws, nuts, spacers, springs, and so on.** Use an old paint tray or a (clean) cat litter box to make it easy to root through the heap in search of a certain part. The junk box can be a real time and money saver.

TIP

Sporting-goods and craft stores have frequent sales on inexpensive tackle boxes and multiple-drawer cases that are perfect for electronic components.

TIP

To get experience working with electronics, build some kits! The instructions help you get the kit built and working. You gain experience with how electronic devices are made and in handling components. Plus, you get a working piece of electronics that you made yourself. See the section on building kits later in this chapter.

Maintaining Your Station

The best thing you can do for your station is to spend a little time doing regular maintenance. Maintenance works for cars, checkbooks, and relationships, so why not ham radio?

Be sure to keep a station notebook (see Chapter 13). Open the notebook whenever you add a piece of equipment, wire a gadget, note a problem, or fix a problem. Over time, the notebook helps you prevent or solve problems, but only if you keep it up to date.

You also need to set aside a little time on a regular basis to inspect, test, and check the individual components that make up the station. Along with the equipment, check the cables, power supplies, wires, ropes, masts, and everything else between the operator and the ionosphere. Check these items when you plan to be off the air so that you don't have to do a panic fix when you want to be on the air. Your equipment and antennas are of no use if they're not working.

You can make routine maintenance easy with a checklist. Start with the following list and customize it for your station:

>> **Check all RF cables, connectors, switches, and grounds.** Make sure all connectors are tight because temperature changes can work them loose. Rotate switches or turn relays on and off to keep contacts clean and check for problems. Look for kinks in or damage to feed lines. Be sure that ground connections are snug and clean.

>> **Test transmitters and amplifiers for full power output on all bands.** Double-check your antennas and RF cabling. Use full power output to check all bands for RF feedback or pickup on microphones, keying lines, or control signals.

>> **Check received noise level (too high or too low) on all bands.** The noise level is a good indication of whether feed lines are in good shape, preamps are working, or you have a new noise source to worry about.

>> **Check SWR on all antennas.** Be especially vigilant for changes in the frequency of minimum SWR, which can indicate connection problems or water getting into the antenna or feed line connectors. Sudden changes in SWR (up or down) mean tuning or feed line problems. SWR is discussed in Chapter 12.

>> **Inspect all antennas and outside feed lines.** Use a pair of binoculars to check the antenna. Look for loose connections; unraveling tape, ties, or twists; damage to cable jackets; and that sort of thing.

>> **Check weatherproofing.** Water and RF don't mix. Check your outdoor connectors and reseal if necessary. There is a very good how-to on sealing RF connectors at `static.dxengineering.com/pdf/WeatherProofingCoax-TechTip.pdf`. Avoid putty and silicone sealant — they make a mess and are really hard to remove and rework. Also, open any enclosures and be sure water (and bugs) are kept out.

>> **Inspect ropes and guy wires.** Get into the habit of checking for tightness and wear whenever you walk by. A branch rubbing on a rope can eventually cause a break. Knots can come loose.

>> **Inspect masts, towers, and antenna mounts.** The best time to find problems is in autumn, before the weather turns cold. Use a wrench to check tower and clamp bolts and nuts. Fight rust on galvanized surfaces with cold galvanizing spray paint. (Don't spray it on aluminum or non-ferrous metals.) In the spring, check again for weather damage.

>> **Vacuum and clean the operating table and equipment; clear away loose papers and magazines.** Sneak those coffee cups back to the kitchen, and recycle the old soft-drink cans. Make sure that all fans and ventilation holes are clean and not blocked.

TIP

I realize that you may not want to haul the vacuum cleaner into the station, but it may be the most valuable piece of maintenance gear you have. Heat is the mortal enemy of electronic components and leads to more failures than any other cause. The dust and crud that settle on radio equipment restrict air flow and act as insulators, keeping equipment hot. High-voltage circuits, such as in vacuum tube gear like an amplifier, attract dust like crazy. Vacuuming removes the dust, wire bits, paper scraps, and other junk before they cause expensive trouble.

As you complete your maintenance, note whether anything needs fixing or replacing and why, if you know. You'll probably get some ideas about improvements or additions to the station, so note those ideas too.

Over time, you'll notice that some things regularly need work. In my mobile station, the antenna mounts need cleaning, vibration loosens connectors, and cables can get pinched or stretched. I'm always on the lookout for these problems.

REMEMBER

If you do routine maintenance three or four times a year, you can dramatically reduce the number of unpleasant surprises you receive.

Overall Troubleshooting

No matter how well you do maintenance, something eventually breaks or fails. Finding the problem quickly is the hallmark of a master, but you can become a good troubleshooter by remembering a few simple rules:

>> **Try not to jump to conclusions.** Work through the problem in an orderly fashion. Write your thoughts down to help focus.

>> **Start with the big picture.** Work your way down to equipment level.

>> **Avoid making assumptions.** Check out everything possible for yourself.

>> **Read the equipment owner's manual, and get a copy of the service manual.** The manufacturer knows the equipment best.

>> **Consult your station notebook.** Look for recent changes or earlier instances of related behavior.

>> **Label or tag "temporary" changes.** It's hard to remember these things and "temporary" often becomes "permanent." Label the affected equipment with colored tape or labels to remind you of a change.

>> **Write down any changes or adjustments you made while troubleshooting so you can reverse them later.** You won't remember everything. Keep a pocket notebook handy to make taking notes easy.

Troubleshooting Your Station

Your station is a system of equipment and antennas. To operate properly, each piece of equipment expects certain signals and settings at each of its connectors and controls. You can trace many station problems to those signals and settings, often without using any test equipment more sophisticated than a voltmeter.

Most station problems fall into two categories: RF and operational. RF problems are things such as high SWR, no signals, and reports of poor signal quality. Operational problems include not turning on (or off) properly, not keying (or keying inappropriately), or no communications between pieces of equipment. Start by assigning the problem to one of these categories. (You may be wrong, but you have to start somewhere.)

It's often helpful to do an Internet search for problems like yours. Pick the key words for your search. For example, searching for *TS-440 display dots* turns up pages of sites describing this well-known problem and how to fix it.

Power problems

Power problems can be obvious (no power), spectacular (failure of the high-voltage power supply), or subtle (AC ripple, slightly low or high voltage, or poor connections). The key is to never take power for granted. Just because the power supply light is on doesn't mean the output is at the right voltage. I've wasted a lot of time due to not checking power, and now I always check the power supplies first. Try these tests to find power problems:

>> **Check to see whether the problem is caused by the equipment, not the power supply.** You can easily isolate obvious and spectacular failures, but don't swap in another supply until you're sure that the problem is, in fact, the power supply. Connecting a power supply to a shorted cable or input can quickly destroy the supply's output circuits. If a circuit breaker keeps tripping or a fuse keeps opening, don't jumper it. Find out why it's opening.

>> **Check for low output voltage.** Low voltage, especially when transmitting, can cause radios to exhibit all sorts of strange behavior. The microprocessor may not function correctly, leading to bizarre displays, loss of external control, and incorrect responses to controls. Low voltage can also result in low power output or poor RF stability (chirpy, drifting, or raspy signals). Check with the supply disconnected and with a light load. Remember that "12 volt" radios usually need 13.8 V or so to really work properly.

>> **Check the supply output using both AC and DC meter ranges.** Ripple on your signal can mean a failing power supply or battery. A DC voltmeter check may be just fine, but power supply outputs need to show less than 100 mV of AC. Watch for intermittent or erratic voltages that indicate voltage regulation or internal connection problems.

>> **If you suspect a poor connection, measure voltage at the load (such as the radio) and work your way back to the supply.** Poor connections in a cable or connector cause the voltage to drop under load. They can be difficult to isolate because they're a problem only with high current, such as when you're transmitting. Voltage may be fine when you're just receiving. Excessive indicator-light dimming is a sure indicator of poor connections or a failing power supply.

WARNING

Working on AC line-powered and 50-volt or higher supplies can be dangerous. Follow safety rules, and get help if you're unsure of your abilities.

TIP

If your USB device is powered from the USB host, be sure that the host can supply power at the required amount of current. Remember that portable USB hubs often don't supply power unless connected to an AC adapter. Similarly, a laptop may be configured not to supply USB power unless its battery charger is working.

RF problems

TIP

Some RF problems occur when RF isn't going where it's supposed to go. These problems generally are caused by a bad or missing cable, connector, or switching device (a switch or relay) that needs to be replaced. Try fixing these problems with the following suggestions:

>> Replace cables and adapters one at a time, if you have spares that you know work.

>> Note which combinations of switching devices and antennas seem to work and which don't. See whether the problem is common to a set or piece of equipment or specific cables.

>> Bypass or remove switches, relays, or filters. Make a note in the station notebook to put the device back in and put a label or sticky note on the equipment to remind you of the change.

>> Measure resistance through antenna feed lines. Take into account whether the antenna feed point has a DC connection across it, such as a tuning network or impedance-matching transformer. Gamma-matched Yagi beams show an open circuit, whereas beta-matched Yagis and quad loops have a few ohms of resistance across the feed point. (*Note:* Recording the normal value of such resistances in the station notebook for comparison when troubleshooting is a good idea.)

Other problems you may come across include "RF hot" microphones and equipment enclosures, and interference to computers or accessories. (You haven't fully lived until you get a little RF burn on your lip from a metal microphone!) Usually, you can fix these problems by bonding equipment together (see Chapter 13). Try these suggestions:

>> Double-check to ensure that the equipment is connected to the station RF bonding bus.

>> Check the shield connections on audio or control cables. These cables are often fragile and can break when flexed or yanked. (You never yank cables, do you?)

>> Coil up an excessively long cable or swap in a shorter one.

>> Add ferrite RF suppression cores to the cables (see the "Ferrites as RFI suppressors" sidebar, later in this chapter).

TECHNICAL STUFF

On the higher HF bands (particularly 21, 24, and 28 MHz), cables and wires begin to look like antennas as their lengths exceed ⅛ wavelength. A 6-foot data cable, for example, is about ³⁄₁₆ wavelength long on 28 MHz and can have a sizable RF voltage at the midpoint, even though both ends are connected to equipment enclosures. If you have RF pickup problems on just one band, try attaching a ¼-wavelength *counterpoise* wire to move the RF hot spot away from the equipment in question. A ¼-wavelength wire left unconnected with high RF voltage at one end can act to reduce voltage at the other end. Attaching the counterpoise to the enclosure of the affected equipment may lower the RF voltage enough to reduce or eliminate the interference. Keep the wire insulated and away from people and equipment at the unconnected end.

Operational problems

Operational problems fall into three categories: power, data, and control. After you determine which type of problem you have, you often come very close to identifying the cause of the problem.

Data problems

Data problems are more and more common in ham stations. Interfaces among computers, radios, and data controllers usually are made with RS-232 or USB connections. Bluetooth connections for audio and text are starting to become common. Internet-connected equipment uses Ethernet or WiFi networks. If you installed new equipment and can't get it to play with your other equipment, five common culprits are to blame:

>> **Data speed:** An incorrect *baud* setting (including the *framing*: the number of start bits, stop bits, and setting for parity) renders links inoperative, even if the wiring is correct. Baud specifies how fast data is sent. Framing specifies the format for each byte of data. These parameters are usually set by a menu or software configuration.

>> **Protocol errors:** Protocol errors generally result from a mismatch in equipment type or version. A program using an Yaesu control protocol can't control a Ten-Tec radio, for example. Be sure that all the equipment involved can use the same protocol or is specified for use with the exact models you have.

>> **Improper wiring configuration:** Be sure that you used the right cables. A null modem RS-232 cable or a crossover network cable may be required.

>> **Port assignment problems:** Use the device management tools of your PC operating system (Device Manager in Windows) to be sure each port is configured the way your software expects. For example, a USB serial port adapter will be assigned to a COMx port ID where *x* is a number indicating which port.

>> **Network problems:** These problems are in a class of their own, but the equipment generally has a configuration or setup procedure that you can perform or review to see whether you have these problems.

If equipment that was communicating properly suddenly fails, you may have a loose cable, or the configuration of the software on one end of the link may have changed. Double-check the communications settings, and inspect the connections carefully.

WARNING

An operating system upgrade can reassign or renumber ports so that your logging or control software can't find them. With everything working the way you want, note all port assignments for future reference in case you have to manually restore or reset them. Software drivers may have to be reinstalled as well.

TIP

USB interfaces go through a process of establishing a connection when the cable is connected. On a computer, icons indicate that the equipment is recognized and working properly (or not). On stand-alone equipment, you may see indicator lights change or icons on a display. Check the user manual, and watch for these changes carefully.

Control problems

Control problems are caused by either the infamous *pilot error* (in other words, you) or actual control input errors.

Pilot error is the easiest, but most embarrassing, type to fix. With all the buttons and switches in the station, I'm amazed that I don't have more problems. Follow these steps to fix your error:

1. Check that all the operating controls are set properly.

Bumping or moving a control by accident is easy. Controls are frequently changed during repairs or testing. Refer to the operator's manual for a list of settings for the various modes. Try doing a control-by-control setup, and don't forget controls on the back panel or under an access panel.

TIP

Speaking from personal experience, before you decide that a radio needs to go to the shop, check every control on the front panel, especially squelch (which can mute the audio), MOX or XMIT (which turns the transmitter on all the time), and Receive Antenna (which makes the receiver sound dead if no receive antenna is attached). If you're really desperate, most radios have the capability to perform a *soft reset,* which restores all factory default settings. A *hard* or *factory reset* also restores the defaults but wipes out the memory settings.

2. **Disconnect every cable from the radio one at a time, except for power and the antenna.**

   Start with the cable that contains signals related to the problem. If the behavior changes for any of the cables, dig into the manual to find out what that cable does. Could any of the signals in that cable cause the problem? Check the cable with an ohmmeter, especially for intermittent shorts or connections, by wiggling the connector while watching the meter or listening to the receiver. Similarly, you can remove all of the cables and reconnect them one at a time, watching for changes in behavior.

3. **If the equipment isn't responding to a control input, such as keying or PTT, you need to simulate the control signal.**

   Most control signals are switch or contact closures between a connector pin and ground or 12V. You can easily simulate a switch closure with . . . a switch! Replace the control cable with a spare connector, and use a clip lead (a wire with small alligator clips on each end) to jumper the pin to the proper voltage. You may want to solder a small switch to the connector with short wires if the pins are close together. Make the connection manually, and see whether the equipment responds properly. If so, something is wrong in the cable or device generating the signal. If not, the problem is in the equipment you're testing.

At this point, you'll probably have isolated the problem to a specific piece of equipment, and your electronics skills can take over. You have a decision to make. If you're experienced in electronics and have the necessary information about the equipment (schematic or service manual), by all means go ahead with your repairs. Otherwise, proceed with caution.

Troubleshooting RF Interference

TIP

If you have problems outside your station, they usually consist of dreaded RF interference (RFI), as in "I can hear you on my telephone!" or "My garage door is going up and down!" Less known, but just as irritating, is the man-bites-dog situation, in which your station receives interference from some other electric or electronic device. Solving these problems can lead you through some real Sherlock Holmes-ian detective work.

Start by browsing the ARRL RFI Information page at www.arrl.org/radio-frequency-interference-rfi. For in-depth information, including diagrams and how-to instructions, read *The ARRL RFI Book,* which covers every common interference problem. Your club library may have a copy. Consult your club experts for assistance. Occasional interference problems are facts of life, and you're not the only ham who experiences them. Draw on the experience and resources of other hams for help.

TIP

If you want to dig in deep, download *RFI, Ferrites, and Common Mode Chokes for Hams* by Jim Brown, K9YC from k9yc.com/publish.htm. Jim's tutorial covers how ferrite works, how to apply it, along with describing many common RFI problems and solutions. There are a number of other useful papers on his website, as well.

Dealing with interference to other equipment

Start by making your own home interference-free. Unless you're a low-power VHF/UHF operator, you likely own at least one appliance that reacts to your transmissions by buzzing, humming, clicking, or doing its best duck imitation when you're speaking. It's acting like a very unselective AM receiver, and your strong signal is being converted to audio, just like the old crystal radio sets did. It's not the ham radio's fault — the appliance is failing to reject your signal — but it's still annoying.

Your goal is to keep your signal out of the appliance so that it doesn't receive the signal. Sounds simple, doesn't it? Start by removing all accessory cords and wires to see whether the problem goes away. If it does, put the cords and wires back one at a time to see which one is acting as the antenna. Power cords and speaker leads are very good antennas and often conduct the RF into the appliance. Wind candidate cables onto a ferrite interference suppression core (see the sidebar "Ferrites as RFI suppressors") close to the appliance to find out whether that cures the problem. You may have to add cores to more than one of the leads, although generally just one or two are sensitive. You can buy component-level Corcom AC line filters from most distributors and MFJ makes AC line filters, as well.

If the device is battery-powered and doesn't have any leads, you probably can't fix the problem, I'm sorry to say. You either have to replace the device or get along with the interference. The manufacturer's website may have some interference cures, or you may find some guidance from ham radio websites or club members. Try searching online for the model number of the appliance and *interference* or *RFI* to see what turns up.

TIP

The stored messages in the archives of the RFI email reflector (lists. contesting.com/_rfi) are good sources of information. You don't have to be a member of the reflector to read them. Start by searching for messages about the type of device you're having trouble with. If you find information about it, narrow your search with additional terms.

The following common devices are often victims of interference:

>> **Cordless telephones:** Older phones that use 47 MHz frequencies are often devastatingly sensitive to strong signals. Luckily, newer phones use 900 MHz and 2.4 or 5.6 GHz radio links and are much less sensitive to your RF. If you come up against one of the 47 MHz units, just replace it with a newer one.

>> **Touch lamps:** These accursed devices respond to nearly any strong signal on any frequency. You can try ferrite cores on the power cord, but results are definitely mixed. Internal modifications are described on the ARRL RFI website. Replacing the lamp may be the easiest option.

>> **TV, video, and audio equipment:** A common path for interference is via the speaker wires, but any of the many connections among pieces of equipment can be picking up RF. Make sure that all the equipment ground terminals are connected by short, stout wires. The ARRL's RFI website is probably your next stop.

>> **Alarm systems:** The many feet of wire strung around the house to the various sensors and switches create a dandy antenna. Unfortunately, the system controller sometimes confuses the RF that these wires pick up for a sensor signal. System installers have factory-recommended interference suppression kits that take care of most problems.

PART 15 DEVICES

Unlicensed devices that use RF signals to operate or communicate are subject to the Federal Communications Commission's Part 15 rules. These rules apply to cordless phones, wireless headphones, garage-door openers, and other such devices. Devices that may radiate RF signals unintentionally, such as computers and videogame consoles, are also subject to Part 15 rules. The rules make a tradeoff: Device owners don't need a license to operate, say, a cordless phone, but they must not interfere with licensed stations (such as ham radio stations), and they have to accept interference from licensed stations. This agreement generally works pretty well except in the strong-transmitter/sensitive-receiver neighborhood of a ham radio station. See the extensive discussion of Part 15 rules at www.arrl.org/part-15-radio-frequency-devices for more details.

By practicing on your own home electronics, you gain valuable experience in diagnosing and fixing interference problems. Also, if a neighbor has problems, you're prepared to deal with the issue. See the nearby sidebar, "Part 15 devices."

Dealing with interference to your equipment

Electric noise is caused primarily by arcing in power lines or equipment, such as motors, heaters, and electric fences. *Electronic noise* is caused by leaking RF signals from consumer appliances and computers operating nearby. Each type has a distinctive *signature*, or characteristic sound. The following list describes the signatures of common sources of electric noise:

>> **Power line:** Steady or intermittent buzzing at 60 Hz or 120 Hz. Very wet or very dry weather may affect interference.

Power-line noise is caused by arcing or corona discharge. *Arcing* can occur around or even inside cracked or dirty insulators. It can also occur when two wires, such as neutral and ground wires, rub together. *Corona discharge* occurs at high-voltage points on sharp objects where the air molecules become ionized and electricity leaks into the atmosphere. The interference is a 120 Hz buzzing noise because the arc or discharge occurs at the peaks of the 60 Hz voltage, which occur twice per cycle.

WARNING

Do not attempt to fix problems with power lines or power poles. *Always* call your power company.

You can assist the power company by locating the faulty equipment. You can track down the noise source with a battery-powered AM radio or VHF/UHF handheld radio with an AM mode (aircraft band works well). If you have a rotatable antenna at home, use it to pinpoint the direction of the noise. (The null off the side of a beam antenna is sharper than the peak of the pattern.) Walk or drive along the power lines in that direction to see whether you can find a location where the noise peaks. I've found several power poles with bad hardware by driving around with the car's AM radio tuned between stations. If you do find a suspect pole, write down any identifying numbers on the pole. Several numbers for the different companies that use the pole may be on it; write them all down. Contact your utility and ask to report interference. You can find a great deal more information about this process on the ARRL RFI web page (www.arrl.org/radio-frequency-interference-rfi).

>> **Industrial equipment:** Sounds like power-line noise but with a more regular pattern, such as motors or heaters that operate on a cycle. Examples in the home include vacuum cleaners, furnace fans, and sewing machines.

- >> **Defective contacts:** Highly erratic buzzes and rasps, emitted by failing thermostats or switches carrying heavy loads. These problems are significant fire hazards in the home, and you need to fix them immediately.

- >> **Dimmers, electronic ballasts, and variable-frequency speed controls:** Low-level noise like power lines that comes and goes as you use lights or motors. Variable-frequency drive (VFD) systems in large appliances like washers or heat pumps can generate a *lot* of noise that is present when the appliance is operating.

- >> **Switching power supplies:** Most miniature power supplies that mount directly on the AC outlet (called *wall warts*) use rapidly switching electronics to convert AC power to low-voltage DC. They are used for low-voltage lighting and all sorts of appliances, computer equipment, and phone chargers. Although these devices are efficient, they can be real noisemakers. Sometimes, a ferrite core on the DC output helps, but the best solution is to replace the supply with a regular linear supply that uses a transformer.

- >> **Battery chargers for scooters and bikes:** These sometimes use switching electronics, too, and they're much harder to fix or replace. You can often get relief by turning them off or unplugging them when the batteries are fully charged.

- >> **Vehicle ignition noise:** Buzzing that varies with engine speed, which is caused by arcing in the ignition system.

- >> **Electric fences:** Regular pop-pop-pop noises at about 1-second intervals. A defective charger can cause these problems, but the noise is usually caused by broken or missing insulators or arcing from the fence wires to weeds, brush, or ground.

Finding an in-home source of electric noise depends on whether the device is in your home or a neighbor's. Tracking down in-home sources can be as simple as recognizing the pattern when the noise is present and recognizing it as the pattern of use for an appliance. You can also turn off your home's circuit breakers one at a time to find the circuit powering the device. Then check each device on that circuit.

TIP

Battery-powered equipment and computer UPS (uninterruptible power supply) systems may not turn completely off via the power switch if internal control circuits are still on in a low-power state. You may have to physically disconnect or remove internal batteries to be sure the device is off. If the noise is coming from outside your home, you have to identify the direction and then start walking or driving with a portable receiver. Review the ARRL RFI website or reference texts for information about how to proceed when the interfering device is on someone else's property.

What about electronic noise? The following list describes the signatures of common sources of electronic noise:

>> **Computers, videogame consoles, and networks:** These devices produce steady or warbling tones on a single frequency that are strongest on HF, but you can also hear them at VHF and UHF. (See the K9YC tutorial for a list of frequencies.)

>> **Cable and power-line modems:** You hear steady or warbling tones or hissing/rasping on the HF bands.

>> **Cable TV leakage:** Cable TV signal leakage at VHF and UHF sounds like hissing noise. Cable channel 12 covers the same frequencies as the 2 meter band. If you have leakage, a strong ham signal can also cause interference on the same channel.

Each type of electronic interference calls for its own set of techniques for finding the source and stopping the unwanted transmission. You're most likely to receive interference from devices in your own home or close by because the signals are weak. If you're sure that the source isn't on your property, you need a portable receiver that can hear the interfering signal.

TECHNICAL
STUFF

FERRITES AS RFI SUPPRESSORS

Ferrite is a magnetic ceramic material that's used as a core in RF inductors and transformers. It's formed into rods, *toroid cores* (circular and rectangular rings), and beads (small toroids made to slip over wires). Ferrite has good magnetic characteristics at RF and is made in different formulations, called *mixes,* that optimize it for different frequency ranges. Ferrites made of Type 31 material work best for suppressing RFI at HF, for example, and Type 43 ferrites work best at VHF.

Winding a cable or wire on a ferrite core or rod creates *impedance* (opposition to AC current) that tends to block RF signals. The more turns on the core, the higher the impedance. Because they're small, you can place ferrite cores very close to the point at which an undesired signal is getting into or out of a piece of equipment. You can secure cores on the cable with a plastic cable tie, tape, or heat-shrink tubing. This technique works particularly well with telephone and power cords. *Split cores* come with a plastic cover that holds the core together, which makes placing the core on a cable or winding turns easy if the cable already has a large connector installed.

The ARRL RFI website has some helpful hints on each type of interference, as well as guidance on how to diplomatically address the problem (because it's not your device). The Overview page of the ARRL RFI website contains excellent material on dealing with and managing interference complaints (both by you and from others). ARRL members have access to the league's technical coordinators and technical information services.

You can eliminate or reduce most types of interference to insignificant levels with careful investigative work and application of the proper interference-suppression techniques. The important thing is to keep your frustration in check and work the problem through.

Building Equipment from a Kit

Building your own gear — even just a simple speaker switch — is a great ham tradition. By putting equipment together yourself, you become familiar with the operation, repair, and maintenance of your existing equipment.

If you're just getting started in electronics, I recommend that you start your building adventures with kits. Although the fondly remembered Heathkit is no longer around, kits are available today from many sources. To find companies selling ham equipment kits, check out the ads in ham magazines. Non-ham vendors such as Adafruit (www.adafruit.com) and Sparkfun (www.sparkfun.com) have many useful kits, accessories, tools, and parts. You can also search online for "ham radio kit" or just add "kit" to the type of gadget you want to build.

Choose simple kits until you're confident about your technique. Kits are great budget-saving ways to add test instruments to your workbench and various gadgets to your radio station. Also, you don't have to do the metalwork, and the finished result looks great.

After you build a few kits, you'll be ready to move up to building a complete radio. Although the Elecraft K3 (www.elecraft.com) is the top-of-the-line radio kit available today, numerous smaller QRP radio kits are available from other vendors.

You can build most kits by using just the maintenance tool kit described at the beginning of this chapter. Concentrate on advancing your soldering skills. Strive to make the completed kit look like a master built it, and take pride in the quality of your work. Read the manual and use the schematic to understand how the kit works. Observe how the kit is put together mechanically, particularly the front-panel displays and controls.

TIP

You may also like to try the kit-like projects at Instructables (www.instructables. com) and Makezine (www.makezine.com). They are a good halfway point between kits and from-scratch projects.

TIP

Encourage your club to sponsor a build-a-kit project as everyone builds and tests the same kit or kits so help is available on the spot!

Building Equipment from Scratch

Building something by starting with a blank piece of paper or a magazine article and then putting it to use in your own station is a real accomplishment. Building from scratch isn't too different from building from a kit, except that you have to make your own kit. Your first electronic project should be a copy of a circuit in a magazine or handbook — one that's known to work and that comes with assembly and test directions. If a blank printed circuit board is available, I recommend ordering one. You might also want to try building an antenna like a dipole or vertical. Then bring your completed project to "homebrew night" for your club!

Imagine that you have to make a kit for someone else based on the instructions, schematic, and list of components. Photocopy the article, and highlight all the instructions. If an assembly drawing is included, enlarge it for guidance. Make extra copies so you can mark them up as you go. Read the article carefully to identify any critical steps. When you get your components together, sort them by type and value, and place them in jars or the cups of an old muffin pan. Keep a notebook handy so that you can take notes for later use. As you build and test the unit and finally put it to use, everything is completely documented.

If you choose to design a circuit from scratch, I salute you! Documenting your work in a notebook is even more important for a project that starts with design. Take care to make your schematics complete and well-labeled. Record whatever calculations you must make so that if you have to revisit some part of the design later, you have a record of how you arrived at the original values. Take a few high-resolution, in-focus, well-lit photos at important milestones of construction. When you finish, record any tests that you make to verify that the equipment works.

REMEMBER

Don't let failure get you down! First designs hardly ever work out exactly right, and sometimes, you even wind up letting all the smoke out of a component or two. If a design doesn't work, figure out why and then move on to the next version. Don't be afraid to ask for help or to try a different angle. Ham radio isn't a job, so keep things fun. After all, it's *amateur* radio!

5

The Part of Tens

Discover ten common types of jargon you'll hear on the air — don't be confused!

Learn ten important technical fundamentals that make ham radio work.

Follow up with ten tips that the masters use on the air every day.

Chapter 17

Ham Radio Jargon — Say What?

L ike any hobby, ham radio involves a fair amount of jargon. To a newcomer (or an experienced ham starting a new activity) using jargon can make it harder to get going. This chapter explains terms you're likely to encounter.

Spoken Q-signals

In theory, these abbreviations are just supposed to be used in Morse operation. In practice, however, hams use spoken versions on voice which can be confusing. The meanings are often a little different than the formal definition, as well. (A full list of common Q-signals can be found in Chapter 8.)

>> **Kyew-are-emm (QRM):** Any kind of interference. *Local QRM* refers to audio noise bothering the speaker: "I'm getting some local QRM from the TV."

>> **Kyew-are-eks (QRX):** A request to stop talking or stand by, "Can you QRX for a minute?"

>> **Kyew-are-zed (QRZ):** What was that call sign? "Zed" is a phonetic for Z as well as its British pronunciation.

>> **Kyew-so (QSO):** "In contact with," as in "I'm in QSO with NØAX right now."

>> **Kyew-ess-ell (QSL):** Often means "I agree!"

>> **Kyew-aitch-bee (QHB):** A fun way to wish a ham operator "Happy Birthday!"

Contesting or Radiosport

In the fast-paced world of a contest, knowing the terms helps you get up to speed and feel at home handing out contacts. A full contest glossary is available from Contest University (CTU) at contestuniversity.com/attachments/ Contesting_Terminology.pdf. CTU is a full day of training and lectures held at the Dayton Hamvention every year.

>> **Exchange:** Information exchanged during a contest contact. For example, "QTH" in an exchange means your location as a multiplier in the contest.

>> **Serial (number):** Sequence number of the contact in the contest for you. The serial number of your 10th contact is 10.

>> **Zone:** Either CQ zone (www.cqww.com/resources.htm) or IARU zone (www.iaru.org/about-us/organisation-and-history/regions/), depending on the contest.

>> **Run:** Stay on one frequency and call CQ.

>> **Search and pounce (S&P):** Tune the band looking (searching) for stations calling CQ and calling (pouncing on) them.

>> **Cabrillo:** A standard format for contest logs submitted to the sponsor by email or web upload. See the sponsor's website for instructions.

Antenna Varieties

You'll hear all kinds of references to antennas — from Chapter 12 you already know *beam, Yagi, dipole,* and so forth. Here are a few more:

>> **Quad:** A type of Yagi beam with elements that are square one-wavelength loops.

>> **J-pole:** Half-wavelength VHF/UHF vertical with the base section giving it a "J" shape (they're easy to build!).

- **EFHW or OFCD:** End-fed half-wave or off-center-fed dipole. Both are just half-wave dipoles fed somewhere other than the center.

- **Log:** Log-periodic antenna that looks like a Yagi but has many elements close together, often slightly V-shaped.

- **Doublet:** Similar to a dipole but not designed to be a resonant half-wavelength on the operating frequency(s).

- **Zepp:** Wire antenna fed at one end, refers to the Zeppelin airships that used this kind of "trailing wire" antenna.

- **Squalo:** Horizontal VHF/UHF loop in the shape of a square.

Feed Lines

There are almost as many terms for feed lines as there are antennas!

- **Heliax:** Trade name for solid-shield coax with the center insulator made from a strip of plastic wound around the center conductor.

- **Foam:** Coaxial cable (either flexible or hard-line) with center insulation made of foamed plastic.

- **Direct burial:** Feed line that can be buried without any protective conduit.

- **Ladder line, window line, twin-lead, open-wire line:** All used as generic terms for parallel-conductor feed line.

- **Balun:** Short for "balanced-unbalanced," allows an unbalanced, coaxial feed line to be connected to a balanced load such as a dipole antenna or parallel-conductor feed line.

Antenna Tuners

Unless you get lucky and your antenna and feed line present the transmitter with an impedance close to 50 ohms, you'll need an antenna tuner. Antenna tuners are referred to by several names:

- **Impedance Matcher:** This is what the antenna tuner actually does. Remember that an antenna tuner doesn't really tune your antenna; it just changes the impedance presented to your transmitter.

>> **Matchbox:** Originally a model name for an E.F. Johnson antenna tuner.

>> **Transmatch:** Name of a tuner design described in a popular 1960s *QST* construction article and became a generic term for tuners.

>> **Balanced tuner:** An antenna tuner designed to be connected to parallel-conductor feed line. It usually includes a balun (see previous section) so that coaxial cable can be used between the tuner and the transmitter.

>> **Auto-tuner:** Microprocessor-controlled antenna tuner that makes adjustments automatically. If your web browser can run Java applets, check out the online tuner simulator by W9CF at `fermi.la.asu.edu/w9cf/tuner/tuner.html`. You can enter impedance values and watch the tuner adjust itself or you can operate it manually.

Repeater Operating

With repeater operation so common, it would be surprising if there wasn't any jargon! Here are some of the more common terms you'll hear and there are many regional variations:

>> **Flutter** or **mobile flutter:** Rapid variations in strength of a mobile station's signal due to reflections as the vehicle moves.

>> **Picket-fencing:** A flutter that sounds like a stick being dragged along a picket fence.

>> **Scratchy:** Intermittent or low-level static in the audio of a weak signal.

>> **Machine:** Reference to the repeater station. *Making the machine* or *hitting the machine* means a signal strong enough to open the repeater's squelch and activate the transmitter to retransmit your signal.

>> **Sub-audible** or **PL:** Low-frequency control tones (see Chapter 8).

>> **Kerchunk:** Pressing the PTT switch momentarily without identifying to see if the signal is heard by the repeater.

>> **Squelch tail:** Time during which the repeater is still transmitting after the input signal has ceased.

>> **Timeout:** Transmit long enough to activate the repeater's shut-down timer, usually about three minutes.

Grid Squares

Earning the VHF/UHF Century Club (VUCC) award requires you to know about grid squares. You can find out everything you need to know about grid squares at www.arrl.org/grid-squares, but here are the common terms:

>> **Grid circling:** Operating while driving around a point at which four grids come together at a corner.

>> **Water grid:** Grid square without any land.

>> **Maidenhead:** Location in England where the grid system was defined, grid squares are part of the Maidenhead Locator System.

>> **Grid field:** 20° (longitude) × 10° (latitude) rectangle identified by just two letters, such as FN, EM, or DM.

>> **Grid square:** 2° (longitude) × 1° (latitude) rectangle identified by two numbers after the grid field designator, such as FN01, EM48, or DM03.

>> **Grid sub-square or locator:** 5' (longitude) × 0.25' (latitude) rectangle identified by two letters after the grid field, such as FN01ah, EM48ss, or DM03pt. (The apostrophe represents degrees.)

>> **Locator:** Any grid reference; field, square, or sub-square.

Interference and Noise

Like drivers and traffic, there are many ways to describe the various disturbances to contacts. Since they often refer to specific problems, knowing the terms can also lead to a solution.

>> **Hum** versus **buzz:** True hum is a low tone at the frequency of the local AC power grid, 60 Hz in the U.S. It is usually caused by magnetic fields from AC wiring or motors. *Buzz* has a sharper, higher tone and is caused by power supplies that rectify the AC power to produce DC power. Sometimes hum is used to refer to any power-related noise. *Line noise* is a type of buzz caused by arcing on the power lines.

>> **Popcorn** or **shot noise:** Sharp, irregular pops and crackles that sound a bit like popcorn popping or shotgun pellets being spilled onto a surface. Often caused by erratic connections.

>> **White noise** and **pink noise:** White noise is random noise over a wide range of frequencies. Pink noise is random noise over the audio range. Both sound like hiss to the human ear.

>> **Buckshot** or **splatter:** Distortion caused by speaking too loudly or over-modulating a transmitter, causing intermittent signals to appear on adjacent channels.

>> **Ignition noise:** Sharp snapping noise that varies with engine speed, caused by the ignition system sparks.

>> **Alternator whine:** Mid- to high-pitched audio tone that varies with engine speed, caused by the vehicle battery charging system.

Connector Parts

There are lots of different types of connectors (see Chapter 15), but many of them have similar parts with similar names:

>> **Plug** and **receptacle:** Plugs have *prongs* or *pins* that extend from the body of the connector. Receptacles have *sockets* recessed into the body of the connector. Plugs are usually installed on the end of cables whereas receptacles are wall- or panel-mounted.

>> **Body** and **shell:** The part of the connector that holds the pins and sockets. Some types of connectors have a shell that can be removed.

>> **Barrel connector:** Connector used to join two cables together. Usually refers to RF connector families such as the common UFH-series PL-258 that joins a PL-259 plug with another PL-259.

>> **Reducer:** Type of adapter that lets a small-diameter cable be used with a plug designed for thicker cable.

>> **Bulkhead connector:** Like a barrel connector but long enough to extend through a thick panel or a wall.

>> **Crimp connector:** RF connector installed by crimping or compressing a sleeve or tube over a coaxial cable shield.

>> **Tip-ring-sleeve (TRS):** The three terminals of a stereo phone plug/jack. Tip is the contact at the very end of the plug. Sleeve refers to the barrel of the plug and ring (if present) is the contact between the tip and sleeve. *TRRS* connectors have a second ring contact between the ring and sleeve.

Solar and Geomagnetic Activity

Events on the Sun have a great deal to do with HF and lower-VHF radio wave propagation here on Earth. Satellites and telescopes combine to give us a good idea of what's happening on the Sun that affects radio propagation. For more information, tutorials, and real-time data and images, check out the Spaceweather website (www.spaceweather.com). The following measurements are used to describe space weather conditions:

>> **Solar flux:** Light energy coming from the Sun as microwaves, visible light, ultraviolet, and X-rays that create the *ionosphere*. Measured in *solar flux units (SFU)* with a minimum value of 65.

>> **A and K indices:** Measures of disturbances of the Earth's geomagnetic field. Higher values indicate greater disruption and generally poorer propagation.

>> **Solar flare:** Sudden, large release of visible light, UV, and X-rays from the surface of the Sun.

>> **Coronal Mass Ejection (CME):** Release of charged particles from the Sun's outer layers; it takes about 36 hours to travel from the Sun to the Earth. The charged particles enter the ionosphere above the Earth's *geomagnetic poles* and help create the *aurora*.

>> **Geomagnetic field:** The Earth's magnetic field, which interacts with both HF radio signals and solar phenomena.

Chapter 18

Technical Fundamentals

N o matter what interests you in ham radio, from ragchewing to equipment design, you'll get more out of the hobby if you have a basic understanding of a few technical details.

TIP

A number of license exam questions involve these technical details. Your license study guides will show you how to use this information.

If you want to dive in a little deeper, *The ARRL Handbook* and *The ARRL Antenna Book* have been reliable technical references for many years. Online you can use the ARRL Technical Information Service (www.arrl.org/technology) that is available to all hams. The appendix contains a radio math supplement that provides some common math formulas you'll encounter in ham radio.

Electrical Units and Symbols

You should know each of the basic electrical units and what they represent:

» **Voltage (volts, V):** The electrical potential between two points, represented as *V, v, E,* or *e* in equations.

» **Current (amperes, A):** The electrical charge flowing in a circuit, represented as *I* or *i* in equations.

>> **Power (watts, W):** The rate at which energy is expended or dissipated, represented as P or p in equations.

>> **Resistance (ohms, Ω):** Opposition to current flow, represented as R or r in equations. Ω is a capital Greek letter omega.

>> **Reactance (ohms, Ω):** Opposition to AC current flow, represented as X in equations.

>> **Impedance (ohms, Ω):** Combination of resistance and reactance, represented as Z in equations.

>> **Conductance (siemens, S):** The inverse of resistance, represented as G or g in equations.

>> **Capacitance (farads, F):** The ability to store energy as an electric field, represented as C in equations.

>> **Inductance (henries, H):** The ability to store energy as a magnetic field, represented as L in equations.

>> **Frequency (hertz, Hz):** The number of complete cycles per second of an AC current, represented as f in equations.

>> **Wavelength:** The distance traveled by a radio wave during the time it takes to complete one full cycle, represented by λ (a lower-case Greek letter lambda) in equations.

Ohm's Law

The most basic relationship in electronics and radio is *Ohm's Law*, which states that the voltage (V, in volts) across a resistance (R, in ohms or Ω) is proportional to the current flowing through it (I, in amperes). Mathematically it looks like this, shown three different ways:

$$V = I \times R \quad I = V / R \quad \text{and } R = V / I$$

The first version is why the voltage across a resistor is sometimes referred to as the *IR drop*. Voltage may also be represented by E.

Power

Another fundamental equation is the *Power Equation* that shows power (P, in watts) being used or dissipated by a device or component is equal to the voltage across it (V, in volts) multiplied by the current through it (I, in amperes): $P = V \times I$. (Voltage can also be represented by E.)

By substituting the Ohm's Law relationships for V and I, we also get:

$$P = V^2 / R \qquad P = I^2 \times R$$

When calculating power in an AC circuit, use the *RMS (root-mean-square)* voltage measured by a voltmeter.

Transmitter *peak envelope power* (PEP) is the power measured at the very highest peak of the RF waveform. If you use an RF *wattmeter*, be sure it is calibrated to show PEP and not average power.

Decibels

Decibels (dB or "dee-bee") are a ratio of any two quantities such as voltage or power, expressed as factors of 10. A change by a factor of 10 (from 10 to 100 or from 1 to 0.1) is a change of 10 dB. Positive values of dB represent an increase, and negative values are a decrease. The following formulas calculate changes in power and voltage as decibels:

$$dB = 10 \log (\text{power ratio})$$

$$dB = 20 \log (\text{voltage ratio})$$

If you memorize these ratio–dB pairs, you can save yourself a lot of calculating:

Power x 2 = 3 dB	Power x 1/2 = –3 dB
Power x 4 = 6 dB	Power x 1/4 = –6 dB
Power x 5 = 7 dB	Power x 1/5 = –7 dB
Power x 8 = 9 dB	Power x 1/8 = –9 dB

Power x 10 = 10 dB	Power x 1/10 = –10 dB
Power x 20 = 13 dB	Power x 1/20 = –13 dB
Power x 50 = 17 dB	Power x 1/50 = –17 dB
Power x 100 = 20 dB	Power x 1/100 = –20 dB

A change of 1 receiver S unit represents approximately 6 dB.

For more about how to calculate and work with decibel values, the ARRL has published my *QST* magazine article "Untangling the Decibel Dilemma" as an online PDF at www.arrl.org/files/file/Education/Untangling_the_Decibel_Dilemma.pdf.

Attenuation, Loss, and Gain

The decibel is used to measure or specify the following quantities:

>> **Attenuation:** A reduction in signal level by a circuit, such as a filter or attenuator, or by the signal traveling through a feed line or through space.

>> **Loss:** A reduction in signal level caused by the signal flowing through a component or feed line.

>> **Gain:** An increase in signal level caused by a circuit, such as an amplifier, or by an antenna focusing signals in a preferred direction.

>> **Effective Radiated Power (ERP):** Includes both loss and gain in a transmitting station so that the final radiated signal can be compared in strength to a system that uses a reference antenna such as a dipole (ERPD) or an isotropic antenna (EIRP). ERP is given in watts and is equal to the *transmitter output power (TPO)* plus any gain created by the antenna system less any losses created by the feed line and any feed line components like filters. Both ERPD and EIRP are common.

Bandwidth

Bandwidth is a range of frequencies over which a circuit or signal behaves in some specified way. Figure 18-1A shows a typical signal occupying a channel. The level of the signal is shown by the heavy black line. Two vertical lines show where the signal level is one-half that of the signal's peak level. The frequency range between the vertical lines is the signal's bandwidth. Similarly, Figure 18-1B illustrates filter bandwidth. The heavy black line shows the filter *response*, meaning how much signal is passed or removed by the filter. This *notch filter* removes signals over a range of frequencies. The range over which the signal coming out of the filter is one-half or less of the signal going in is the filter's bandwidth. The bandwidths of other filters and signals are measured similarly.

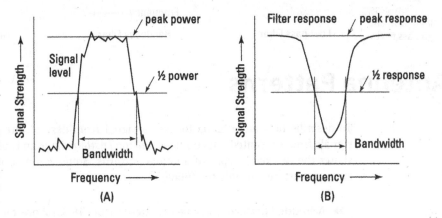

FIGURE 18-1:
(A) Signal
bandwidth
example and (B) a
filter bandwidth
example.

Filters

A circuit that intentionally increases or decreases a signal's strength based on its frequency is a *filter*. (There are other types of filters but this is the most common type.) Ham radio uses a lot of filters! Most are of four common types:

>> **Low-pass filter (LPF):** Attenuates signal level above a *cut-off frequency*. Generally used to remove unwanted harmonics or signals above the desired range.

>> **High-pass filter (HPF):** Attenuates signal level below the cut-off frequency. A high-pass filter used to remove hum or buzz and sub-audible signals from audio is called a *low-cut filter*.

>> **Bandpass filter (BPF):** Attenuates signal level above and below a specified range of frequencies (the *passband*). Can also be thought of as a combination of high- and low-pass filters.

>> **Band-stop** or **notch filter:** Attenuates signals over a narrow range of frequencies, shown in Figure 18-2.

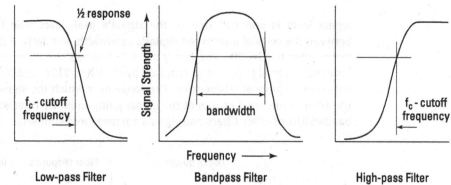

FIGURE 18-2:
Filter types, showing the responses of low-pass, bandpass, and high-pass filters.

½ response

Signal Strength

f_c - cutoff frequency

Low-pass Filter

bandwidth

Frequency

Bandpass Filter

f_c - cutoff frequency

High-pass Filter

Antenna Patterns

To describe how an antenna focuses a signal from desired directions or rejects signals from unwanted directions, an *antenna pattern* diagram is used. Figure 18-3 shows the two basic types of antenna patterns. Some common terms related to antenna patterns include the following:

>> **Azimuthal pattern:** This type of pattern (Figure 18-3a) shows the relative signal strength radiated by the antenna in all horizontal directions. It is as if you were looking down on the antenna at the center of the chart. The distance from the center to the solid line shows how strong the signal is or isn't.

>> **Elevation pattern:** Figure 18-3b shows an antenna pattern from the side, showing how well it radiates at all vertical angles above the ground.

>> **Null:** A point of minimum radiation in the antenna pattern.

>> **Lobe:** The region of the antenna pattern between nulls.

>> **Front-to-back ratio (F/B):** The ratio of signal strengths directly to the front of the antenna (0° in Figure 18-3a) to directly off the back of the antenna (180°).

>> **Front-to-side ratio (F/S):** The ratio of signal strengths directly to the front of the antenna (0° in Fig 18-3a) to the signal strength 90° away from the front of the antenna (90° and 270°).

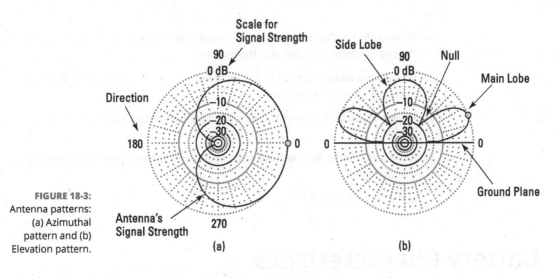

FIGURE 18-3:
Antenna patterns:
(a) Azimuthal
pattern and (b)
Elevation pattern.

Standing Wave Ratio (SWR)

You don't have to be a radio engineer to use SWR as an indicator for tuning and troubleshooting antenna systems. It basically represents power bouncing around in your feed line until it is radiated by the antenna or dissipated as heat. Lower SWR is good (1:1 is the best), but you don't have to get too worked up over it. Most transmitters are perfectly happy with SWR values of up to 2:1. For moderate SWR, you can use an antenna tuner. Big changes in or very high values of SWR can indicate a bad connection or other problem.

As SWR increases, so does the amount of signal lost in the feed line since some is lost on each trip back and forth. The additional feed line loss caused by SWR can become significant on the upper HF bands and at VHF/UHF and microwave frequencies.

TIP

To learn all about SWR in an easy-to-understand presentation, watch the excellent AT&T Archives video "Similarities in Wave Behavior" on YouTube at www.youtube.com/watch?v=DovunOx1Y1k&t=4s. Don't let the black-and-white presentation put you off — it's one of the clearest introductions to SWR and related topics I've ever seen!

You'll encounter the following terms when discussing SWR and how to measure it:

>> **Standing wave:** A pattern of voltage and current in a feed line caused by power reflected from an antenna or load

>> **SWR bridge:** A common accessory used to measure SWR directly

>> **Mismatch:** A condition where the impedance of the antenna isn't the same as the characteristic impedance of the feed line

>> **Directional wattmeter:** An RF power meter that can measure power flowing in each direction in a feed line

>> **Forward power, P_f:** Power traveling from the transmitter to the antenna or load

>> **Reflected power, P_r:** Power reflected from a mismatched antenna toward the transmitter

$$SWR = (P_f + P_r) / (P_f - P_r)$$

Battery Characteristics

There sure are a lot of batteries! You need to know which ones your radio needs as well as how (and whether) to charge them. Start by learning a few terms for when you start shopping:

>> **Primary:** Rechargeable batteries such as lead-acid or Li-ion. *Secondary* batteries are not rechargeable, such as regular alkaline cells.

>> **Ampere-hours (A-hr):** Battery *energy capacity*. Since a battery's output voltage is fairly constant, the time a battery can supply a certain current before voltage begins to drop tells you how much energy the battery can supply.

>> **Energy density:** The amount of stored battery energy per unit of weight (pound or kilogram) or unit of volume.

>> **Charge profile:** The sequence of steps that a charger can take for a specific type of battery.

>> **Battery chemistry:** The types of chemicals that make up a battery and determine both its output voltage and energy density.

For complete information on batteries, battery chargers, and other related topics, check out Battery University (batteryuniversity.com).

Satellite Tracking

You don't have to be a rocket scientist to know when a satellite is visible to your radio signals and where it will be in the sky. There are terrific *tracking* programs available from AMSAT (www.amsat.org) and other sources. (Websites like

Heavens Above [www.heavens-above.com] show you where the satellites are at any given time, as well.)

TIP

Heavens Above shows when the ISS is visible at your location. It's an impressive sight!

To use tracking software, you need a set of data for each satellite called its *Keplerian elements*. These values describe the satellite's orbit with enough precision for you to known when and where to aim an antenna at the satellite. Here is a set of "Keps" for recently launched amateur satellite AO-91:

```
AO-91
1 43016U 17073D 17324.81992359 .00000855 00000-0 73563-4 0 9991
2 43016 97.6901 259.8749 0259544 229.6727 128.1552 14.77757019 356
```

This is the *two-line NASA format*. Each piece of data is defined as:

```
KEY: A-CATALOGNUM B-EPOCHTIME C-DECAY D-ELSETNUM E-INCLINATION F-RAAN
G-ECCENTRICITY H-ARGPERIGEE I-MNANOM J-MNMOTION K-ORBITNUM Z-CHECKSUM
```

Many tracking software packages can acquire this information online automatically without you having to type it in.

You'll need to be familiar with the following terms to know when to begin listening for the satellite and how to adjust your receiving frequency as the satellite moves past your location:

>> **AOS:** Acquisition of signal. When the satellite's signal is receivable at your location.

>> **LOS:** Loss of signal. When the satellite's signal is no longer receivable at your location.

>> **Doppler (shift):** The change in a satellite signal's frequency caused by its motion. When the satellite is moving toward you, its signal appears to be slightly higher in frequency, and the reverse is true when the satellite is moving away.

Chapter **19**

Tips for Masters

Your ham radio license is really a license to learn. Take advantage of every opportunity, including learning from your mistakes. (You'll have plenty!) Each problem or goof is also a lesson.

Masters got to be masters by starting as new hams just like you, then making one improvement at a time, day in and day out. You may think that ham radio veterans surely have stores of secret knowledge — knowledge that makes them the masters of all they survey. Certainly, the veterans have worked hard to gain experience and expertise, but they also rely on simple principles that work in many situations. You can use these principles, too.

Listening to Everything

Masters get more out of listening and monitoring than anyone else because they've learned the value of doing it. Every minute you spend listening is a minute learning and a minute closer to being a master, whether it's as a net control, a top contest operator, setting up a balloon tracker, or just giving out directions to the club meeting. Listen (or watch) a skilled operator and learn how.

Learning How It Works

Operating a radio and building an efficient, effective station are much easier if you know how the equipment works. Even if you're not terribly tech-savvy, take the time to get familiar with the basics of electronics and how your equipment functions. You will be much more effective if you learn the effects of controls and their adjustments. Learn how to make simple repairs to keep your station on the air.

Following the Protocol

Use the expected terms and give information in the form and order in which it is expected. When calling another station, follow "Gift Tag Order – To then From:" Start with that station's call sign to alert that operator, then give your call once or twice as necessary. Use the recommended phonetics that others in your group prefer. In a competition, exchange your information in the same order published by the sponsor. When checking into a net, follow the procedures set by the net control station.

Keeping Your Axe Sharp

When asked what he would do if he had eight hours to cut down a tree, Abe Lincoln replied that he would spend the first six hours sharpening his axe.

If you have battery-powered equipment, be sure that the batteries are charged and fresh. Make sure fuel for a generator is fresh. Lay out your "go kit" from time to time so that you are sure it's all there when you need it. Test your station's basic operation from time to time on all bands and modes. Keep your equipment and skills sharp. When they're needed on the air, you'll be ready.

Practice to Make Perfect

Even a sharp axe gets dull if it isn't used. Get on the air regularly, keeping in touch with conditions. An experienced operator knows what stations are active, from where, and when, as well as when important nets and on-the-air events take place. Even if you know the procedures by heart, check in to your local net each week. Take advantage of contests or special events to exercise your skills and make sure your equipment is working.

Make operating your radio station a natural and comfortable activity by keeping yourself in shape with regular radio exercise.

Paying Attention to Detail

Masters know that the little things are what make the difference between 100 percent and 90 percent performance — or even between being on the air and off the air. The most expensive station isn't worth a nickel if it doesn't work properly when you need it. Waterproofing that connector completely or having your CQ sound just right really pays off in the long run. Masters are on the radio for the long run.

Knowing What You Don't Know

Take a tip from Mark Twain, who warned, "It ain't what you don't know that gets you into trouble. It's what you know for sure that just ain't so." If you get something wrong, don't be too proud to admit it. Find out the right way; track down the correct fact. People make their worst mistakes by ignoring the truth.

REMEMBER

Radio waves and electricity don't care about pride. A master isn't afraid to say, "I don't know."

Maintaining Radio Discipline

When you are performing public service, whether in an emergency or not, practice your radio discipline: Know and follow the rules of the operation, follow the instructions of a net control station, transmit only when authorized and necessary, use plain language, and pay attention so you are ready to respond. Strive to make your operating crisp and clear so that anyone can understand. It will pay off when every second counts.

Make Small Improvements Continuously

Any improvement in the path between stations should not be neglected. Anything that makes your signal easier to understand — 1 dB (decibel) less noise received, 1 dB better audio quality, 1 dB stronger transmitted signal — makes the contact easier. Keep improving your station and your operating in small increments and you'll get a lot more out of ham radio. In Japan, the art of continuous small improvements is called *kaizen*. It took Honda and Toyota from fringe players to world-leading status, and it will do the same for you!

Help Others and Accept Help from Others

Sooner or later, you will encounter operators needing assistance. If they ask for help, offer your services. They may not be aware there is a problem, such as with poor audio, a distorted signal, or erratic operation. New operators may not know the right way or time to call another station. Before informing them of the problem, ask yourself how you would want to learn about a problem with your station. When describing the problem, be polite and be as clear as you can in your description.

When other operators tell you that you have a problem, don't get mad or embarrassed. Thank them for bringing the problem to your attention and make them feel good about helping you. Ask them to help you troubleshoot. Ham radio is all about helping each other, on and off the air.

Index

About the Author

H. Ward Silver, ham radio call sign NØAX, has been a ham since 1972, when he earned his Novice license and became WNØGQP. His experiences in ham radio led to a 20-year career as an electrical engineer, designing microprocessor-based products and medical devices. In 2000, he began a second career as a teacher and writer, receiving the Bill Orr Technical Writing Award in both 2003 and 2016. In 2008, he was recognized as the Dayton Hamvention's Ham of the Year. Finally, in 2015, he became a member of the CQ Contesting Hall of Fame.

Silver is lead editor of the two primary amateur radio technical references, both published by the American Radio Relay League: *The ARRL Handbook* and *The ARRL Antenna Book*. He has authored all three ARRL licensing study guides, and for 15 years wrote the popular monthly column "Hands-On Radio," now available in three volumes of experiments from the ARRL. He has written two other Wiley titles as well: *Two-Way Radios and Scanners For Dummies* and *Circuitbuilding Do-It-Yourself For Dummies,* as well as the ham radio detective mystery *Ray Tracy: Zone of Iniquity* (published on www.lulu.com). His most recent book is the tutorial *Grounding and Bonding for the Radio Amateur* (www.arrl.org/shop/Grounding-and-Bonding-for-the-Radio-Amateur).

On the air, he enjoys contacting faraway (DX) stations, competing in radiosport competitions, building antennas, and participating in his local club and emergency communications team. He is a founder of the World Radiosport Team Championship (www.wrtc.info) and president of the Yasme Foundation (www.yasme.org). Outside ham radio, he plays the mandolin; dabbles in digital photography; and enjoys biking, camping, canoeing, and kayaking. Occasionally, he finds time to sleep.

Dedication

As always, my work is dedicated to my wife, Ellen. Her kindness, encouragement, and understanding enable me to write with confidence. Thank you. I also owe a tip of the laser pointer to the cats, the venerable Dander and young Electra, and to Bennie the dog, all of whom keep me on a tight leash.

Author's Acknowledgments

This book has been enriched by numerous individuals and organizations that contributed photos, drawings, tables, and various bits of hard-won knowledge and wisdom. In addition, they've made their knowledge freely available on websites and blogs and videos in the best traditions of ham radio.

It's particularly important to acknowledge the generosity of the American Radio Relay League (ARRL; www.arrl.org), both in making material available to me for use in the book and in maintaining an enormous website of amateur radio know-how, significant portions of which are freely available to members and nonmembers alike. The loyal support of its members — please consider joining if you're not a member already — is an important reason why the organization is available to facilitate, represent, and defend amateur radio.

Publisher's Acknowledgments

Acquisitions Editor: Kelsey Baird

Project Editor: Christopher Morris

Copy Editor: Christopher Morris

Technical Editor: Kirk Kleinschmidt

Proofreader: Debbye Butler

Production Editor: Mohammed Zafar Ali

Cover Image: ©2021 of IC-705 courtesy of Icom America Inc.